T0297638

Intelligence Emerging

Intelligence Emerging

Adaptivity and Search in Evolving Neural Systems

Keith L. Downing

The MIT Press
Cambridge, Massachusetts
London, England

This book was set in Palatino by diacriTech, Chennai.

Library of Congress Cataloging-in-Publication Data
Downing, Keith L.
Intelligence emerging: adaptivity and search in evolving neural systems / Keith L. Downing.
    pages cm
Includes bibliographical references and index.
ISBN 978-0-262-02913-1 (hardcover: alk. paper), 978-0-262-53684-4 (paperback: alk. paper)
1. Neural networks (Computer science). 2. Machine learning. 3. Experiential learning. 4. Genetic algorithms. 5. Neural networks (Computer science) 6. Adaptive computing systems. I. Title.
QA76.87.D69 2015
006.3'2—dc23
                                                                                    2014045996

To Nancy, Marvin, Målfrid, Neva, Asta, and Jasem

Contents

Preface

In the late 1970s, the computer scientist Douglas Lenat designed an artificial intelligence (AI) system that, when given very basic mathematical concepts, was able to discover important relations such as Goldbach's conjecture (i.e., every even integer greater than 2 is the sum of two primes). He improved and generalized the system to invent surprising new VLSI chip designs and then to craft a winning strategy for the 1981 U.S. national championships of a popular role-playing game (involving the command of a fleet of vessels). That discovery forced a rule change for the 1982 contest, which Lenat's program also won, with yet another unique twist: sinking some of its own ships!

In the early 1990s, Gerald Tesauro adapted a few popular AI machine learning techniques to play the game of backgammon. That yielded several surprising new strategies that enabled the program to play at an expert level. A few of these strategies even changed conventional backgammon theory so that now, among experts, one of the more popular opening moves has been replaced by the AI-devised alternative.

In the mid-1990s, Karl Sims used simulated evolution of brains and bodies to produce creatures that moved about on land and in water. Some looked extremely elegant and natural, while others were amusingly odd in appearance and behavior but successful nonetheless. When evolved to compete against one another in a simple game—be the first to grab a small object placed midway between two combatants—most of the top individuals acquired long, agile arms to quickly snap up the target. But one successful individual had a more unconventional strategy: it rushed to the opponent, pushed it away, and then returned to grab the prize with its short arms.

An even odder family of automated discoveries is the antennas that have been developed at NASA since 2006. The asymmetric bent and twisted wires, discovered by AI's evolutionary algorithms, look nothing like the products of human engineering but perform exceptionally well; NASA now deploys them on space missions.

To surprise, much more than *to reason*, is the feature of AI systems that has intrigued me throughout a 30-year career of teaching and research in the field. Oh, sure, an AI program deducing that Socrates is most likely dead by now—given his birth date and a few properly encoded facts about human longevity—can still keep the front row of my AI class bubbling with enthusiasm. But to get every head off the desk, it takes stories of AI programs that adapt to changing situations, modify themselves, or discover interesting facts, patterns, or designs, often in ways that catch a human observer completely off guard. These systems display extraordinary creativity. Besides achieving results like those just mentioned, they have made discoveries in areas such as furniture, art, music, electronic circuits, data compression, protein sequencing, and controllers (e.g., for mobile robots, race cars, rockets, and biochemical plants). These discoveries impress with their performance while baffling with their unorthodoxy. Socrates and his mortality really are quite mundane in comparison.

Though AI (and technology) skeptics claim that computers can only act as we program them—no real surprises—the fact is that we now give some AI programs a fixed (usually quite small) set of primitive components and operations along with a computational carte blanche to combine them in numerous ways. These mix-and-match possibilities are so great that with respect to the final products, the computer essentially chooses from a menu with more options than there are sand grains on earth or atoms in the universe (approximately 10^{80}). In all those combinations are surely many that no human has ever contemplated. Computers surprise us all the time. Rather than grumble about this unpredictability (or the inability to formally prove the correctness or rationality of the computer's decisions), many AI researchers revel in their machine's creativity and work to channel it into useful artifacts.

Another argument that tries to rain on the AI surprise party is that if we know the primitives and the combining processes, then in theory we can predict the outcome of any computer run. However, the computer may actually perform trillions of combining operations in a way that cannot be compressed or succinctly summarized to allow the human brain to predict the result (via a much smaller set of logical inferences). In reality, the only intelligence capable of predicting the computer's output (in any reasonable amount of time, such as a human lifetime) is an artificial intelligence: another computer.

Surprise plays a major role in the field of complex adaptive systems (CASs), where complex structures arise from (often simple) interactions between (numerous) components or agents. In many cases, the only way to accurately predict the outcome is to build detailed computer models and simulate all those interactions; simpler alternative models for processing by our limited mental machinery are in short supply. For example, we may predict patterns of rush hour traffic (a classic CAS) quite successfully using logic and a few simple rules, such as when the Main Street bridge clogs up, drivers exit early onto Palm Avenue. But when an oil truck breaks down on the Palm Avenue exit ramp during rush hour, predicting the ensuing traffic flows (along numerous side roads) may evade human inference and require a complex multiagent (automobile) simulation, the bread and butter of CAS research.

This book is motivated by considerable evidence that beneath our logical, rational, and often quite predictable cognitive activities run numerous complex adaptive systems, all creative and exploratory but subject to various forms of selection that tend to keep things running smoothly. Still, there is always room for surprise, particularly when we need to adapt to new and changing conditions on time scales from milliseconds to millennia. Just as AI constrains and molds the computer's creative leaps into practical designs, Darwinian evolution has tweaked and tuned these natural CASs into a repertoire that systematically coerces relatively unpredictable trial-and-error behavior into the structures and dynamics of cognition.

In these pages I investigate many of those biological CASs, and the computational tools that they have inspired, for the purpose of fostering appreciation of the emergent nature of both natural and artificial intelligence and exploring additional natural mechanisms that could strongly impact AI research. These CASs occupy three distinct time scales: evolutionary, developmental, and learning, whose relevance to AI is explored in this book.

Emergence and search underlie much more of cognition than modern science often acknowledges, and they enable intelligent behavior to strike the proper balance between stability and adaptability. I hope the reader will begin to appreciate this perspective on intelligence.

I have had the good fortune of working with many exceptional AI researchers in several institutions in the United States and Europe and with a wide variety of AI techniques. I started out in classic Good Old Fashioned AI (GOFAI) and transitioned gradually into biologically inspired AI (Bio-AI). Since 1997 I have taught a university course in Bio-AI. Much of this book stems from that course and from my desire to get my perspective written down as an accompaniment to my lecture slides, which are heavy on diagrams but light on text. I had hoped that my classroom performance could draw students out of their dorm rooms, but I could not match the convenience of online education: all the course slides were available on a Web page. So my eagerness to thoroughly document my views on Bio-AI dovetails nicely with the students' need for extra sleep.

Writing a book was not an obvious next step, since there are several good Bio-AI books already on the market. But I perceived a significant flaw in my Bio-AI course. The introductory lecture was a showcase of emergence—it focused on the artificial life (ALife) aspects of AI—but after a sojourn into the lives of ants, termites, flocking birds, and frustrated freeway commuters, this theme vanished as we marched through the algorithmic details of evolutionary algorithms (EAs) and artificial neural networks (ANNs). I realized that my Bio-AI book needed to bring emergence to the forefront and keep it there. So I began to look at EAs and ANNs from the emergent perspective and found more and more instances of emergence in studies of intelligence.

An aha! moment on the jogging trail made me realize that *it really is all about search*. Formal search methods lie at the heart of the field. Just about everything in AI—when coded for the benefit of our machines, including automated logical reasoning itself—embodies search. Issues of how these encodings (representations) facilitate efficient search are also a pivotal aspect of AI.

In evolution, development, and synaptic fine-tuning of brains, search shows up as well, often as the underlying mechanism of the emergent result. Finding these connections among search, representation, emergence, evolution, and neural processing helped dissolve some of the mental barriers I had placed between GOFAI and Bio-AI. Still, emergence (and a belief in its ability to produce useful results) remains a major difference between the disciples of GOFAI and of Bio-AI. Whatever the reader's philosophical AI leanings, I hope this book encourages appreciation of these connections.

This book is geared toward people with interests in cognitive science and machine intelligence. It is written with third- and fourth-year college students in mind, particularly those who have had an introductory course in AI. Basic AI terms are explained here, but some previous acquaintance with AI will be helpful. The book can serve as a supplement to standard textbooks in a broad range of AI and CAS courses.

Introductory chapters on five core concepts—emergent phenomena, formal search processes, representational issues in Bio-AI, ANNs, and EAs—are followed by intermediate chapters that delve deeper into search, representation, and emergence in ANNs, EAs, and evolving brains. Then, advanced chapters on evolving artificial neural networks (EANNs) and information-theoretic approaches to assessing emergence in neural systems combine many of the earlier topics to provide some perspective on the future of Bio-AI.

The introductory and intermediate chapters contain thin cross-sections of several very broad fields, slices in which search, representation, and emergence are prominent. The advanced chapters have considerably more breadth and depth while keeping these three core concepts in view. However, no chapter offers a complete overview or review of an area. Many evolutionary, neuroscientific, and connectionist concepts come into play, but the full breadth of these fields does not.

In the preface to *The Computational Beauty of Nature* (2000), Gary Flake writes that producing his book was a labor of love because he was able to convert interesting concepts into code. It is true that nothing in AI is completely fulfilling without a running program, but—though many of the concepts and algorithms described here have been implemented by myself and others, often with great success—I make little attempt to delve into actual programming specifics. Unfortunately, the scope of this book is a little too broad for any one AI programming platform. There are surely devils in the details whose exposure will come only from a complete implementation. This is very often the case with emergent processes: predicting their outcomes is nearly impossible. Someone has to write the code, but in some cases neither I nor anyone else has yet done so.

I hope this book engages some of your own neurons, that it challenges you to think a little differently about what constitutes thinking in humans and machines.

Acknowledgments

At the age of 88, my grandfather, Kenneth L. Downing, a never-really-retired Methodist minister, published his first book, *Holy Gossip*, an extremely liberal, commonsense explanation of some of the more miraculous events in the Bible. The book was banned in church libraries across the United States. We were all very proud of him for sticking with his convictions and braving the inevitable storm of criticism. It planted a seed in several of us grandchildren: maybe we too could write something, someday.

At the age of 39, I was promoted to professor in the Norwegian academic system, at which point one of those who assessed me, Wolfgang Banzhaf, proclaimed that it was time to write a book. That was 14 years ago, but the message did stick. I am very grateful for his encouragement back then, and more recently.

At the age of 18, I met a wonderful person, Eric Allgaier, while we were both students and teammates at Bucknell University. Eric gave me all sorts of advice throughout college, but more recently, in our adult years, he has repeatedly prodded me about writing a book. I thank him dearly for that. Eric knows me well enough to understand that I absolutely HAD to have that push.

At the age of 0, I was born into a family that didn't waste much time worrying about impressions, limitations or *spilled milk* (as my grandmother, Gino, used to say). Life is about opportunities, and my mother, Nancy, and father, Marvin, did their best to keep me focused on them, despite my tendency to buck the family tradition and worry about everything. In particular, Nancy exudes a spirit of unsinkability that has always been an inspiration. Without her constant reminder to keep my chin up and eyes ahead, I would have barely survived grade school, let alone grad school. From Marvin, I got the core of my sense of humor; from Nancy, I get the constant reminder to use it!

Writing a book is one of those endeavors to which the label *CBA* certainly applies: conceive, believe, achieve. The people mentioned above are all about the C's and B's. Mark Bedau, another B, administered the deciding shove at an ALife conference banquet in 2012, somewhere high atop the football field at Michigan

State University. There, totally giddy over the chance to see the inside of a Big-10 stadium, I dropped my guard and spouted off my wild book idea. Mark's very positive response buoyed me for months as I worked to flesh out the details, and, in general, convince myself that I could actually carry through with it all.

Once over that hump, the rest, the actual achievement, was a walk in the park—a very LONG walk, but one that was overwhelmingly pleasant. And for that, I must thank all of the people at MIT Press.

It began with Marie Lufkin Lee and Marc Lowenthal, both of whom have helped with untold practical matters and, most importantly, offered much-needed encouragement during the extensive review process. The person who knows this book almost as intimately as its author is Alice Cheyer, whose copy editing exceeds all superlatives. It both comforted and amazed me to witness such thoroughness. Finally, Katherine Almeida has logged her share of hours into sewing all of these chapters together and helping me to meet the high production standards of MIT Press. Colleagues have asked the highly rhetorical question of whether I'd ever want to go through this process again; with the help and guidance that I've received from MIT Press, I can confidently say "Yes!"

I received feedback on technical content from a number of colleagues in Europe and the United States. Pauline Haddow (NTNU) worked through several chapters and provided a whole host of helpful suggestions, along with general encouragement. Kathleen Freeman (University of Oregon), Mico Perales (Oberlin), and Jim Torreson (Oslo) also gave loads of general advice and commentary, while Michael Loose (Oberlin) helped minimize the gap between my AI-tainted understanding of the brain and conventional neuroscience. Andrea Soltoggio (Loughborough University) provided invaluable advice on reinforcement learning.

Like many academics, I'm not a real conventional thinker, and this book certainly reflects many of my personal quirks. It is neither a textbook nor a popular science offering, but a hybrid grounded in my own eclectic understanding and appreciation of science. As Gene Luks (University of Oregon), my mathematics and computer-science professor once told me (while tending to the many bird houses in his back yard), you never know what information or experience will one day come in handy. Discount nothing. Not long after receiving that advice, I had the aha! moment that lead to my computer-science PhD thesis ...while reading a *Scientific American* article about snakes!

My PhD advisor, Sarah Douglas (University of Oregon), has always assured me that the middle ground between AI and the natural sciences is a basket worth putting many eggs into, despite the insecurities one constantly feels about not being expert enough in any of the specialty areas. My many hours spent in her office, discussing a wide spectrum of cognitive science, are still some of the most stimulating and challenging intellectual exchanges of my career.

Steve Fickas (University of Oregon) has never met a problem he doesn't like, and, more precisely, a pair of problems or solutions that he cannot (somehow) find interesting connections between. That refreshing, open-minded attitude, spanning multiple academic fields, encouraged many of us students to think outside the box, and it launched many of us into multidisciplinary work.

I owe a deep gratitude to Bucknell University and a handful of exceptional professors who got me started down this road so many years ago. David Ray (mathematics), Al Henry (computer science) and Andrea Halpern (psychology) were three of the finest. Michael Ward has always been my professorial role model. The calm joy with which he coated a wall of blackboards with a mathematical proof inspired me to choose academia over industry at a young age. Whenever I stand in front of a room full of students, a vision of Mike floats somewhere between my conscious and subconscious. I hope that, to some degree, I too have been able to infect students with my (less than calm) enthusiasm.

Here in Norway, I am grateful for the working environment in the computer science department (IDI) of the Norwegian University of Science and Technology (NTNU). These conditions make it possible for professors to strike a healthy balance between teaching and research. I am particularly thankful to Guttorm Sindre, Jon Atle Gulla, and Maria Letizia Jaccheri, who, during their tenures as department head, have been generously supportive of my book-writing effort, recognizing this too as an important form of research and scholarship.

Here at IDI, the hard work of many people—including Birgit Sørgård, Eivind Voldhagen, Kristin Karlsen, Unni Kringtrø, Kai Dragland, and Rolf Foss—makes the lives of us professors nearly utopian. Berit Hellan and Ellen Solberg, thanks for reminding me to laugh. Anne Berit Dahl, thanks for reminding me about everything else!

My longtime colleague Agnar Aamodt also deserves special mention for both helping me to secure an academic position and for a perpetual wisdom that has ferried me across plenty of rough water. Our overlapping, but distinct, backgrounds and interests make every AI discussion an invigorating one.

A significant chunk of this book was written at Oberlin College, during my sabbatical and other visits. Oberlin has become a second home for our family, and for that opportunity, I am deeply grateful to the Oberlin neuroscience department and three of its leaders who have all welcomed me with open arms: Janice Thornton, Lynne Bianchi, and Catherine McCormick. Thank you, and go Yeomen!

Tom Ziemke and Skövde University have given me a place to hang my hat on several occasions. A chapter or two of this book was written during one of those "research vacations," and many of the underlying ideas emerged during other visits. Tom is simply a brilliant cognitive scientist, and I appreciate the assistance that he has given me throughout my career.

My family has been a continuous source of support during this long process. Our children, Neva, Asta, and Jasem are so busy with their own activities that I often wonder if they miss me much in the evenings, when I often hide away in a dark corner of the house and write. Lately, I think they've primarily viewed me as a marginalized competitor for the television, treadmill, and car keys. But my comeback is imminent; watch out! In their eyes, a book carries infinitely more status than a journal article, so if I ever walk into a bookstore and find this on the science shelves, I hope they are there to share in the celebration.

The love and encouragement from my wife, Målfrid Rustad, has always been my greatest source of stability and strength. I tend to worry too much about too many (little) things, so it's always soothing to see a don't-worry-be-happy attitude in a person who, on a daily basis, deals with orders of magnitude more stress than I do—and in a manner that has made her very popular with her friends and co-workers. She tends to look beyond barriers to focus on possibilities. Of all the wonderful gifts that I have been given, none compares to the opportunity to share a life with Målfrid and our children.

1 Introduction

1.1 Exploration and Emergence

From the comfort and safety of a soft rubber floor, I have watched my children grapple with climbing walls. Vertical explorers, spider people, they silently struggle upward toward the victory bell, with legs extending, retracting, then extending again, weight shifting and hands clutching for a centimeter or two of ledge. They blend tentative, decisive, reversible, and unforgiving movements, making progress that often requires time-lapsed photography to appreciate.

Though some ascents appear methodical, even well-planned, others seem more like vertical random walks, with moves being *tossed out there* to see what gives; the climber is just winging it, spitballing, or flying by the seat of her pants. This book claims that the rudiments of cognition work in a similar manner: by myriad micro- and macroprocesses relying on little more than persistence and dumb luck to find their own footholds, their own stability. And from these numerous local acts of random exploration, the phenomenon of intelligence arises.

Emergence, the formation of global patterns from solely local interactions, is a frequent and fascinating theme in popular science books, and for good reason. There is something very intriguing and *pure* about a collection of simple agents that indirectly *team up* to produce something complex, without the slightest hint of central control but relying heavily on other central mechanisms from control theory: positive and negative feedback.

The examples are many. Social insects alone provide dozens, from the building of mounds, hives, and nests to role and task allocation to the formation of optimal routes to food sources (Holldobler and Wilson, 2009) (figure 1.1). The expansive, branching, highly dynamic pathways generated by raiding army ants (Camazine et al., 2001) justify the term *superorganism* even though the individual organisms are nearly blind, relying on pheromone signals for guidance. These chemicals are typically emitted when ants leave the nest (bivouac), thus marking the route for

Figure 1.1
Ants have a surprisingly common presence in the artificial intelligence community (Dorigo et al., 1996; Bonabeau et al., 1999; Simon, 1996; *AI Challenge*) with their most memorable performance coming in the *Ant Fugue* of the Pulitzer prize–winning book *Gödel, Escher, Bach* (Hofstadter, 1979). There, Hodstadter draws an important analogy between the manner in which the (complex) colony emerges from the (comparatively simple) behavior of individual ants, and the way our mental *self* emerges from the collective activity of many billions of neurons. This book continues along the trail blazed by Hofstadter, aided by 35 additional years of neuroscience progress.

return trips, and when raiding ants find prey, attack it, and cart the corpses back to the bivouac. At the front of a raiding swarm, ants move slowly, tentatively, in search of prey. When ants succeed, their pheromones attract (many) others, and portions of the front converge quickly upon the area of high prey density; and the more ants that succeed in finding prey, the more pheromones get emitted, and the more reserves invade to continue the conquest. Conversely, ants that venture a few centimeters ahead of the front without encountering prey or pheromones quickly rebound back to the front before beginning another scouting expedition. Via the trial-and-error movement of numerous ants supplemented with a pheromone-mediated collective memory, the emerging superorganism efficiently forages large swaths of land.

The positive and negative feedbacks inherent in emergent ant foraging patterns have parallels in the formation of the body's capillary networks, albeit via different chemical signals (Wolpert et al., 2002). The low oxygen levels in undernourished tissues induce the expression of a particular gene (*Vegf*), which stimulates vascularization and enhances oxygenation. Thus, the circulatory pathway branches and grows to efficiently match the spatial pattern of oxygen demand, just as the ant swarm distributes to match prey density.

Not to be outdone by ants or blood vessels, networks in the brain arise from multiple forms of distributed search. First, neurons migrate to their proper locations, along glial pathways, by using filopodia (slim cytoplasmic spikes) that repeatedly extend and retract in search of stable attachment sites (Wolpert et al., 2002), much like the limbs of climbing-wall enthusiasts. Next, in attempts to find proper recipients for their electrochemical signals, neurons sprout axons that explore surrounding extracellular spaces in search of compatible chemical markers (Sanes et al., 2011). Once again, progress stems from repeated extensions and retractions of the axonal growth cone, with the resulting network tailored to the sizes, chemical tags, and activity levels of the intertwined neural populations.

The functional differentiation of neural regions also exhibits considerable emergence, based on patterns of perceptual stimulation and the nascent task demands. For example, the mammalian visual cortex is replete with ocular dominance columns: alternating neural neighborhoods that preferentially respond to stimuli from one eye or the other but (generally) not both. These show up as stripes (the emergent global pattern) in stained regions of cortex. Pre- and postnatal processes combine to form these segregated regions (Kandel et al., 2000), with axonal migration driven by chemical gradients producing minor innervation biases: individual neurons may have slightly more incoming axons from one eye than the other. After birth, the asynchronous stimulation of each eye (which, in the real world, is much more common than synchronous stimulation) produces heterogeneous activity patterns in the visual cortex, with no sizable regions dedicated to one eye or the other. However, one additional local mechanism quickly accentuates any minor, randomly generated asymmetries: young axons that succeed in stimulating a target neuron often sprout extra branches, thus allowing them to connect to and stimulate neighboring neurons. Via parallel runs of this process, millions of times over, the cortex gradually separates into distinct columns dedicated to one eye or the other.

This established fact of neuroscience closely resembles other, more classic, examples of emergent grouping and segregation found in the popular literature. For instance, ants use only local cues to both aggregate corpses and sort larvae into groups based on their size (Bonabeau et al., 1999). As one moves further along the intelligence spectrum, simulation studies (Ball, 2004) show that strongly segregated cities can emerge naturally when nonprejudiced citizens have the occasional ability to relocate, along with a simple preference for neighborhoods in which at least a few residents are similar to themselves.

Finally, every fan of emergence is familiar with the synchronous flashing of fireflies, popularized by the book *Sync* (Strogatz, 2003). Among humans, similar entrainment occurs at concerts, plays, and sporting events, where entire crowds may clap or chant in unison, not because anyone leads the way but simply because of the strong effect of neighboring stimuli upon one's own sound-producing rhythm (Ball, 2004). Once again, these traditional examples have direct parallels in the brain, where the emergent synchrony of neural activation stems from the dominating influence of afferent/upstream activity and plays a key role in everything from perceptual binding (of diverse stimuli) to sequence learning to task switching between memory formation and recall (Buzsáki, 2006).

So the brain and cognition are a potpourri of emergence that is no less spectacular than termite mounds, flocking birds, raiding ants, and crowdsourcing. However, unlike many of these textbook examples, intelligence does not arise from one (or even a few) basic processes. Hence, it does not fit neatly into a popular science chapter on the topic. Rather, intelligence is emergent (and driven by trial-and-error search) across many temporal and spatial scales, thus confounding the investigation—and motivating book-length discussions of the topic.

This book examines instances of search and emergence—at several levels of neural organization and across several temporal granularities of adaptivity—in order to deepen the understanding of cognition and improve the ability to automate it in artificial intelligence (AI) systems. The goal is not a complete picture of natural intelligence, nor a new style of AI system that outperforms all others, but rather a deeper inquiry into some of the principles that motivated the newer, biology-based perspective of AI, and some additional biological mechanisms that could prove useful to AI.

1.2   Intelligence

Have you ever tried to define *intelligence*? Using a dictionary for this chore often leads to a confusing, ultimately circular, pathway among words such as *knowledge, thought, understanding, reasoning,* and *rationality.* So pick your favorite route through this maze of ill-defined concepts, or simply settle on a working definition that helps frame your own investigations into the matter. My personal favorite has evolved into this:

> Doing the right thing at the right time, as judged by an outside human observer.

This is a very behavioristic, black box definition: it evaluates intelligence based on the *result* of a process (such as thought), not the process itself. Of course, since scientists only vaguely understand the process, it is hard to include its details in the definition of intelligence; we are essentially on the outside of this black box, struggling to catch a glimpse of its inner workings but coming away with only a few hints.

So this definition may be a convenient cop-out for one hoping to keep the cans of philosophical worms closed up tight. Maybe concepts such as thought and rationality will eventually have well-accepted, formal, and noncircular definitions upon which a definition of intelligence might stand. But maybe not. It could well be that everything just below the surface of an intelligent act is actually quite confusing: there is no logical, deterministic explanation of how the agent's situation produces its action. The process may be stochastic—not random but not fully deterministic either. It may even include hints (or large doses) of chaos.

Intelligence may *emerge* from all this stochasticity in ways that we will eventually understand but can never fully predict. This might lead to an improvement of the earlier definition:

> The activation of neural processes $P_1, P_2, \ldots, P_n$, yielding a cumulative result that evokes an action that when assessed by an external human observer, appears to be the correct response to the current situation.

So what are $P_1, P_2, \ldots, P_n$? This is the gist of my interaction with neuroscientists: I joke about wanting them to list the key neural mechanisms that underlie their view of intelligence, and then I will promise to scurry back to my AI lab, never to bother them again.

Unfortunately, I am too old to wait any longer for *the list*, and I have spent several years digging around the neuroscience literature in hopes of assembling it myself. Alas, the digging continues, but a few gems have come to my attention. Some are straight from experimental neuroscience, others involve the fruits of computational neuroscience, and still others stem from artificial intelligence. None give the complete picture, and some may (despite their current status) turn out to be either patently false scientifically or completely useless for the engineering of automated intelligences. But all provide interesting explanations of how various low-level aspects of intelligence arise from processes that bear little resemblance to logic but exhibit high degrees of emergence and trial-and-error search.

In fact, this investigation gives grounds for the claim that *the adaptivity that is so crucial to advanced intelligence is the emergent result of many serial, parallel, hierarchical, and heterarchical trial-and-error search processes.* Since trial and error is the antithesis of *intelligent search*, it is easy to conclude that the activities at just a level or two below the *intelligent act* are comparably unintelligent. They may ooze with interesting biology but have no apparent ties to the psychology of rationality.

Have you ever tried to recall the name of, say, a celebrity? You know that you know it, and you have some vague idea of the beginning letters or the sound of the name, but you cannot spit it out. Then, an hour later, while doing some completely unrelated task (or just relaxing), out pops the celebrity's name, seemingly from nowhere. Few would dispute memory's pivotal role in intelligence, but equally few could explain this tip-of-the-tongue scenario without invoking something at least bordering on emergence; and in general, many everyday mental pursuits seem to involve it. The study of intelligence from an emergent perspective includes investigations into the neural bases of memory and learning, of which this book includes several.

However, the mystery goes deeper than this. $P_1, P_2, \ldots, P_n$ may not be enough to capture the full emergent character of intelligence. Biologists are fond of saying that no living systems (or their parts) can be completely understood without considering their evolutionary origins; and in fact, it is often the case that biological designs make no sense whatsoever—they would earn failing marks as engineering school projects—unless one takes evolution into account.

Darwinian evolution is frequently cast as a search process—grounded in variation, inheritance, and selection—from which highly fit genotypes emerge. Phenotypes then arise from genotypes through developmental processes that are, quite likely, the most intricate and impressive examples of emergence known to modern science. Many of the mechanisms of development double as maintenance processes during postnatal life and as facilitators of the neural plasticity that underlies lifetime learning. Learning also exhibits considerable emergence and trial-and-error search, and oddly enough, the results of learning can even affect the course of evolution. There is no clean separation between these processes, no straight line of causality.

So just as definitions of intelligence are typically circular, the story of intelligence emerging is profoundly cyclic. The tale involves several adaptive loops that appear somewhat concentric but are also hopelessly intertwined, thus adding to the mystery but accentuating the fascination when pieces begin to fall into place, when links among the cycles reveal themselves. The next section discusses these adaptive mechanisms in a little more detail, and later chapters of the book dig much deeper.

1.3   Adaptivity

One of the key differences between primitive and complex life-forms is the degree to which they can adapt to changing conditions: more sophisticated and intelligent organisms can handle a wider range of new and unforeseen situations. Starfish adapt to limb loss by growing new ones; chameleons adapt to changing backgrounds by altering their skin color; many mammals adapt to seasonal temperature variation by changing the thickness of their coats; and of course, humans adapt to a gauntlet of daily challenges via a mixture of off-the-shelf and exceptionally creative solutions.

From a longer time perspective, species adapt to changing environments via evolution; and the nature of today's animal intelligences surely reflects many of these past worlds that the brains and bodies of the time had to navigate. Characteristics of many of those brains not only linger in the design of our own but are predominant features, with only a few small (but architecturally encompassing) tweaks needed to boost general mammalian intelligence to that of primates and humans.

Between the time scales of minute-to-minute daily activity and evolutionary progression lies that of development, a process spanning days, weeks, months, and even years. At this scale nature offers some of the most impressive examples of emergence, as complete organisms self-organize in a gradually unfolding cellular dance. However, the adaptive nature of development often goes unnoticed, particularly during the early prenatal stages, since the environment (e.g., womb) has evolved to be reasonably stable. Furthermore, DNA clearly exerts considerable control over development, giving the (false) impression that nature doggedly follows the DNA script to build a being. However, DNA hardly encodes all the details of the full-blown organism; rather, it serves as a recipe for a multitude of adaptive cellular mechanisms that, together, will produce the fully formed individual. Many of those processes will continue to work in the background during postnatal life, often supplanting repeated minute-to-minute regulatory behavior with longer-term adaptive change, as described in Bateson's classic work (1979) on the economy of flexibility.

These three levels of adaptivity are commonly described as *phylogenetic* (P), *ontogenetic* (O), and *epigenetic* (E), or less formally, *evolution*, *development*, and *learning*, respectively. Artificial systems that employ elements of each often bear the POE label.[1] This book argues that POE capabilities are critical for the attainment of intelligence by emergent means, whether the system is biological or artificial, and the following section on AI reveals a growing appreciation for the role of emergence in producing thinking machines. Though this newer approach to AI, grounded more in biology than psychology, does take a completely different angle on the intelligence issue, there is no reason to ignore classic AI while embracing the emergent perspective.

Although the classic AI view of formal logic as fundamental for intelligence is somewhat outdated, the conceptual tools popularized by logic-based and other knowledge-intensive approaches to AI, namely, search and representation, deserve careful consideration in alternative AI paradigms. In particular, the P, O, and E of POE both embody and arise from search processes, some of which have only recently come to light via empirical discoveries and theoretical advances. Thus, search and the representations that enable it become fundamental aspects of many approaches to AI, including those grounded in the power of emergence and the perspective of complex systems.

1.4   I Am, Therefore I Think

In the late 1950s the field of artificial intelligence began in a blaze of glory. Within a decade or so of its inception, computers were solving geometry and physics problems at a college freshman level, playing chess like regional champions, diagnosing serious illnesses on par with expert physicians, and designing complex VLSI circuits. No problem was too complex, but, as AI researchers discovered in the 1980s, many were too simple.

Indeed, the capabilities that humans take for granted, our basic sensorimotor skills such as walking, climbing, and grasping for objects, turned out to be orders of magnitude more difficult to program than backgammon, bridge, and biochemical analysis. By the mid 1980s AI researchers realized that a serious shortcoming in their systems was none other than common sense. AI systems behaved like idiot savants, producing exceptional results on a wide range of situations but floundering miserably on cases that demanded basic intuitions about the world, intuitions that most humans have acquired by their second birthday.

Many attempts were made to force-feed this common sense into AI systems, in much the same manner and using similar knowledge representation formats as had been successfully used to load expert rules of thumb into AI systems. In fact, the whole AI subfield of qualitative reasoning (QR) (Forbus, 1997) was dedicated to this aim. Although QR produced many useful paradigms whose applications range from intelligent tutoring to plant monitoring to automobile and Mars rover diagnosis, it did little to fortify AI systems with that broad base of general knowledge needed to transform idiot savants into well respected (and trusted) gurus.

As AI faced this disappointing truth in the 1980s, a related but diametrically opposed field began to take root: artificial life (ALife) (Langton, 1989). Although ALife researchers primarily sought to understand the life process at a level far removed from that of neuroscience and psychology, the basic philosophy had immediate implications and inspiration for AI, although only a few AI researchers took note (R. A. Brooks, 1999; Steels, 2003; Beer and Gallagher, 1992; Pfeifer and Scheier, 1999). The essential transferable concepts from ALife to AI were situatedness and embodiment. Although often trivial in their detail, the vast majority of ALife systems consist of simulated organisms that reside in environments (situatedness) and have a body (embodiment) whose survival depends upon a fruitful interaction with those surroundings. This ALife-inspired route to AI bears acronyms such as SEAI (situated and embodied AI) and Bio-AI (biologically inspired AI), which this book uses interchangeably.

Classic AI systems, often called GOFAI (Good Old-Fashioned AI) systems, assume away all environmental and bodily factors to focus on cognition in a vacuum. This works well for chess but fails consistently in robotics. As GOFAI researchers found out, general abstract reasoning systems do not plug-and-play with any set of sensors and motors. General common sense exists not in a platform-independent piece of software but in a behavioral repertoire that is finely tuned to the structure and dynamics of both body and environment. And although some aspects of this repertoire are easily explained in everyday terms and rules of thumb, many are the unique province of biology and engineering. Logics, so common to GOFAI, are of little utility, but many of ALife's kernel concepts—emergence, feedback, competition, and cooperation—form the backbone of this low road to understanding intelligence (Downing, 2004).

For that handful of AI researchers who saw ALife as more than cute, abstract simulations of self-organization—rather, as a more fundamentally sound approach to cognition—the motivating thesis can be approximated as follows:

> Complex intelligence is better understood and more successfully embodied in artifacts by working up from low-level sensorimotor agents than by working down from abstract cognitive mechanisms of rationality such as deduction, induction, and means-ends analysis.

Essentially, Bio-AI researchers believe that GOFAI's holy grail, common sense, comes only via the learned experiences of a body in a world. There are significant limits to how much knowledge one body (a teacher or an expert system designer) can transfer to another (a student or an expert system), and with common sense, these limits are very stringent. Whereas "I think, therefore I am" might have been an appropriate slogan for GOFAI, its converse more aptly summarizes Bio-AI. That is, by living, we acquire common sense, which then supports more complex reasoning.

Clark (2001) uses the term *cognitive incrementalism* to denote this general bootstrapping of intelligence:

> This is the idea that you do indeed get full-blown, human cognition by gradually adding bells and whistles to basic (embodied, embedded) strategies of relating to the present at hand. (135)

Hans Moravec, a renowned roboticist, draws an interesting parallel between the evolution of living organisms and that of computers (Moravec, 1999). As summarized in table 1.1, animals have always had the ability to sense and act but have gradually evolved advanced cognitive and then formal, explicit calculation capacities, whereas the evolution of computers has gone in the opposite direction, from their World War II roots as industrial-strength calculators to advanced automated reasoning during the heyday of GOFAI to the more recent appearance of relatively sophisticated autonomous robots.

Table 1.1
Comparison of advances in animal evolution versus computer capabilities

	Living Organisms	Computers
Sense and act	10,000,000	25
Reason	100,000	40
Calculate	1,000	60

Note: Each number denotes the approximate number of years that the (natural or artificial) system has possessed a respectable level of the given faculty.

Bio-AI must recognize that sensing-and-acting organisms cannot (in all probability) simply evolve an independent *reasoning unit* or *calculating module* in a single generation. The relatively homogeneous nature of the human brain (in terms of the basic electrochemical properties of its neurons, the structure of its cortical columns, and so on) and reasonably tight integration of its regions indicate that any newly evolved region would both have similar neural machinery as that of the preexisting sense-and-act areas, and be required to communicate with those areas. Hence, any evolutionary brain improvements would be enabled and constrained by previously evolved sensorimotor mechanisms.

The grand challenge to cognitive incrementalism may come from the work of Lakoff and Núñez (2000), who explain mathematical reasoning, both simple and complex, as an extension of our sensorimotor understanding of the world. The neuroscientific grounding of their theory is weak, but the metaphorical ties between embedded and embodied action on the one hand and mathematical concepts on the other are striking. By linking everyday sensing and acting to one of humanity's most abstract cognitive endeavors, the authors implicitly motivate a Turing-type challenge for Bio-AI: build a sense-and-act robot that evolves the ability to do mathematics.

Ignoring the obvious difficulty of this challenge, the general Bio-AI philosophy and its bottom-up approach to knowledge acquisition hold some promise. After all, one can hardly deny the importance of firsthand experience in learning simple facts of life such as that wet things can be slippery, sharp things can cut, and loud sounds can warn of an unpleasant immediate future. However, the cruel realities of engineering raise major obstacles, as all roboticists know.

Whereas GOFAI began with a divergent radiation of impressive applications that displayed many forms of (shallow) intelligence, Bio-AI seems to have converged on a menagerie of wall-following robots, all of which have very deep, functional (albeit implicit) understandings of their own body and domain: a barren floor surrounded by walls. To date, there are no biologically inspired robots that display transferable common sense. That is, many systems behave intelligently and exhibit minimally cognitive behaviors, such as perceptual classification, attentional focusing, and so on (Beer, 2003), but the common sense is so tightly embedded in reactive routines (or controllers with simple notions of internal state) that it evades reuse for other tasks. So far, sensing and acting have not produced commonsense scaffolding for cognitive activities, such as the planning of motor sequences.

After 20 years of Bio-AI, cognitive scientists probably expect more. GOFAI adherents can arguably write off Bio-AI as overly optimistic biological envy in the same way that an AI *scruffy* might criticize a *neat's* logical approaches to intelligence as mere mathematical envy.[2] Still, the fact remains that GOFAI was built on a computational foundation that was rock-solid for building useful engineering tools but flimsy and misleading as a model of intelligence. Bio-AI has yielded a few useful artifacts, such as floor-sweeping and lawn-mowing robots, and a host of insectlike intelligences, but the functional cornerstone (normally neural networks or other systems of distributed processors) maps much more readily to the basis of animal behavior than do GOFAI's theorem provers, frames, and semantic networks. Bio-AI is clearly paying the price for building a proper biologically rooted foundation, but as GOFAI's failures in cognitive science indicate, a field's initial progress is not a clear indicator of its ultimate success.

Regardless of the rather weak state of the art in Bio-AI vis-à-vis its ultimate goals, the bottom-up approach to machine intelligence holds a promise that deserves continued exploration. Of course, it is the path followed by nature, and since natural evolution took hundreds of millions of years to go from simple multicellular organisms to humans, one cannot expect a synthetic progression to happen overnight, no matter how many computers are crunching away on the task. Still, a royal helping of skepticism is in order. After all, many aspects of natural intelligence may only be contingencies of the past 3 billion years of life on earth, not vital principles of cognition. For instance, mass, volume, and general structural constraints imposed on the brain and cranium—by the process of childbirth, the energetic demands of signal-processing cells, the conversion of DNA recipes into fully functioning organisms, and the common need to funnel visual and auditory information through a mere pair of receptors—seem only peripherally related to general issues of artificial intelligence.

The idiosyncrasies of the human brain should therefore be viewed as fascinating mechanisms for achieving intelligence but certainly not as prerequisites for all forms of sophisticated thought. The author's confidence in the bottom-up approach stems not from these large-scale aspects of neuroanatomy but from the general intuition that a neural substrate, or some similar network of relatively simple, interconnected processors, is most effective for the representations and information processing that underlie intelligence. The brains of extant organisms are simply enlightening illustrations of anatomical scaffolding that supports the emergence of mind from the interactions of basic computational units. Obviously, brains provide many insights into intelligence, but many GOFAI researchers have ignored this.

Still, over a half century of GOFAI research has shed considerable light upon issues of mind and machine. Bio-AI followers who dismiss these results as *merely* engineering or as grounded in the *weaker* sciences of psychology, sociology, and economics, are disregarding huge conceptual resources. For example, one of the early battle cries of Bio-AI was *reasoning without representation* (R. A. Brooks, 1999), a nice reductionistic idea—and one supported by a host of slick examples in simple domains—but one that has proven exceptionally difficult to realize for complex tasks. Traditional knowledge representation (KR), as formulated by GOFAI workers over the years, may not directly apply to bottom-up approaches to intelligence, but KR concepts still offer considerable leverage. Similarly, search, a fundamental process for most of GOFAI, has extensive and diverse connections to emergent intelligence, though the particular algorithms often differ between GOFAI and Bio-AI.

1.5   Evolving Neural Networks

This book champions the use of evolving artificial neural networks (EANNs) as an emergent route to artificial intelligence. EANNs are a combination of evolutionary algorithms (EAs), a search technique motivated by Darwinian evolution, and artificial neural networks (ANNs), a representation and reasoning framework based loosely on neurons and the brain. So the biological inspiration for EANNs comes directly from the brain and its evolution, but the conceptual inspiration comes from GOFAI, Bio-AI, and fields such as dynamic systems theory and complex adaptive systems (CAS), where emergence has special significance. The CAS literature abounds with abstract references to intelligence as yet another emergent phenomenon, but rarely with much elaboration.

In the brain the same type of signal (an action potential) can represent anything from the feel of leather, the taste of watermelon, the sound of Grandpa's voice, and the motor movements necessary to swing a baseball bat to abstract concepts such as baseball's infield fly rule, center domination in chess, and a Hausdorff space in topology. It all depends upon which neurons in which regions are exchanging the signals. Thus, simulated neural mechanisms should possess the needed generality to govern sensorimotor and high-level cognitive behavior, and to support the multiple levels of adaptivity that enable the latter to eventually emerge from the former.

In fact, the commitment to a neural mechanism seems to demand an evolutionary design process. Essentially, the basic structure of brains is beyond the design capabilities of standard engineering. Brains tend to have modules, but these are very tightly interconnected, with tens or hundreds of thousands of projections between one another. This violates most principles of engineering design, particularly software engineering, which prefers well-encapsulated modules with a limited number of signaling pathways. Although neuroscientists often characterize biological neural networks as box-and-arrow diagrams, they are well aware of the extreme complexity of connections both within and between the boxes, and the strong contribution of this spaghetti wiring to intelligence.

Although duplicating the complexity of an entire mammalian brain seems far-fetched (at least today), the basic pattern of highly interconnected neural modules is feasible on a smaller scale, in artificial neural networks with thousands (or even millions) of neurons instead of 100 billion. So Bio-AI can set its sights on large, but not necessarily life-size, neural networks and attempt to find useful topologies in this intermediate-size class. For such networks, designing useful connection patterns by hand seems daunting. And even if the topology is handmade, the weights between nodes are nearly impossible to hand code and must either be acquired by experience via supervised learning algorithms or discovered by search techniques, where simulated evolution is one of the most popular for the job.

Evolutionary algorithms can function as a replacement for (or another level of adaptation on top of) the ANN learning algorithms. They permit the exploration of a wide design space of topologies. In addition, ANNs with recurrent connections (from downstream neurons back to upstream neurons) are so difficult to train with standard learning techniques that many researchers use EAs instead to evolve proper weight vectors. Since real brains exhibit extremely high recurrency, which many neuroscientists believe to be a critical foundation of cognitive processes such as attention and learning, synthetic brains probably need this topological trait as well.

In short, to find proper topologies and weight vectors for complex neural networks, evolutionary algorithms are extremely useful. Again, this may not be a coincidence, since natural evolution takes much of the honor for the amazing complexity of our own brains. Hence, to appreciate Bio-AI's prospects of achieving high-level intelligence, familiarity with ANNs and EAs seems prerequisite; they are ubiquitous in the field. Several of this book's chapters introduce these concepts and expound upon their contributions to emergent intelligence.

1.6   Deciphering Knowledge and Gauging Complexity

A huge advantage of logical and other declarative approaches to knowledge representation lies in the ease with which humans can analyze and interpret not only the results of the system but the knowledge content and inference process as well. The system's intelligence is easily explained in human-friendly concepts, since these often compose the set of primitive terms for the representation.

Unfortunately, the secrets of success in neural systems are much less transparent. Though the overt behavior of a neural system—for example, in terms of the motions taken by an ANN-controlled robot—is readily comprehended, the actual *reasoning* done by the net yields few explicit clues, merely large vectors of activation values and synaptic weights. These activity and strength readings typically arise from simple local behaviors such as signal spreading and Hebbian learning (described later), but the emergent knowledge/competence of the network resides in the global patterns of both firing activity and synaptic influence. Additional tools are required to find and evaluate these nonlocal structures, which often span both space and time.

Information theory (Shannon, 1948; Cover and Thomas, 1991; MacKay, 2003) has become a popular framework for detecting global patterns in complex networks of simple, locally-interacting components. It helps pinpoint interactions of a semicausal (or at least correlational) nature, wherein the patterns in one spatiotemporal network region can predict those in another. This basic recognition of relations between patterns above the local level yields valuable insights into network functionality. It now becomes possible (at least in theory) to prove that the thought of Grandma, though distributed across the activation values of many neurons, has a strong tendency to evoke thoughts of her popular oatmeal cookies.

Summing over the many pattern-based interactions supported by neuron populations, researchers can now formalize the complexity of entire networks. These metrics help inform comparative neuroanatomical studies across species and across developmental stages, giving formal, mathematical accounts of, for example, why humans handle symbol processing better than chimps, or why creativity often declines with age. Thus, information theory allows the fruits of emergent POE processes to be recognized, and their complexity appreciated.

Furthermore, just as the action potential serves as a common currency for information exchange between all parts of the brain, whether perceptual, motor, or deeply cognitive, the concept of information applies not only to the entire neural network but to the body and environment as well. Hence, information theory formalizes the entire scope of situated and embodied intelligence, allowing an evaluation of emergent complexity in the complete supersystem. This book examines several of the information-theoretic metrics that can help scientists recognize the emergent complexity of intelligence.

1.7   Simon's Great Ant

Today, much of AI involves building sophisticated engineering systems, regardless of their relation to human intelligence. However, in its infancy, AI was tied very tightly to fields such as cognitive science. In fact, two of the (small group of) early pioneers of both disciplines were the same Carnegie Mellon researchers: Allen Newell and Herbert Simon. Their psychological and computational studies laid the groundwork for both fields. They worked extensively with logic-based reasoning systems (Newell and Simon, 1972), which were highly regarded models of natural intelligence.

A philosophical cornerstone of Newell and Simon's work was the *physical symbol system hypothesis* (PSSH), which states that physical symbol systems (e.g., logical theorem provers) provide both necessary and sufficient bases for intelligent behavior. Though the sufficiency half of PSSH seems obvious—many physical symbol systems exhibit intelligent behavior—the claim of necessity raises many eyebrows; few people believe that an intelligent system *must* have a PSSH as its foundation.[3] As further proof of his academic versatility, Simon also authored one of the most influential books on emergent intelligence, *Sciences of the Artificial* (1996), another publication that helped popularize ants in the AI community.

Anyone who has studied or implemented logical (or more generally, symbolic) reasoning systems appreciates their complexity. In fact, they are quite a bit harder to code than many types of ANNs. Yet, Simon views the complexity of human behavior as the work of a complex environment forcing the moves of a simple (but presumably symbolic) controller (i.e., brain), in much the same way that a (relatively dumb) ant traverses an intricate route because of the complexity of the surrounding terrain. Clearly, reconciling the *necessity* half of the PSSH with the wandering-ant view of intelligence requires a deeper explanation. How can anyone claim that an intelligent system requires symbol processing if it is simply being shoved around by its environment?

Simon rescues the argument by including an agent's memories as part of the environment; his research into human expertise shows that proficiency at mental tasks primarily depends upon a huge, well-indexed store of experiences, not upon a powerful search procedure. Constructing and efficiently accessing those memories would then be the major job of the symbol-processing system. In short, trial and error is fine as long as the choices of the trial stage are far from random but are rather based on well-indexed memories of previous problem-solving episodes. In fact, Simon (1996) equates problem solving with selective trial and error:

> All that we have learned ... points to the same conclusion: that human problem solving, from the most blundering to the most insightful, involves nothing more than varying mixtures of trial and error and selectivity. The selectivity derives from various rules of thumb, or heuristics, that suggest which paths should be tried first and which leads are promising. (195)

This insight extends quite naturally to all three levels of adaptivity, since evolution, development, and learning all exhibit strong signs of trial-and-error search

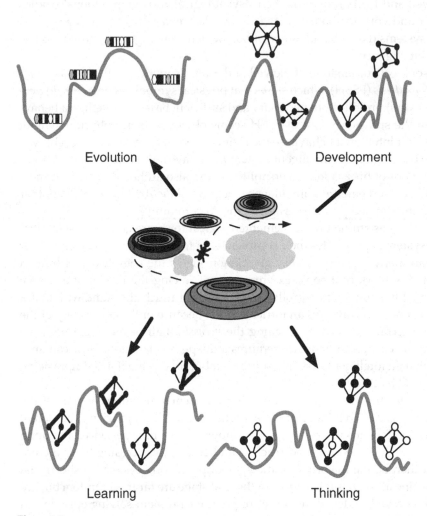

Figure 1.2
Four different types of search that underlie emergent intelligence, all driven by landscapes formed by
the system's history. (*Top left*) Evolution involves search in the space of possible genomes, with landscape height mirroring the fitness of the produced phenotype. (*Top right*) Brain development searches a
space of possible configurations (of the resulting neural network), all supported by the same genome,
but only some of which are fully compatible with the ontological environment. (*Bottom left*) Learning
consists of search among the space of synaptic-strength combinations (depicted by connector lines of
varying thickness, representing strength) for the network topology produced by development. Here,
the environment consists of inputs from the body and external world. (*Bottom right*) Thought invokes a
search process among combinations of neural activation states (drawn as networks of filled and open
circles to depict high and low activity, respectively), with a landscape formed by the synaptic strengths
produced by learning.

molded by overriding selective mechanisms, many of which stem from historical contingencies. Just as a species' phylogeny strongly biases the viability of evolutionary change—since all new structures must properly interface with the old ones—an organism's lifetime experiences filter and modify ideas that arise during activities such as memory retrieval and problem solving. Furthermore, many aspects of development involve *structural search* wherein conduits, such as microtubules, axons, and capillaries, grow in search of preexisting targets, whose presence fortifies (and absence extinguishes or redirects) these extensions.

Figure 1.2 summarizes four aspects of emergent intelligence: the three primary adaptive levels plus the basic act of thinking. Like the ant on the beach, all involve search processes that are strongly influenced by an environment. But unlike frictionless planes in physics, these search processes provide little enlightenment when viewed as blank slates: the *context* of emergent intelligence normally matters; it forms the landscape for search, tightly constraining each process and thus strongly biasing the emergent result. As shown in the figure, the product of one search process often forms the context for the next. The genome—evolved amidst a landscape formed by both the past history of DNA sequences and the present ecosystem in which the organism must survive—combines with the developmental environment (e.g., the womb) to mold the emerging neural topology. That topology, along with external sensory and internal bodily signals, forms the context for learning via synaptic modification; and the trillions (in the human brain) of finely tuned synapses tightly constrain neurons in their search for the stable firing patterns that support conscious and subconscious thought.

1.8   Search in Carbon and Silicon

As a simple example of the parallels between search in classic AI and in biology, consider the K Queens problem, a staple in AI texts (Russell and Norvig, 2009). The goal is to place K queens on a $K \times K$ chessboard such that no two queens attack one another. The two main classes of search algorithms for solving this and many other problems are incremental and local, also known as partial-solution and complete-solution methods, respectively.

In incremental search queens are placed on the board one by one. Whenever a newly placed queen attacks another, the placement is modified. If all possible placements for the queen create attacks, then the algorithm backtracks to a previously placed queen and tries to reposition it. Failure to do so results in further backtracking. This combination of sequential placement of queens intermixed with backtracking and repositioning eventually produces a complete solution: an attack-free cell choice for all K queens. An animation of this process shows a vine growing from the start state (representing a chess board without queens) at the top of the screen to deeper sites (where depth mirrors the number of queens on the board), then retracting to higher points, then plunging again in a different direction. This combination of upward and downward movement of the vine's tip continues until a state with K nonattacking queens is found, somewhere near the bottom of the screen.

This animation closely resembles real movies of axonal growth cones searching for target neurons. In these fascinating glimpes of neural network formation, immature neurons emit many neurites, which grow in all directions. Each extends, then retracts, then reorients and extends again. In this case, there are many vines. However, only those finding the most promising paths to target neurons extend large distances (a few millimeters) and eventually form synapses with the targets. Initially, many synapses form, but as the brain matures, many are pruned, leaving only the essential connections. There is no master plan for the network's structure, just a general ability to explore by growth, retraction, and regrowth.

The second major search paradigm, local search, involves the (rather immediate) generation of complete solution attempts followed by a simple loop among two actions: (1) testing the solution(s) for proximity to the goal, and (2) tweaking the solution(s), often in random or only mildly intelligent ways. For instance, a local search approach to K Queens begins by randomly placing K queens on the board, often one per row. The number of attacking queens then provides an estimate of the solution's quality. Tweaking entails moving one or a few queens, normally those involved in attacks. Amazingly enough, for problems such as K Queens, if tweaking involves either random repositioning or weak heuristics (such as moving a queen to the column in its current row that causes the least number of attacks), the local search algorithm can regularly find solutions, even for K in the millions (whereas the incremental method has problems with more than 25 or so queens). Just tossing (solutions out there) and tweaking can go a long way.

The activity of neural networks during recall gives this same local search impression. Driven by parallel updates of each node, complete patterns of neural activity, often spanning an entire network, repeatedly morph into new forms and gradually transition to states that are most compatible with the constraints imposed by the network's existing synaptic strengths, which are strongly determined by previous experiences. Thus, past knowledge exerts a form of selection upon the evolving activity patterns, sculpting them into a form that matches the constraints. This knowledge supports an otherwise *unintelligent* search process, akin to the wandering of a dumb ant along a complex terrain, which itself is analogous to the intricate collection of synaptic strengths.

1.9   Traveling Light

In summary, the relations between AI and biological intelligence have taken many forms over the years. Early AI systems were inspired by many psychological findings, and popular AI concepts (e.g., rationality, satisficing and heuristics) and tools (the von Neumann computer in general and logical reasoning systems in particular) significantly impacted theories of human cognition. Today, a newer brand of AI, motivated by neuroscience and evolutionary theory, provides a more plausible model of natural intelligence while also producing a good many sophisticated problem-solving systems that rely heavily upon stochasticity and emergence, in stark contrast to the deterministic, proof-driven techniques of GOFAI.

Surprisingly, two of GOFAI's cornerstone concepts, search and representation, remain quite useful for understanding the emergent nature of cognition and for cultivating digital intelligences. This book promotes that perspective by examining these two concepts and their relationships to the phylogenetic, ontogenetic, and epigenetic forms of adaptivity that underlie emergent intelligence. A full appreciation of these interactions will hopefully bring science just a little bit closer to *the list*.

So buckle up for a brief journey along the exciting but bumpy low road to intelligence. There is no need to pack a lot of philosophical baggage associated with the definitions of intelligence, knowledge, rationality, and so on. This is a vacation from all that. Instead, the focus shifts squarely to neurons, under the assumption that whatever goes on down at that level plays a crucial role in intelligence. The second assumption is that multilevel adaptivity is the hallmark of intelligence. The main goal of our journey is to illuminate the contributions of neural systems to that adaptivity using a set of conceptual tools from GOFAI and Bio-AI: search, representation, emergence, and complexity. These take up little room in a suitcase, but the reader must decide if they are fitting attire for one so noble as the king of human faculties.

2 Emergence

2.1 Complex Adaptive Systems

At various points in our lives, we come to the realization that neither our parents nor our teachers know everything, that people in charge make all kinds of mistakes, and that many powerful entities decidedly do not have the best interest of the people or the planet at heart. In short, we learn that the world is not a well-oiled machine.

For the most part, things normally just work out, though no laws can guarantee it. For sure, there are plenty of laws, but there is no master plan to maximize much of anything: happiness, economic growth, traffic flow, environmental sustainability, and so on. In reality, the world works because of a complex combination of imposed constraints (e.g., traffic laws) and individual interactions (e.g., drivers swerving to avoid those who occasionally forget the laws).

From all of this, large-scale patterns and systems emerge, ones that are rarely considered optimal or well-oiled but that generally work well for at least a sizable fraction of the people. The eye-opening fact that the world works without excessive global manipulation and without many guarantees may initially disappoint us, but the understanding of how predominantly local interactions produce sophisticated global results can turn childhood dismay into adult fascination.

The field of *complex systems* or *complex adaptive systems* (CAS) specializes in these intriguing feats of self-organization in a variety of systems as diverse as the planet itself: rain forests, termite mounds, traffic jams, epidemics, global economies, climate, the Internet, immune systems, politics, migrations, stampedes, stock markets, bacterial colonies, retail distribution chains, brains, languages, metabolic pathways, and ancient civilizations.

As described by various authors (M. Mitchell, 2009; Holland, 1995; Kauffman, 1993), complex systems typically consist of many interacting components (or agents) and exhibit several of the following properties:

- *Distributed control* Nothing oversees or manages the components and their interactions.

- *Synergy* The sophistry or complexity of the gestalt greatly exceeds the summed complexities of the components.

- *Emergence (self-organization)* The global structure arises solely from the local interactions, often via positive feedback.

- *Autopoiesis* The global structure is maintained by local interactions, often via negative feedback.

- *Dissipation* The system is far from equilibrium (and thus in a low-entropy state) yet stable.

- *Adaptivity* The system can modify its own structure or behavior in order to maintain other aspects of itself.

Precise definitions of *complexity* abound but have little formal agreement. Within particular disciplines, such as computer science or physics, formalizations of the term aid in intradisciplinary assessments, but overarching characterizations, for example, to compare the complexities of a sorting algorithm to that of a steam engine, do not exist and seem somewhat ludicrous.

However, a few cross-disciplinary generalizations are in order. Complexity rarely depends upon the number of components or their individual intricacies, but rather on their interaction protocols. Though individual humans seem infinitely complex on their own, the mind-boggling complexity of society stems not from the many billions of individuals but from the billions (or trillions) of interactions among them, many of which are (even in complete isolation) highly unpredictable.

Typically, complex systems defy the prediction of global behavior from cursory analyses of components and their interactions. Surprises are commonplace, and in many cases, the only way to avoid them is to observe the system itself or run very large-scale simulations. In fact, Bedau (2008) defines a *weakly emergent* system in terms of *explanatory incompressibility*: the only way to ascertain the system's behavior is to run a complete simulation of it; no shortcuts, such as slick heuristics or solutions to analytical models, will do the trick.

Finally, *emergence* and *self-organization* are the two words most synonymous with complex systems, while many of the other adjectives, such as *nonlinear* and *autopoietic*, have straightforward ties to these anchor terms. For example, emergence and its inherent surprise often result from synergistic interactions, where the global sophistication greatly exceeds the sum of the individual agent complexities; and this superadditivity is the hallmark of a nonlinear system. Similarly, self-organization and autopoiesis are often co-occurring complements: self-assembly and self-maintenance typically require one another.

A common element in discussions of emergence is that the self-organizing global structure puts constraints upon the local structure, thus biasing component interactions in ways not immediately predicted by the local behaviors. This leads to a useful working definition of emergence:

The transition of a multicomponent system into a state containing high-level properties and relationships

- *whose causal origins involve a nonlinear (and often unpredictable) combination of the low-level components' properties and relationships, and*
- *that often add further constraint's to the low-level interactions.*

For example, convection cooling of water on a pond's surface produces vertical flows wherein warm water masses rise, are cooled, and then sink. These large-scale flows emerge from the local properties of water molecules that lose energy and are simply displaced from the surface by higher-energy molecules from warmer water. As small vertical streams form, their molecules may collide with nearby molecules, imparting vertical momentum and gradually broadening the stream. This constitutes positive feedback and leads to the rapid cooling of water.

Since water is most dense at 4°C, any cooling below that temperature produces lower-density surface water, which then halts the convection stream and reduces cooling to a (much slower) conductive process. Below 0°C, ice formation begins and further reduces the cooling rate of the remaining liquid. Thus, the local properties of water (i.e., its maximum-density and freezing points) support the emergence of the light upper layer and then ice, both of which have a negative feedback upon the vertical streams and thus upon water temperature, which they help stabilize.

In sum, one emergent structure, a convection current, accelerates the cooling rate via positive feedback. The other emergent structure, ice, provides negative feedback to maintain water temperatures (near freezing) despite much colder air temperatures.

An often-cited example of emergence comes from the renowned neuroscientist Roger Sperry (1991). As shown in figure 2.1, the round wheel pattern enables a rolling motion, which imparts circular momentum upon the individual wheel molecules. Though an elegant example of top-down constraint, this fails to illustrate the actual self-organization of the global wheel form.

Tumbleweed more convincingly exemplifies emergence. As the plant dries out and dies, it disengages from the root and is easily blown around. Being dry but still malleable, the dead plant attains a rounder form (i.e., global pattern) via repeated

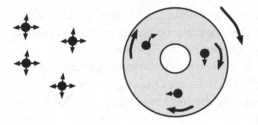

Figure 2.1
Roger Sperry's example of a wheel and the top-down effects of global structure upon the motions of individual molecules (dark circles with emanating arrows), which indicate their degrees of translational freedom. (*Left*) Molecules in isolation can move in many directions, but (*right*) inside the rotating wheel, their options narrow to those induced by the enclosing system.

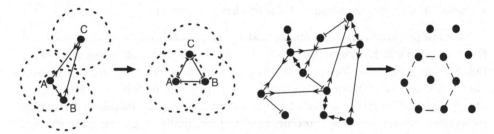

Figure 2.2
Emergence of a hexagonal lattice from simple attraction-repulsion behavior among agents (dots). Dashed circles (of radius R about each agent) denote borders between attraction and repulsion. (*Left*) All interactions between three agents are shown, with attraction (repulsion) indicated by arrows pointing to the middle (end) of the line. The attracting/repelling agents form triangles in which the resultant force vector on each agent vanishes (drawn as a line with flat ends). (*Right*) Sketch of some (not all) interaction forces among a larger group of agents, which eventually lead to a hexagonal lattice.

contact with the ground and the bending and breaking of individual stems (i.e., local changes). Eventually, the accumulated deformations round the weed to the point of being able to roll and impart top-down rotational momentum upon its molecules. Upon reaching a wetter area, the tumbleweed absorbs water and physically opens up, thus imparting another top-down momentum upon its molecules.

As a more practical example of emergence, consider the work of Spears et al. (2004), in which purely local information guides the movements of agents (representing military vehicles), which then self-organize into target global patterns, such as hexagonal and square lattices. As shown in figure 2.2, a hexagonal pattern emerges when all agents follow the simple rules of (1) repelling all agents located within a radius of R units of themselves while (2) attracting all agents outside of that circle. For instance, on the left of the figure, agent B is pushed to the right by A and pulled upward by C, yielding a resultant force vector that moves it

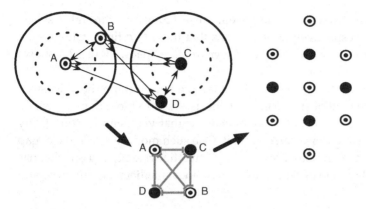

Figure 2.3
Emergence of a square lattice in a system involving two agent types, differentiated by their spin: either
up (solid black dot) or down (black-and-white dot). (*Top left*) Two concentric circles surround each agent
and define the borders of attraction versus repulsion for agents having the same (solid outer circle) and
different (dashed inner circle) spin type. For example, A and B repel one another, since B lies within a
$\sqrt{2}R$ distance (solid circle) of A; whereas A and D attract one another, since D lies outside the R radius
(dashed circle) of A. (*Bottom*) The resultant forces on each agent eventually lead to the formation of
squares, and with many agents (*top right*), a square lattice.

in a northeasterly direction, while agent C is pulled to the southwest, and agent A
to the north. Together, these forces drive the agents into a stable triangular pattern.
When many agents interact, collections of the resulting triangles form hexagons.

As shown in figure 2.3, the authors achieve the emergence of square lattices by
adding a binary spin attribute to each agent. The preferred separation distance
for opposite-spinning agents is R, and $\sqrt{2}R$ when the spins match. Squares then
emerge among groups of four agents, two of each type, with alternating types
along the perimeter (with edges of length R) and like types in opposite corners
(a diagonal distance $\sqrt{2}R$ apart). In addition, by allowing agents to occasionally
change spin, they facilitate an ingenious method of self-repair (Spears et al., 2004).

2.2 Chaos and Complexity

Complex systems are frequently characterized as those exhibiting a mixture of
order and disorder; and the *complex regime* of a dynamic system is often defined
as *the edge of chaos*: an area in state space between order and disorder/chaos. So
understanding complexity requires understanding chaos as well.

The trademark of a chaotic system is *sensitive dependence upon initial conditions*
(SDIC). Assume two separate runs of a system, one in which it begins in state s_0
and another in s_0*; these two states are distinct but extremely (even infinitesimally)

close in state space. The two runs of the system cause a transition through T states, ending in s_T and s_{T*}, respectively. SDIC entails that these two final states are *not* close: the two trajectories of the system *diverge* in state space. Figure 2.4 shows SDIC for a fictitious two-variable system.

The problem with chaotic systems is that they are nearly unpredictable, since the precise value of the initial state can rarely be assessed to more than a few significant digits. The mathematical beauty of chaos is that very simple, completely deterministic, nonlinear systems can exhibit it. One such model is the logistic map of equation (2.1), one of a few classic examples of chaotic systems, and one that has applications in a wide range of fields, such as ecology, medicine, economics, and neural computation.

$$x_{t+1} = rx_t(1 - x_t). \tag{2.1}$$

The gist of the model is simple: the next value of x, (x_{t+1}), is the product of a rate constant, r, the current value of x, (x_t), and 1 minus that value. x_0 is assumed to be between 0 and 1, and the model insures that x never goes outside those bounds. The choice of r strongly influences the trajectory of x in the following manner:

- $r < 3$ The system quickly transitions to a value that remains stable from timestep to timestep, a *point attractor*. Starting values near one another all transition to this attractor.

- $3 \leq r \leq 3.57$ The system transitions to an infinite loop between k different values, a *periodic attractor* of period k. When $r = 3$, that period is 2. Then, as r increases beyond 3, the system goes through various splits or *bifurcation points*,

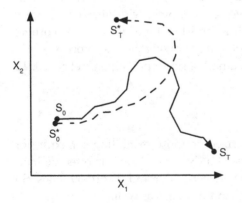

Figure 2.4
Simple illustration of SDIC in a system composed of two variables: X_1 and X_2. Though initial states s_0 and s_0* are very close, the trajectories from each quickly diverge.

where k doubles. The frequency of occurrence of these bifurcations increases as r approaches 3.57. Thus, k doubles at an ever-increasing rate, and the cycles become harder to detect, since they involve more and more states. Again, nearby starting values tend to gravitate toward the same periodic attractor.

- $r > 3.57$ The system never settles into a discernible point or periodic attractor, and nearby starting points no longer follow similar trajectories: SDIC arises.

In short, small values of r produce an ordered state, whereas those above 3.57 produce chaos. To illustrate SDIC in this state, figure 2.5 shows two trajectories that begin with a mere one-millionth separation but gradually diverge. Both trajectories are nonrepeating.

Values of r near 3.57 represent the *complex regime* or *edge of chaos*. Prominent researchers such as Langton (1989), one of the founders of ALife, have shown that within this regime, a system is best able to perform computation: the storage, transmission, and modification of information. Others, such as Kauffman (1993), another ALife pioneer, indicate that systems often transition (or evolve) naturally to this regime.

This natural transition is best exemplified by the hypothetical sandpile of Bak et al. (1988). Imagine dropping single sand grains onto a pile, always from the same point above the pile, and always just one grain at a time: the external perturbation to the pile is both small and constant. Initially, each grain will just fall to the table, but eventually the pile will develop a peak with steep slopes. At this point, the addition of each new grain can potentially cause avalanches of various sizes (i.e., displacing varying numbers of sand grains). But when the sizes of those avalanches are plotted, it turns out that the vast majority of perturbations cause small avalanches, while an exponentially decreasing number of larger avalanches occur. Still, some very large avalanches do occur. This exponentially

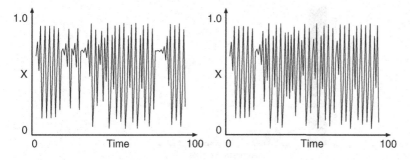

Figure 2.5
Two applications of the logistic map, equation (2.1), with $r = 3.9$ and initial values differing by only one-millionth. The trajectories diverge after approximately 30 timesteps.

declining frequency curve is known as a *power law distribution*, and it is an often-cited signature of a system in a *critical state*.

Figure 2.6 gives the plot of a typical power law curve, along with a cartoon example of a distribution of avalanche sizes for a sandpile as described by Bak et al. (1998). For this and many other systems formally deemed *complex*, the frequency distribution of values for some important system variable (x) is the basis for the power law distribution expressed in equation (2.2). For example, if x denotes the size of a sandpile avalanche, then cx^k is the frequency of occurrence of avalanches of that size.

$$f(x) = cx^k. \tag{2.2}$$

As long as the perturbations are small and constant, the occurrence of a power law distribution over the *internal effects* of the disturbances is often grounds for the following type of claim:

> Most perturbations lead to small changes in the system, indicating that it is pre-dominantly *stable* or *ordered*. But occasionally, a small perturbation can produce a large change, indicating a more *unordered*, or *chaotic*, situation. The mixture of the two, order and chaos, with the former dominating, is a strong indicator of com-plexity in the system.

In addition to formalizing the critical state, Bak and his colleagues argued that many systems will naturally transition to that state of their own accord, with only minor external assistance (such as the dropping of sand grains onto a pile). They are emergent in the same way that a convection current emerges from the internal physicochemical dynamics of water, combined with an external temperature per-turbation. The term *self-organized criticality* (SOC) (Bak et al., 1988) expresses both the emergent and edge-of-chaos aspects of this important phenomenon.

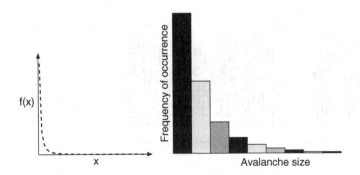

Figure 2.6
(*Left*) Plot of the function $f(x) = cx^k$ with $c = 1$ and $k = -2$, a standard example of a power law curve. (*Right*) Sketch of a power law distribution for the avalanches of a typical sandpile.

A common mathematical construct for assessing order, chaos, and complexity is the *Lyapunov exponent*, λ, in equation (2.3), where $\triangle x_0$ and $\triangle x_t$ are the initial and final separations, respectively, between two trajectories of a system.

$$\triangle x_t = \triangle x_0 e^{\lambda t}. \tag{2.3}$$

Intuitively, when $\lambda < 0$, the trajectories converge; the system is ordered. Similarly, when $\lambda > 0$, the systems diverge, indicating SDIC and chaos. The complex regime is that (often narrow) band of parameter space where $\lambda = 0$ and trajectories neither diverge nor converge. For systems with large state space vectors, this would seem to be a nontrivial band to find, although influential complexity theorists such as Kauffman (1993) support Bak's view that many systems will almost inevitably approach this edge-of-chaos area.

Throughout this book are examples of systems that appear to inhabit the complex regime as evidenced by a mixture of orderly and chaotic behavior, where order dominates but chaos has just enough presence to support the changes that underlie adaptation.

2.3   Levels of Emergence

Some of the more thorough philosophical investigations into emergence come from Terrence Deacon (2003; 2012) who describes three different levels, ranging from basic multiscalarity to those invoking notions of *memory/representation* and *self*.

Related to these levels are *synchronic* and *diachronic* aspects of emergence, the latter involving global pattern formation over time, and the former, changes of spatial scale without temporal progression. For example, temperature emerges synchronically at the macro level from energetic molecular interactions at the micro level. This emergence does not unfold over time but merely becomes evident to the outside observer in the shift of spatial perspective and equipment from the microscope to the thermometer. Conversely, pond freezing exhibits diachronic emergence through the temporal sequence of conduction, convection, conduction, and ice formation.

Deacon's three levels or *orders* of emergence are more elaborate. First-order emergence is primarily synchronic and involves *aggregates* of low-level interactions that produce higher-level properties. Temperature from molecular interactions, color from molecular light absorption properties, and weight from atomic mass are all examples of first-order emergence. This type of emergence is often devoid of surprise, with high-level properties being easily predicted from a statistical analysis of the underlying micro properties and events. Stochastic perturbations of first-order systems tend to *average out* at the macro level, as opposed to being magnified by positive feedbacks; and the global patterns exert no obvious constraints upon local dynamics.

Although this is a fairly mundane type of emergence—essentially a base case—it exhibits some degree of explanatory complexity in that global patterns do not follow from local interactions by simple causal arguments. Tools such as statistical mechanics, which average over the micro events, are needed for the logical derivation of global effects from aggregated local causes.

With second-order emergence, the fun begins. In these scenarios, as sketched in figure 2.7, stochastic microlevel perturbations (asymmetries) can become amplified into global patterns in a diachronic manner. A *recurrent causal architecture* also comes into play as these global structures bias local behaviors. These systems are both self-organizing and autopoietic (self-maintaining), and they involve a significant mix of bottom-up and top-down causality, i.e., local interactions affect the global pattern, and vice versa. These systems are *dissipative*, i.e., far from equilibrium with material and energy flowing through them, but the global structure (the low-entropy formation) is *recirculated*, amplified, and often eventually stabilized. This stability of form engenders notions of self: some hallmark of the system or system class that is created and maintained. Most examples of emergence in the popular complex systems literature are of this type.[1]

Figure 2.7 paints an abstract picture of second-order emergence, which is nontrivial to visualize. At the top, notice that individual components begin with many degrees of freedom. This entails a large *search space* of possible actions, a space that gradually constricts as the global pattern emerges and exerts top-down influence upon the local components. With each component restricted to only a few possible actions, the global pattern (a loop in this case) becomes evident. Although purely physical activities, such as the Brownian movement of molecules, rarely evoke this notion of search, when system components are cellular or multicellular the metaphor begins to feel more appropriate.

In third-order systems the emergent high-level structures are imprinted into another structure: the *memory* or *representation*, which can function as a *seed* for another round of second-order emergence, producing another, similar high-level structure. Essentially, this facilitates reproduction and evolution of the self. This seeding is a *re-presentation* of the memory to the lower-level dynamics, which catalyzes second-order emergence.

Third-order emergence involves a synergy between self-organization and representation. The latter implicitly houses a memory of the self-organizational process, which, in turn, can create new representations. These representations may be disseminated and subjected to differential selection pressures (via different microlevel environments) and the resulting heterogeneous replication frequencies.

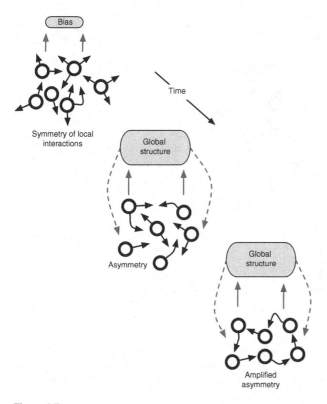

Figure 2.7
Essence of second-order emergence. (*Top*) From predominantly unbiased (symmetric) local interactions, a small amount of bias arises randomly. (*Middle*) Positive feedback accentuates this bias, forming a global pattern that provides significant constraint upon the low-level behaviors (i.e., increasing asymmetry). (*Bottom*) Negative feedbacks eventually kick in to stabilize the local dynamics and global structure. Black arrows denote degrees of behavioral freedom of low-level components, while lighter arrows indicate bottom-up or top-down influence. The global structure and bias labels only indicate *conceptual* additions to the situation; physically, these structures reside within the aggregation of the low-level components.

Thus, third-order emergence facilitates evolution by natural selection, where the representation is analogous to DNA.

Furthermore, any adaptations of the system to the environment will involve changes to the dynamics, some of which may be reflected in the emergent global pattern and thus encoded in the representation. In this way, the representation becomes an implicit memory of a history of environments (via the adaptations necessary to survive in them). For example, any animal's genome serves as a history of its adaptations, a reflection of a long chain of salient environmental factors—ambiguous, noisy, and woefully incomplete, but a history nonetheless.

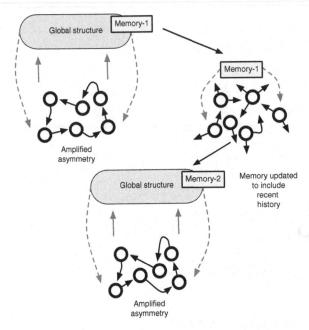

Figure 2.8
Essence of the re-presentation process underlying Deacon's third-order emergence, where a memory structure arises as an additional result of second-order emergence (*top*). When placed in a new environment (*middle*), this memory serves as a seed for regenerating the global pattern while potentially incurring changes to itself (*bottom*).

Figure 2.8 illustrates the basic process of third-order emergence, wherein the representation is built in parallel with the global structure. Then, when transferred to an unordered environment, i.e., one where the local interactions have few constraints, the memory provides a bias leading to both the rejuvenation of a high-order pattern and the biasing of the low-level dynamics. Since this new situation may not be identical to the original, the representation may also change to more accurately reflect the new scenario.

The representation acts as *information* in the formal, information-theoretic sense that when added to a system of relatively unconstrained (high-entropy) local interactions, it provides a strong bias, thereby promoting structure and reducing entropy; information content equates with induced entropy reduction (see chapter 13).

2.3.1   Deacon's Autogen
Deacon (2012) illustrates second- and third-order emergence with an abstract biochemical system called an *autogen*, which resembles a classic early model of

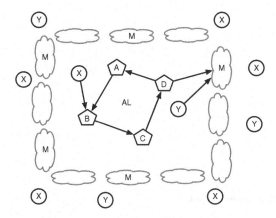

Figure 2.9
Basic structure of Deacon's autogen, with an autocatalytic loop (AL) that requires raw materials
(X and Y), and is surrounded by a membrane, whose building blocks (M) are produced by AL. Arrows
from one symbol to another indicate that the former produces the latter; multiple input arrows into a
product (e.g., B) from reactants (e.g., A and X) imply that B requires both A and X for its production.

autopoiesis (Varela et al., 1974). As shown in figure 2.9, an autogen consists of an
autocatalytic loop (AL) surrounded by a membrane. The loop and membrane are
symbiotic, since (1) a by-product of the loop is membrane building blocks (M), and
(2) the membrane prevents chemical constituents of the loop from dissipating into
the environment. Though many components of the AL are produced from other
AL constituents, a few *raw material* chemicals (e.g., X and Y) are also required to
perpetuate operation.

The autogen's structure emerges through a second-order process in which various compounds come into contact and begin producing other compounds, including M and other molecules of the AL. Positive feedbacks include (1) the ability of
M molecules to partly restrict the movement of the AL compounds, thus helping
to prevent their immediate dispersal, and (2) the eventual formation of the complete AL loop, allowing compounds A, B, C, and D to produce one another (in the
presence of X) and M (using Y).

Once the complete membrane and AL have formed, negative feedbacks help to
stabilize the structure. For example, since X and Y are only consumed (not produced) by the AL, their internal concentrations will wane without periodic resupply, which turns out to be an emergent property of the autogen. As the AL loop
begins to slow down, production of M cannot keep pace with natural deteriora-
tion of the membrane, and gaps appear (figure 2.10). These allow X and Y to flow
into the cell, thereby restimulating the AL, producing more M, and mending the
membrane.

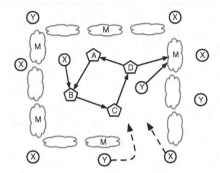

Figure 2.10
Resupply of X and Y via gaps in the deteriorating autogen membrane

In short, the positive feedbacks involving AL and the membrane lead to self-organizing autogen formation, while the negative feedbacks insure autopoiesis.

Furthermore, if the membrane splits in two places, creating two pieces, then each piece may partly surround enough of A, B, C, D, X, and Y to have a private AL, which will then generate a complete new membrane for each, thus producing two copies of the original autogen. So reproduction can also be a second-order emergent phenomenon.

Conceivably, an AL could become modified, for example, by replacing compound D with E. If this produced a more efficient AL, possibly one requiring fever raw materials to produce the same amount of membrane, then it might have a survival advantage in terms of having less frequent membrane tears. Of course, more tears could lead to faster reproduction, so an inefficient AL might actually be a more prolific copier. Either way, the rudiments of an evolutionary process are already evident in this second-order emergent system.

Autogens exhibit third-order emergence with the addition of a representation in the form of a chemical template. In general, chemical reactions are enhanced by mechanisms that increase the proximity of their reactants. For example, by elevating the odds that A and X molecules meet (figure 2.11), production of B should increase. If molecules A* and X* exist and bind to A and X, respectively, and if A* normally binds to X*, then the presence of an A*-X* compound should promote A-X proximity and B production. Thus, A* and X* serve as building blocks for a template that (1) promotes AL formation, maintenance, and productivity, and (2) embodies a memory of the AL such that the infusion of templates into a well-mixed chemical soup can accelerate autogen emergence. The information embodied by the template greatly biases the low-level dynamics of the chemical soup, producing structures (i.e., autogens) much faster than they would normally arise.

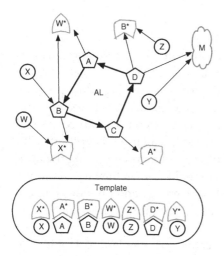

Figure 2.11
Use of an autogenerated template to promote spatial proximity of the reactants of the autocatalytic loop (AL). Reactants (e.g., A and X) come into contact when binding with their respective template complements (A* and X*), thus catalyzing the production of B.

Third-order emergence requires that the representation emerge as well. Although many of the building blocks may arise naturally in the chemical milieu, there is a decided selective advantage to any autogen whose AL produces some or all of them. For example, as shown in figure 2.11, the AL produces four of the seven template blocks: A*, B*, W*, and X*. Clearly then, autogens that produce templates should have an increasing frequency in the population compared to purely second-order *species*, and each template modification that improves AL efficiency should give a selective advantage to these representations.

Templates, like other representations, are structural embodiments of the dynamic processes that underlie self. They play the informational role of catalyzing activities that reduce entropy. And (as has probably not escaped your notice) they behave a lot like DNA. Indeed, Deacon's third-order emergence introduces the genotype-phenotype distinction.

Deacon's three levels of emergence provide useful guidance as one investigates emerging intelligence across another three levels, those of adaptivity: evolution, development, and learning. As shown in later chapters, the levels of emergence and those of adaptivity frequently overlap. For example, development and learning exhibit considerable second-order emergence, whereas Deacon's third level applies not only to evolution but, in some cases, to learning.

2.4   Emergence and Search

Have you ever wondered why migrating birds often fly in a V formation? This is
a classic case of emergence that begins with an important fact of physics: when a
bird flaps its wings downward, it pushes air downward, creating a low pressure
region above one of higher pressure (because of the added air). A split second after
this happens (as the bird moves forward), this pressure difference produces a small
updraft. A second bird, trailing behind and to the side, can use this updraft to lift
its own wing, thus saving some energy. In addition, if birds prefer seeing the horizon ahead of them instead of the backside of another bird, then they will naturally
seek out locations in the moving flock that both save energy and provide an unobstructed view. When all birds have these motivations, the group self-organizes into
a flying wedge.

As illustrated in figure 2.12, from the perspective of a single bird looking for
an optimal location, the initial, unorganized phase affords many options which
constitute a broad, unconstrained, private search space for the bird. However, as
the global V pattern emerges, it embodies a top-down constraint, operationalized as a restriction of the bird's search space. So whereas the bird can randomly choose among many options in the unorganized state, its moves become
forced as the wedge materializes. The global pattern manifests a lot of local
constraint.

As shown in figure 2.13, this transition from an open to a restricted search space
is even more clear for schooling fish, which conserve energy by swimming directly
behind one another and appear to follow a few local behavioral guidelines: head

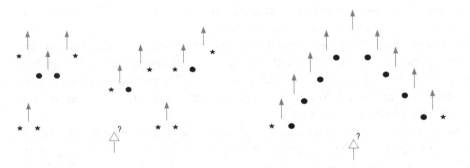

Figure 2.12
Emergence of a V pattern in flying birds (arrows), each of which creates pockets of uplifting air behind
and to each side. Stars denote unused (by a trailing bird) pockets with an unobstructed view; circles
represent unused pockets with a poor view. (*Left*) Prior to V formation, many unobstructed pockets
exist, which provides many options for the focal bird (open arrow). (*Right*) In the V the only optimal
options for the focal bird are at the edges, and these promote extension of the V, a positive feedback.

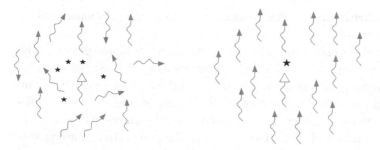

Figure 2.13
Emergence of schooling in fish. (*Left*) In the unorganized state a focal fish (open arrow) has many move-ment options (stars). (*Right*) The organized state offers few degrees of freedom, since moving anywhere but the starred location could produce a collision.

in the same direction as your neighbors while avoiding collisions. The same basic idea pertains to human scenarios as well, such as swimming in crowded pools, driving in congested freeway traffic, and finding your favorite concession stand at halftime of a football game. The important point is that decisions that may appear quite intelligent involve no sophisticated knowledge or planning. They are merely reactions to an emerging set of local constraints, which often make the correct movement choice obvious. In short, as situations accumulate constraints, even the most intelligent of beings exhibits little more problem-solving wizardry than the ant on the beach (Simon, 1996).

Emergence in systems of decision-making agents can therefore be viewed as an interaction among parallel search processes. The initial trial-and-error dynamics of each such process enable portions of the system to randomly jiggle into small-scale configurations—for example, short diagonal lines and small wedges among the migrating birds—where positive feedbacks commence, thus forcing the moves of birds near these patterns. This allows the small patterns to grow into large-scale patterns, which then exert natural restrictions on every individual search process. Without the trial and error, there is no jiggle, and the system can quickly stagnate in a highly ineffective state. Imagine what would happen if every bird, fish or, human simply chose a movement sequence ahead of time and stuck to it, come hell or high water.

From an observer's global viewpoint, the group as a whole might appear to be searching for an efficient pattern, though no overarching goals or instructions guide such a transition. Still, the system gives the impression of trying different alternatives and eventually becoming *attracted* toward particular stable configura-tions. A similar interpretation holds when we observe populations of noncognitive agents, such as neurons. Each has a repertoire of possible behaviors that becomes constrained by the behaviors of other neurons, but the neuron obviously has no awareness of these options.

Despite the lack of cognizant deliberation by molecules or neurons and by the aggregates that they form, the *search perspective* provides useful insights and predictions when analyzing such systems. Of course, at the highest levels, the brain itself performs cognitive acts, but the choices that organisms make are not explicitly about neural patterns, i.e., there are not specific target patterns of neural firing toward which organisms explicitly strive. Yet, from the perspective of an outside observer examining patterns of neural firing, the system can appear to naturally transition to certain attractive states, cycles, or semirecurring patterns. Just as in physics, where systems are often described as *seeking a low-energy state*, the search metaphor provides useful explanatory leverage in any population of interacting entities, each of which has several degrees of freedom. Then, in cases of emergence, these degrees may change as global patterns form and evolve, giving the outward appearance of a system seeking a particular ordered state.

In emergent systems individual agents make local choices, thus affecting a small subset of the global situation. However, unlike many localized design decisions in AI search (see chapter 3), those in emergent systems are not tested in isolation but *in vivo*: in direct interaction with other system components. Through an understanding of the interaction topologies and behavioral repertoires of the agents as well as the changing constraints, one can begin to attach probabilities to the agents' actions and to differentiate between situations in which strong constraints *force* agents' choices and those in which agents possess a legitimate freedom to explore.

It turns out that this exploratory freedom of individual agents is the foundation of natural adaptivity—a foundation composed of an extremely malleable substrate, one that tends to bend without breaking. The ability of living systems to modify themselves to accommodate external change relies heavily on the ability of their components to exhibit fairly explicit trial-and-error behavior. As the basis of adaptivity, parallel search among multiple agents is pivotal in the emergent complexity of life and intelligence.

For example, the development of the nervous system is a highly adaptive process in which brain hooks up to body, not by carrying out a genetically specified topology-building plan (i.e., connect neuron A to neuron B, then B to C and D) but by individual neurons performing various search processes, such as sending out axons in search of dendritic targets. Those that find a dendrite to synapse onto may survive, while failure to connect can spell death. Similarly, during cell mitosis, a pair of centromeres send out exploratory microtubules *searching* for chromosomes with which to attach. Those that hook up will form part of the mitotic spindle; the rest will wither away. This is exploration, trial-and-error search within individual cells, that enables the centrosomes to find all the chromosomes, wherever they reside. The cell adapts to random chromosome locations because the centrosomes can search/explore in the same way that the nervous system adapts to varied locations of muscle cells via the exploratory axons of motor neurons.

2.5   Emergent Intelligence

Although many different properties of complex systems have already been discussed, Melanie Mitchell (2006) provides another four (all motivated by a network perspective) that seem very relevant for the emergence of intelligence:

- Global information is encoded as patterns over the component states.
- Stochasticity abounds and is often magnified by positive feedback.
- Parallel exploration and exploitation at the level of individual agents is ubiquitous.
- Top-down and bottom-up interactions co-occur continuously.

The brain exhibits these in many different ways; a single example of each (with more in later chapters) should illustrate the strong conceptual bonds. First, neural representations are commonly believed to involve population coding: concepts such as tumbleweed are encoded by large combinations of active neurons, not the individual cells alone. Second, random neural activations occur continuously in the brain, and these often employ positive feedback to elicit large-scale firing patterns, such as the synchronous activity common to brain oscillations like theta and gamma waves, which are critical for memory formation and retrieval (Buzsáki, 2006). Third, during development, neurons grow axons that independently explore the environment for other neurons with which to connect. In this search for efferents, links to already well-connected neurons have a selective advantage, since this exploitation of proven networkers enhances the new neuron's chances of participating in active circuits and thus surviving. Finally, the brain's interpretation of sensory information relies on a healthy mix of bottom-up (sensory-driven) signals that interact with top-down (expectation-driven) signals from higher cortical regions.

Most popular accounts of complex systems, such as that of Melanie Mitchell (2009), include the brain in their list of examples, with intelligence portrayed as an emergent result of billions of interacting neurons. Unfortunately, intelligence is a rather vague concept, and certainly not one easily described by a particular global pattern, although, scientists have attempted to define concepts such as consciousness in terms of information-theoretic metrics and the neural activation patterns that satisfy them (see chapter 13). However, a detailed analysis of the many neural structures and mechanisms (along with their origins) that underlie intelligent behavior reveals a parade of self-organizing systems. These span the three primary levels of adaptation, and at each level emergence relies heavily upon a search process.

From an evolutionary perspective, brains have emerged along a million-year time scale via the gradual expansion of a very distant ancestral genotype (by duplication and modification of successful genetic components). Evolution has devised ingenious developmental recipes for converting genotypes to phenotypes. Although evolution is often characterized as random, the concept of selection pressure and its ability to steer evolution via genotypic adaptations suffices to convince that evolution has the ability to craft complex organisms that purely random processes could never achieve. In short, evolutionary search is highly biased, favoring some genetic and phenotypic variations over others.

Still, many biologists view the processes by which variations arise (prior to selection) as random, with selection then separating the wheat from the chaff. However, the popular new theory of *facilitated variation* (Kirschner and Gerhart, 2005) indicates quite the opposite: the processes that generate variations are also highly biased. Essentially, millions of years of evolution have hit upon complex combinations of genes and versatile genotype-phenotype relations that greatly improve the odds that random mutations will produce viable phenotypes, thus giving selection a very high-quality pool of applicants. It seems that the DNA representations of natural third-order emergence are extremely informative and robust. Clarifying the source of this robustness can enhance our general understanding and appreciation for evolution as a designer, and suggest feasible strategies for artificial evolutionary searches for intelligent systems. This permits an interpretation of the evolutionary search process as, in fact, quite well informed and strongly biased to choose routes that history has preselected by virtue of the constraints gradually molded into the genotype-phenotype mapping.

Developmental processes themselves exhibit considerable self-organization, as a multitude of embryonic cells send and receive chemical signals, many of which strongly affect behavior and lead to the formation of global structures. As discussed later, exploratory search, involving moving molecules and cells along with their connective *tubing* and *wiring*, plays out in many parts of the body and brain. Without this developmental search and the *grow-to-fit* assembly paradigm that it engenders, the genome would become more a blueprint than a recipe and lose most of its robustness along the way. An examination of the second-order emergence in developmental search helps explain how instances of apparently improbable simultaneous coevolution of many traits are actually the modification of one trait followed by a host of phenotypic adaptations to that perturbation, all made possible by the flexible search processes implicitly encoded by the genome.

Development is one of the most impressive emergent phenomena in existence, as the complex global structure of the organism arises from an intricate interplay among a wide variety of cells.

Postnatal modifications are often considered examples of phenotypic *plasticity*, with classic examples being the increase of muscle mass via weight lifting, the increase of red blood cell count by moving to a higher altitude, and the modification of synaptic strengths in the brain (often equated with learning). Whereas development lays down the main components and their connections, plasticity/learning involves smaller, more local tweaks to that infrastructure. In the brain development involves the formation of large populations of neurons, along with the extensive axonal and dendritic networks linking them, whereas learning typically involves synaptic formation or tuning but may include the addition of new neurons to established populations (and their linking up to others), although only a few areas of the mammalian brain appear to support postnatal neurogenesis (Shors, 2009).

Learning also has many emergent aspects, again resulting from processes often characterized as search. For example, synaptic tuning in supervised-learning scenarios is commonly modeled as a gradient-descent search for a vector of synaptic weights that reduces the total error of a neural network's input-output mapping. Classic AI supervised-learning algorithms are not particularly emergent, because of the presence of externally provided control signals and complex algorithms with a highly nonlocal flavor. However, some brain circuitry exhibits more decentralized supervised learning from which the mature, fully functioning network gradually arises. Other brain networks display emergent reinforcement learning involving global chemical broadcasts but certainly nothing resembling global "hands of god" steering the process (as in classic AI methods of reinforcement learning). Finally, unsupervised learning is a common neuroanatomy-modifying mechanism capable of crafting complex wiring patterns that faithfully mesh with the real world's sensory statistics, and all from purely local interactions. Here, both natural and artificial versions of the algorithm exhibit considerable self-organization. In addition, the dynamic progression of firing states in these fine-tuned networks—commonly viewed as the essence of thought—is often modeled as search in the space of neural activation patterns.

Compared to the products of development, the self-organized, well-tuned network produced by learning is less impressive for its physical differences from the original (simpler, less-organized) state but more recognized for its accentuated information content, behavioral intricacy, or functional ability. In many cases, to fully appreciate the emergence within a learning system, one needs mathematics (e.g., statistics and information theory) instead of a microscope or phylogenetic tree.

Indeed, intelligence is emergent, and just as imperfect as Mom, Dad, tax laws, and the United Nations. Still, the neural machinery is oiled well enough to perform many impressive feats. The full extent of this phenomenon goes well beyond examples of sophisticated coordination arising from individual decisions, such as the ability of five basketball players to weave intricate patterns from individual movement principles, or of jazz musicians to produce lively entertainment with band members whom they've never met before, or of fireflies to blink in sync. The true beauty of emergent intelligence is spread across the three adaptive mechanisms and Deacon's three levels. And it's all about search and neurons.

3 Search: The Core of AI

3.1 Design Search

Imagine that you have been given the task of creating an origami sculpture of your best friend for her upcoming birthday party. Knowing nothing of this technique, you begin an *information search* among your friends, at the public library, on the Internet. Since no source will have the precise folding instructions for the birthday girl's face (unless you have very well-known friends), your primary goal for this search process is a set of general origami principles and techniques. In your pursuit of knowledge, many of the intermediate points are both source and pointer: friends refer to other friends, books reference other books, and Web pages link to other sites. Still, many of these sources could end up being the final target of your search: a friend may be an origami expert and agree to do the whole job himself, or a book may have all the essential folds for producing faces. And although this seems obvious, each source is an *existing, complete entity*, though it may only provide partial information toward your goal. In short, your search involves a transition among well-indexed sources (via the pointers), all of which are preexisting, and any of which could turn out to be *the* solution in and of itself (completely independent of the other sources along your search path).

Assume that you have gathered and digested all the relevant materials and are sitting before a large sheet of paper. A second, quite different process now begins: *design search*. You fold the paper many times, notice a discrepancy between the intermediate result and a never-to-be-violated principle from an obscure Japanese guidebook, and need to undo the three most recent folds. From that point, you set off along another path of bending and mutilation. Each intermediate creation is far from complete and may hint as to what succeeding folds might bring but certainly does not, on its own, point your fingers in the right direction (although an origami master may indeed get such tactile clues). Rather, you rely on heuristics (rules of thumb) based on general principles (gleaned from your information search) to give some indication of whether a given configuration is a promising stepping stone on the path to a final solution. In general, the intermediate configurations do not exist a priori: you create them and then either build further upon them, undo parts of them, or eventually recognize one of them as the final design goal.

This type of design search occurs in a space of partial solutions: most stops along the way cannot provide a respectable solution as is. Other design search protocols work with full solutions as their primitive units: each point in the virtual search space constitutes a complete (albeit often suboptimal) solution to the problem. For example, you might (methodically or randomly) craft long sequences of paper-folding operations and just try them out in their entirety without thinking too much about the consequences until the sequence finishes. Then, you would evaluate the final product, decide whether it matches the goal (i.e., your friend's face). If not, you would modify some (any) operations of the sequence, start with a new piece of paper, and carry out a new sequence to yield a new solution/attempt, evaluate that one, and then finish, or modify and repeat. Here, although each intermediate state does represent a full solution, it gives only implicit hints (guided by heuristics) as to the next move, i.e., fruitful changes to the folding sequence.

Terms such as *search* and *search engine* make the technology section headlines all the time. This is predominantly information search, the undisputed power of the Web and the process that has changed the way people work and think. Artificial intelligence is all about design search in virtual spaces of both complete and partial solutions. It, too, has changed the way people work and think (about thinking, in fact), and it also plays a huge role in the Internet.

This chapter is about design search and some of its many formalizations that have conquered challenging GOFAI problems over the past several decades. In addition, these same formalisms offer useful perspectives for understanding biology, Bio-AI, and emergent intelligence.

3.2   Classic AI Search Techniques

Now imagine a different problem: finding efficient connection patterns among a collection of sites. Figure 3.1 depicts the layout of four buildings that might compose, for example, the science complex of a small college. Assume that the college wishes to interconnect the four buildings with fiber optics in a manner guaranteeing that the computers in any building can send signals to those in another building using only the cable of this new network. Furthermore, since fiber optic cables are not cheap, the college wants to minimize the total amount of cable.

As figure 3.1 shows, some solutions utilize less cable than others; those using the least possible amount of cable are known as *minimal spanning trees* of the network, and tasks of this sort are known as *minimal spanning tree* (MST) problems. Formally, an MST instance involves a graph (G) that consists of a set V of vertices

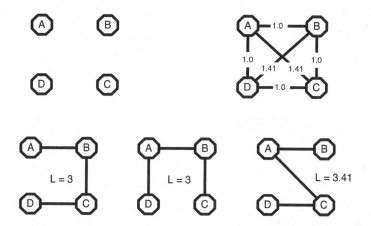

Figure 3.1
(*Top left*) Layout of buildings A–D. (*Top right*) Distances between buildings. (*Bottom*) Three possible cable topologies for connecting the four buildings, with total cable length L.

(points, sites, locations) and a set E of edges (connections between the vertices), with each edge having a corresponding weight (typically representing a cost or distance associated with linking or traveling between the two vertices). The goal is to find a subset E* of E that both minimizes the total weight and provides a *fully connected graph*: there is a path along edges of E* between any two vertices of V. Finding solutions to MST problems requires only polynomial time, namely, a run-time proportional to $|E| \times \log(|E|)$, where $|E|$ is the size of E. For example, $|E| = 6$ in figure 3.1 (top right).

The MST problem can easily be couched as a search task in a space of partial solutions, each of which consists of V and a subset of E. As shown in figure 3.2, MST search problems of this form begin with a single *root node* of a search graph (top of figure) having no edges. Movement in search space involves applying state-transforming *search operators*, which in this case consist of the possible addition of single edges to the graph. When search states consist of partial solutions, the application of operators is known as *node expansion*, since the state associated with the parent search tree node is typically supplemented in some way (in the child node), thus moving the state closer to a completed solution. Furthermore, parent nodes typically sprout many child nodes—one for each possible operator application—so the search tree appears to *expand* or branch out from each parent node.

The root node of figure 3.2 can expand into 15 possible child nodes, since in an unrestricted topology 15 bidirectional edges exist for six vertices, but only three child nodes are shown. The search procedure may then decide to expand the second of these children in three different ways (though 14 are possible), and the third of these children may be expanded using two (of the 13) edge-adding operators.

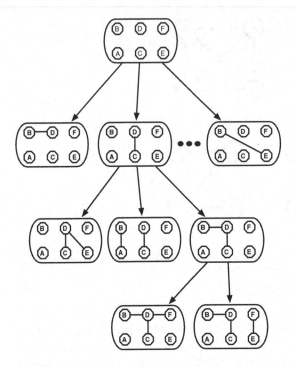

Figure 3.2
Search space exploration for a six-vertex minimum spanning tree (MST) problem, where search states are the complete vertex set along with a subset of edges. The only search operators are the addition of single edges to the graph.

Searching *intelligently* involves

1. initially choosing a potent set of search operators,

2. deciding which of the *open* (i.e., unexpanded) nodes to next expand, and

3. deciding which of the operators to apply to that node.

A well-designed search procedure for a given task will find a good (or optimal) solution without expanding too many nodes. In other words, a good search algorithm can start with a partial solution and complete it relatively quickly without exploring too much of the (often vast) space of possible partial solutions.

The MST problem has two well-known solution/search procedures, both of which run in polynomial time: Prim's algorithm and Kruskal's algorithm. The former begins with any empty set of *chosen edges*, E*, and a single *component* C consisting of a randomly chosen vertex v_0 in V. It then adds to E* the lowest-weight, unchosen edge that connects v_0 to any other vertex, v_1. Next, v_1 is added to C, and the process continues by adding the lowest-weight, unchosen edge connecting any member of C to any nonmember of C, which then joins C when that edge is added to E*. The loop continues until all vertices of V are in C. E* is then guaranteed to be a minimal spanning tree.

Kruskal's algorithm begins with the complete list of edges, E, sorted by ascending weight values, smallest-weighted edges first, and with | V | clusters, each containing one of the vertices of V. It then reads through the sorted edges until it finds one (e) that connects two vertices that are *in different clusters*. It then adds e to E* and merges the clusters that e connects. The algorithm repeats (walking further along the sorted-edge list, choosing edges, and merging clusters) until only one (large) cluster exists, indicating that all vertices are connected to one another. Again, E* is guaranteed to be an MST.

Compared to the abstract search process of figure 3.2, both Prim's and Kruskal's procedures appear extremely intelligent, since neither expands any nodes that are not along a path to the final solution. Granted, many edges are considered for addition to the partial graph, but none but the smallest component-enlarging (Prim) or cluster-joining (Kruskal) edge is added at each step. Partial solutions involving suboptimal networks are never actually generated, and certainly never added to a list of nodes that need to *compete* to be next in line for expansion. In essence, the search tree is one long vine, and MST is hardly considered a search problem at all.

Although the shortest path between two points is a straight line—and thus the MST of a two-vertex graph is just the edge between those two vertices—the step up to three or more vertices complicates things. The minimum total length of connections between three vertices is not always a two-element subset of the three-edge set, e.g., the edges A-B and B-C for vertex set {A, B, C}. In some cases, a shorter MST can be found by adding strategically located vertices to V. This relaxation of the constraints of the original MST problem (which restricted edge and vertex choices to those in V and E) expands the space of possible designs to topologies such as those of figure 3.3, two of which use less cable than the optimal solutions to the corresponding four-vertex MST problem of figure 3.1.

An MST task that permits the addition of new vertices is known as a *Steiner tree problem* (STP) (Bern and Graham, 1989; Gilbert and Pollak, 1968), which has a long history in the area of NP-completeness (Garey and Johnson, 1979), the study of problem types for which intelligent solutions cannot be guaranteed. Though new, improved solutions to special types of STP problems continually crop up in the operations research community (Winter and Zachariasen, 1997; Winter, 1987), no general-purpose, silver-bullet solutions (akin to Prim's and Kruskal's algorithms for MST) exist. Hence, STP remains a legitimate search problem to which concerns of node expansion and operator selection still apply.

3.2.1   Best-First Search

AI search techniques are geared toward these NP-complete problems; easier problems are typically solved by problem-specific techniques, whereas the AI methods
offer useful generality in design-search domains where the available problem-
specific expertise is either sparse or limited to supporting partial information or
ad hoc rules only.

AI search methods try to exploit this partial or ad hoc knowledge of the problem domain in order to make intelligent moves in search space. The core of this
knowledge consists of a rough estimate of the distance from a given state to the
goal. This is known as heuristic information, computed by a *heuristic function* that
takes a search state as input and produces a distance-to-goal approximation.

Search that exploits heuristics is often called *best-first search*. It is typically
applied to search spaces in which solutions are gradually pieced together by
expanding states via operator applications. It contrasts with *depth-first* and *breadth-
first* search, which ignore *nearness-to-goal* estimates and simply expand states based
on a predefined, knowledge-free, protocol. In depth-first search a host of operators are applied to a single state, in series, until either a solution or a failure state
is reached. In case of the latter, search backtracks a level or more up the search
tree before attempting another plunge to a solution or a dead end. By contrast,
in breadth-first search all immediate successors of a single state (those achievable by one operator application) are generated before more distant neighbors are
considered.

In terms of search trees, breadth-first search involves a methodical downward
expansion of the tree, with each horizontal level produced before the next, whereas
depth-first search begins with the expansion of a single vine from the root to a leaf,
followed by repeated plunges to leaf levels from intermediate nodes.

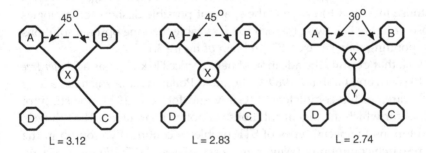

Figure 3.3
Alternative Steiner trees for connecting the original four sites (A–D), with total cable length L

Breadth-first search has the advantage of never overlooking a state, because of its methodical *sweep* of search space, but it suffers the drawback of taking a long time to get to solutions, since these typically reside at significant depths of the tree. Depth-first search can quickly find a solution sometimes, but it can also waste a lot of time exploring false leads.

Best-first search combines the advantages of depth- and breadth-first variants by permitting either type of expansion pattern, depending upon heuristic information (along with exact information about the distance from the root to the current state). Thus, a best-first search might plunge halfway to the leaf before realizing that the current avenue of exploration (i.e., vine) is less promising than an unexpanded node higher up in the tree. At a later time, it might return to the abandoned path as other states reveal themselves to be less useful than expected. At any rate, a best-first search is not a simple series of plunges or level expansions but a heuristic-based movement along the *horizon* of the tree, i.e., the set of unexpanded states, in search of promising nodes. Figure 3.4 illustrates the different expansion patterns for the three search types.

The classic best-first search procedure is the *A* algorithm*, designed by Hart et al. (1968). The heart of A* is the following equation for evaluating a state, *s*, in the search space:

$$f(s) = g(s) + h(s). \tag{3.1}$$

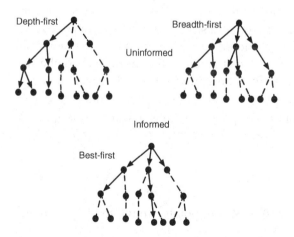

Figure 3.4
Intermediate progress of three common tree-based search techniques: depth-first, breadth-first, and best-first. Dashed lines denote unexplored paths in search space; arrows represent explored areas. The children of each node represent states reachable via one operation applied to the parent state.

Here, $g(s)$ denotes the cost or distance of getting from the root of the search tree to s, while $h(s)$ represents the heuristic: an estimate of the distance from s to a goal state. Thus, $f(s)$ represents the total expected cost of a solution path going from the root through s to a goal. In a nutshell, A* works by expanding the node with the lowest $f(s)$ value, i.e., the node whose state has the lowest combination of known (g) and expected remaining (h) cost.

A* applies well to problems in which

- an initial state, s_0, is given;
- there exists a well-defined, finite set of operators, π, for mapping states into other states;
- at least one goal state, s_g, can be completely described;
- the problem requires a minimum-cost solution, which usually implies that as few operators as possible should be used. In cases where operators have varying costs, the emphasis shifts to exploiting the lower-cost operators when possible.

For the Steiner tree problem, one key piece of knowledge involves the addition of new vertices into the graph: they should only be placed at the *Steiner point* of the triangle formed by three of the preexisting vertices. As shown in figure 3.5, the Steiner point of a triangle ABC resides along its longest side, creates an angle of 120 degrees or more with the two vertices of that side, and gives the minimum total distance from itself to A, B, and C.

By judiciously adding Steiner points (P) to *neighbor* triples (i.e., points closer to one another than to most of the other vertices), a spanning tree can often be found whose total edge weight significantly undercuts that of the MST over only the original vertices. As Bern and Graham (1989) describe, the maximum number of Steiner points is $|V| - 2$, where V is the set of original vertices. However, there are $T = \binom{|V|}{3} = \frac{|V|(|V|-1)(|V|-2)}{6}$ possible triples of points to consider. Thus, the entire search space of possible choices of Steiner points contains Υ states, a compound factorial function of $|V|$:

$$\Upsilon = \sum_{i=0}^{|V|-2} \binom{\binom{|V|}{3}}{i}. \tag{3.2}$$

That is, for each possible number (i) of Steiner points, from 0 to $|V| - 2$, there are $\binom{T}{i}$ ways to pick them.

A few values of Υ for corresponding $|V|$ are shown in table 3.1. For a mere 20-vertex problem, Υ is larger than the number of sand grains on earth. This astronomical size precludes the use of brute force methods, such as breadth-first search, that exhaustively evaluate all possibilities. The constraint imposed by Steiner points reduces the search space from infinite to finite, but still exponential in $|V|$.

Table 3.1
Sizes of vertex sets | V | and corresponding number of states (Υ) in a Steiner tree problem search space

$\lvert V \rvert$	Υ
3	2
4	11
5	176
10	903601306070
20	14667964292635252338493201886625720221593
30	3289176205897046956039094265615683389796991668344203259641620294011355216

Still, these numbers are only rough estimates. For one thing, many STP algorithms include data structures that keep track of the proximity of vertices to one another. Only triples of *nearby* vertices are selected as the basis for Steiner points, although the ease of discerning relative nearness could vary considerably with the problem instance. As another confounding factor, some members of these triples may be Steiner points themselves, thus increasing the full number of choices from $\binom{\lvert V \rvert}{3}$ to $\binom{\lvert V \rvert + \lvert S \rvert}{3}$.

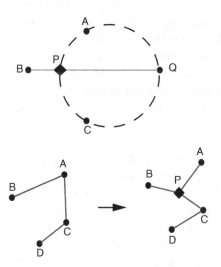

Figure 3.5
(*Top*) Construction of Steiner point for the vertices A, B, and C involves finding AC, the longest side of triangle ABC, then drawing an equilateral triangle AQC opposite to point B (Bern and Graham; 1989). The intersection of line BQ and the circle that circumscribes AQC is the Steiner point (P). If angle ABC $\geq$ 120 degrees, then B is the Steiner point. (*Bottom*) The Steiner point enables a shorter spanning tree of points A–D.

3.2.2   Steiner Trees Using Best-First Search

Best-first search provides a feasible, general-purpose alternative for solving STPs. It cannot guarantee finding a minimal Steiner tree, but the problem fits nicely into the A* protocol and thus serves as an illustrative example of the algorithm. Namely, (1) the state consisting of all vertices and no edges is an intuitive starting point , s_0, (2) the problem requires a minimum-cost solution, (3) a well-defined operator set exists, consisting of adding edges or Steiner points to an existing, partial solution, and (4) a goal state is formally described as one in which $| E* | = | V | + | S | - 1$ (i.e., the number of chosen edges is 1 less than the number of original (V) plus Steiner (S) vertices), and only one vertex cluster remains (using Kruskal's algorithm).

Let A-STP denote a standard version of A* applied to STP, where each node in the A* search tree mirrors a state of the search space, and each such state has a representation housing a partial solution to the problem. Each partial solution is a snapshot of an intermediate stage of the overall computation, which is a mixture of Kruskal's algorithm and Steiner point generation. This state consists of a list of chosen edges (those participating in the spanning tree so far), a list of remaining edges (sorted in ascending order by weight), and a list of Steiner points.[1]

An arc between a parent and child node in the A* tree then signals that the child node's state is achieved by applying a search operator to the parent node's state. Parent nodes are expanded either by adding 1 to J of the next edges recommended by Kruskal's algorithm to a parent state or by adding 1 to K new Steiner points (where J and K are user-chosen parameters). The arc cost for the former operators is simply the summed weight of all new edges, while Steiner point additions incur no arc cost. However, the addition of Steiner points can easily increase the estimated distance to the goal state, the heuristic (h) value, since the number of required edges increases with the addition of each node. As is standard with A*, the g value for any node (n) is the sum of the arc costs from the root down to n.

In devising a heuristic function, $h(n)$, for STP, note that the total number of edges in the final tree should be $| V | + | S | - 1$, so the number of additional edges is $R = (| V | + | S | - 1) - | E* |$, where E* is the current set of chosen edges. The cost of these edges is unclear, but a reasonable underestimate (i.e., lower bound) is the summed cost of the next R edges on the sorted edge list. As it turns out, A* works best when $h(n)$ never overestimates the actual distance-to-goal, as discussed more thoroughly in AI textbooks (Russell and Norvig, 2009; Rich, 1983).

Figure 3.6 sketches the intermediate progress of an A* run on a six-point STP. The vertical bars and horizontal triangles give a rough estimate of arc costs and $h(n)$ values, respectively, while the contents of each node display a wide range of partial states. The algorithm randomly selects triples for Steiner point creation, but

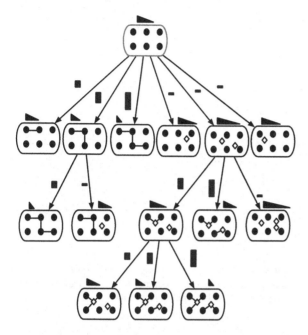

Figure 3.6
Intermediate stage of an A* run on a six-point Steiner tree problem. The node expansion operators are
to add the next 1, 2, or 3 edges using Kruskal's algorithm or to add 1 or 2 new Steiner points (open
diamonds). Vertical bars on search tree arcs denote relative arc costs; horizontal triangles above each
node represent relative heuristic (*h*) values. For clarity, not all operators are applied to each expanded
node.

as indicated by the combinatorics preceding equation (3.2), all such triples cannot
feasibly be tested for problems of nontrivial size. Hence, node expansion involves
only a relatively small number of all possible subsets of triples. In addition, the
algorithm does keep track of neighborhood relations between all vertices, thus
insuring that Steiner triples only contain vertices located relatively close to one
another.

Figure 3.7 displays several Steiner trees found by quick runs of A-STP, where
only 14–16 nodes are expanded for each run. The same method works well for
STPs involving a few hundred vertices, but larger problems involving thousands
of vertices call for special-purpose algorithms. These typically use sophisticated
geometric scaffolding to keep track of neighboring point triples. However, the
point of this example is not to compare and contrast these specialized methods
for STP but rather to illustrate the essence of search among a space of partial solutions using an example problem domain, STP, which happens to have interesting
ties to neural development.

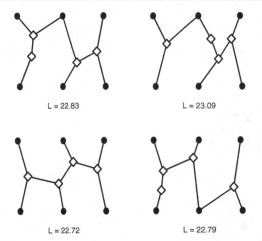

L = 22.83   L = 23.09

L = 22.72   L = 22.79

Figure 3.7
Best-found Steiner trees for four independent runs of A-STP, showing original vertices of the ladder (circles), Steiner points (diamonds), and total length L of each tree. The minimum spanning tree (MST) of the six original points is 24 units in length. The tree at the bottom left is optimal.

3.2.3   Search among Whole Solutions

Whereas the breadth-first, depth-first, and best-first search techniques all work with partial solutions—the most promising of which become further elaborated as search continues—a whole host of alternative approaches, often called *local methods*, deal primarily with whole solutions that are gradually modified in pursuit of improvement. These approaches may work with a single focal solution or a collection of them; and their modification procedure can involve merely random tweaks of current solution components or purposeful changes based on domain-specific knowledge.

Most local search methods follow the same basic scheme:

1. Begin with a set or population (P) containing M *individuals*, each of which constitutes a complete (though probably not optimal) solution.

2. Repeat:

   a. Use an objective function (F) to evaluate each individual in P.

   b. If an individual $p* \in P$ produces an optimal value for F, return p^* and halt.

   c. Produce N children of solutions in P.

   d. Update P by including some or all children and removing some or all previous solutions in P.

Until a predefined maximum number of iterations have been performed.

3. Return the best (though not necessarily optimal) solution found.

Algorithms then vary as to the value of M, the manner in which children are produced, and the criteria for adding children to P (and removing existing members).

For example, in standard *hill climbing* $M = 1$ children are produced by modifying only one feature of p_1 (the only member of P) at a time, and the child with the best evaluation replaces p_1. Note that many children can be produced on each step, since p_1 may have many features, each of which may take on several possible values. One of the drawbacks of hill climbing is that search halts if no children can improve upon p_1, which may not be a global optimum in the search space, only a local optimum.

In *simulated annealing* (SA) again $M = 1$, and children are produced by a host of random modifications to p_1, but now there is a probability p_T that the best child can replace p_1 *even if its evaluation is inferior to* $F(p_1)$. This allows search to continue past local optima by jiggling over to points in search space that, though inferior to their parent, may serve as vital bridges to regions containing a global optimum. In simulated annealing a temperature (T) parameter controls the amount of jiggle such that p_T is directly proportional to T. SA runs typically begin with a high T value that gradually declines. Thus, SA explores more routes that may not appear too promising (i.e., takes more chances) early on but then settles into a hill-climbing mode as T and p_T decrease.

In *beam search* $M > 1$, and the best M children simply replace the current population. Alternatively, in *stochastic beam search* (SBS) the M new members of P are chosen stochastically from the child set, where the child's evaluation affects the probability that it will be chosen. Still, children with poor evaluations can get lucky (and highly rated children get unlucky) during the selection process.

Evolutionary algorithms (EA) are popular relatives of stochastic beam search. They normally have $M > 1$ and produce children via local modifications (*mutations*) to members of P, but many EA variants include a genetic operator akin to recombination (crossover) in biology. This entails the combination of two or more members of P to produce a child, which may vary considerably from each parent and thus represent a point in search space quite distant from the parents. This is the bold new improvement upon stochastic beam search that EAs provide, but many practitioners use it only sparingly. An even bolder supplement to SBS comes from the EA subfield of *genetic programming* (GP), where solutions are no longer vectors of values but complete computer programs. GP adds whole new dimensions to automated design search.

3.2.4   Local Search for Steiner Trees

Returning to the Steiner tree problem, note that if we know the correct set of Steiner points (S^*) with which to supplement a vertex set, then generating the Steiner tree is a simple two-step process:

1. $V^* \leftarrow V + S^*$.
2. Apply Prim's or Kruskal's algorithm to find the MST of V^*.

In short, there is no need to gradually add in the Steiner points to a developing spanning tree. Of course, knowing S^* ahead of time must include knowing any Steiner points for triplets that also include Steiner points. This anticipation of Steiner points by other Steiner points must be baked into S^*. This sounds a bit far-fetched but is relatively straightforward to implement in a local search method.

A simple representation for applying any (of the many) local search methods to STP is a list of Steiner points, each encoded as a triple. For example, the vector representation of equation (3.3) encodes groups of three point indices (where points are labeled from 0 to $|V| - 1$).

$$A_S = \{[p_1^1, p_2^1, p_3^1], [p_1^2, p_2^2, p_3^2], \ldots, [p_1^k, p_2^k, p_3^k]\}. \tag{3.3}$$

As mentioned earlier, the maximum number of Steiner points is proven to be $|V| - 2$, so for any STP, let $k = |V| - 2$ when dimensioning A_S.

Since these vectors may be randomly generated or modified by the search algorithm, the possibility for creating nonsensical triples certainly exists. An STP-specialized decoding process can relieve this problem.

To decode the triple $[p_1^i, p_2^i, p_3^i]$ into a potential Steiner point, begin by converting each index:

$$p_j^i \leftarrow p_j^i \bmod (|V| + |S_{i-1}|), \tag{3.4}$$

where $|S_{i-1}|$ is the size of the set of valid Steiner points generated by the decoding of the first $i-1$ triples of A_S. This insures that each index refers to an existing vertex, either an original point or a newly created Steiner point.

Once converted, a triple (t^i) can still be meaningless if it

- includes duplicate integers (and thus fails to specify three unique points), or
- codes for three vertices whose Steiner point is one of those three vertices, or
- duplicates another triple t^j ($j < i$) in A_S.

Hence, even though A_S has length $k = |V| - 2$, it may produce fewer than k *meaningful* Steiner points. In this way, A_S can encode from 0 to k meaningful triples. Thus, the presence of meaningless triples fortuitously provides needed flexibility to this representation, since prior to running the algorithm, we only know that between 0 and k Steiner points are needed.

A slight modification to A_S can improve search efficiency by allowing triples to be easily ignored or included based on the *flip of a switch*, i.e., the flag bit f_i. Random modifications to these flags then permit speedy transitions in search focus between longer and shorter vectors.

$$A_S = \{[f^1, p_1^1, p_2^1, p_3^1], [f^2, p_1^2, p_2^2, p_3^2], \ldots, [f^k, p_1^k, p_2^k, p_3^k]\}. \tag{3.5}$$

Two basic operators for modifying A_S are sufficient to enable movement about the Steiner tree search space:

- Flip a flag bit from 1 to 0, or vice versa.
- Replace any integer of a triple by a randomly generated positive integer.

As discussed, all local search methods rely upon an objective function (F) to evaluate individual solutions. These functions are a vital prerequisite to success in design search, and though they can be difficult to formulate for some problem domains, STP seems to require nothing more than a simple comparison of the MST of the original vertices V to the MST of V ∪ S. Hence, a useful objective function for STP is

$$F(i) = \frac{W_V - W_{V \cup S_i}}{W_V}, \tag{3.6}$$

where W_V is the total MST edge weight over the original V vertices, and $W_{V \cup S_i}$ is the MST of V supplemented with the Steiner points of solution i (S_i). $F(i)$ is therefore the fractional reduction in the MST when solution i's Steiner points are included. For most STPs, the optimal value of $F(i)$ is 0.13 or less.

The entire process involved in evaluating an individual is summarized in figure 3.8. First, a sequence of quads are converted into potentially viable triples of points, some of which correspond to meaningful Steiner points (S). A well-defined, deterministic process (Kruskal's algorithm) is then used to generate a minimum spanning tree of V ∪ S. Then, an even simpler deterministic process is employed to sum up the edge weights of that Steiner tree and compare it to the summed edge weight of the MST over V, yielding an assessment of the tree, which, in turn, serves as an evaluation of the original vector of quads. This evaluation then helps the search process move in the direction of more promising solutions.

Given (1) a representation for complete solutions, (2) a set of operators for manipulating representations, (3) a means of translating a representation into a meaningful spanning tree, and (4) an objective function, the STP problem is now properly packaged for consumption by any of the local search methods that have been described (and many more). Figure 3.9 shows a collection of Steiner trees

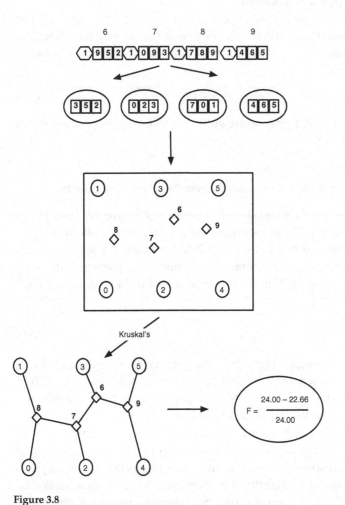

Figure 3.8
Translation and evaluation of a parameter vector during local search. The vector (*top*) consists of four quads, as formalized in equation (3.5). All quads with active (1) flags have their indices scaled relative to the vertex set's size (which varies from 6 to 10 during the process). The integer above each quad indicates this size in effect when the given quad is decoded. Each scaled triple deemed meaningful is converted into a Steiner point (diamond). Kruskal's algorithm is then used to find the MST of the original plus Steiner vertices (assuming that all and only direct connections between any two points constitute the edge set). The total weights of the two MSTs (with and without the Steiner points) are then compared to yield an objective function evaluation.

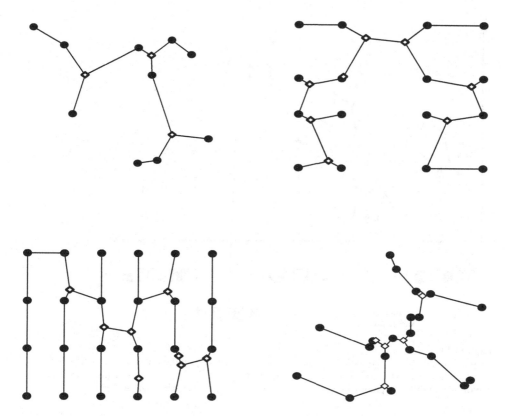

Figure 3.9
Steiner trees found by EA-STP using a population of 20 individuals and run for 200 generations. Circles denote the original vertices; diamonds represent Steiner points. Note that EA–STP creates and connects Steiner points using semi-independent processes, so some Steiner points have only two neighbors in the graph, not three from their source triple.

found by EA-STP, an evolutionary algorithm using this STP packaging. Similar solutions arise from the incremental solution-building approach of A-STP.

As shown in figure 3.10, the search space is often depicted as a k-dimensional *landscape* in which, together, the first $k - 1$ dimensions signify the representation of an individual solution, and the kth dimension denotes its evaluation. Search constitutes movement in that landscape via modifications to the individual representations. The search algorithm cannot see the entire landscape; if it could, the process would more closely resemble map-based navigation than actual search. It only has *spotty* information: the evaluations of some individuals.

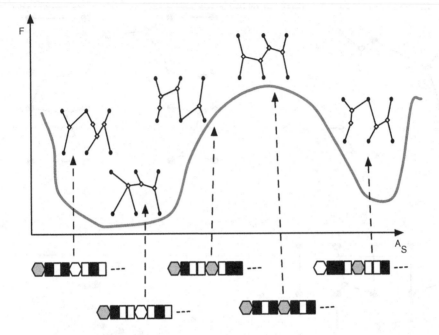

Figure 3.10
Search space of Steiner tree solutions. Each parameter vector (A_S) on the x-axis translates (dashed line) into a Steiner tree, whose evaluation is plotted on the y-axis. Local search mechanisms involve the exploration of different parameter vectors in a manner strongly biased by their evaluations.

The misleading nature of this two-dimensional depiction of a k-dimensional space cannot be overemphasized. In a real search space each solution can have thousands or millions of potential children/neighbors. Generating and evaluating all of them can carry extreme computational demands. The hallmark of an efficient search algorithm is the ability to find a global optimum while evaluating only a small subset of the entire search space.

Figure 3.11 shows a commonly used *footprint* of the progression through search space of an EA looking for a Steiner tree similar to one of those in figure 3.9. Here, the x-axis denotes the generation of the EA (i.e., the round of population-wide evaluation), and the y-axis represents the evaluation/fitness of the best individual of that generation. This plot is typical of an EA run, with stasis punctuated by progress: the algorithm finds an individual that cannot be improved upon for many generations. However, beneath each fitness plateau lies a population undergoing constant modification; but the best fitness in each generation cannot beat that found at the plateau's inception. Often, exploration of these *neutral landscapes* eventually leads to improvement when the algorithm finds a positive slope at one of the many edges of the plateau.

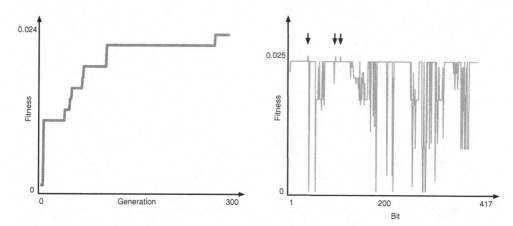

Figure 3.11
(*Left*) Best-of-generation fitness progression for EA-STP using 300 generations and the objective function of equation (3.6). (*Right*) The fitness values of all solutions in the immediate neighborhood of the best individual (B) of generation 300; their genotypes differ from B's by only one of 417 bits. The *x*-axis indicates the deviant bit with the resulting fitness plotted on the *y*-axis. Note that the three minipeaks (small arrows) rising from plateaus represent improvements over B, whose fitness is 0.0236, the level of each broad plateau.

Figure 3.11 (right) illustrates both the size and neutral regions of fitness landscapes. The genotypes in this example consist of 417 bits, since both flags and integers rely on a binary encoding in EA-STP. To get an impression of the local topography, 417 neighbors are generated—each differing from the original in exactly one of the 417 bits—and evaluated, with each fitness plotted in the figure. This gives some indication of the complexity of search in these landscapes. In this case, ascending one of the minipeaks requires generating one of the three advantageous bit mutations (of 417 possibilities). These are quite good odds compared to many local search problems.

3.3   Steiner Brains

Neuronal arbors (for both axons and dendrites) have been shown to display Steiner trees (Cherniak et al., 1999) resulting from self-organizing fluid dynamic effects during brain development. These trees minimize total cable volume, not length, which makes sense in networks with heterogeneous branch thickness.

Consider the simple graphs of figure 3.12. If edge thickness is constant throughout the network, then the four-node Steiner tree at the top of the figure minimizes both channel length and volume. However, if, as in most natural flow systems, the downstream distributing branches are thinner than upstream conduits, then

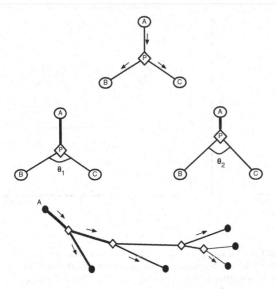

Figure 3.12
(*Top*) A Steiner tree for transporting material from node A to nodes B and C, with edges representing three-dimensional tubes. (*Middle left*) The same tree but with a thicker upper edge. (*Middle right*) An alternative Steiner tree that minimizes total edge *volume*, not length. (*Bottom*) A network built using principles for minimizing edge volume.

the standard Steiner tree no longer minimizes volume (middle left). Moving the branch point (P) closer to the supplying node A produces a volume-minimizing Steiner tree (middle right).

Figure 3.12 (bottom) depicts a flow system reminiscent of an axonal segment. In these networks the goal is to supply the target nodes (black circles) with material from the upper left node (A) using an optimal network structure. Cherniak et al. (1999) found that neuronal arbors of this sort come within a few percentage points of minimizing total conduit volume, while they often deviate by 20%–50% from length- or surface-area-minimizing networks over the same set of target nodes. Thus, the search processes involved in brain development actually produce Steiner trees.

Though neural network topologies self-organize from an intricate growth- and-signaling process involving axons and their targets (Sanes et al., 2011), the Steiner tree structure seems to emerge from simple hydrodynamics. In fact, signal gradients typically interfere with Steiner formations. In a nutshell, hydrodynamic principles determine the optimal thickness of branch points, such that

$$P^k = C_1^k + C_2^k, \tag{3.7}$$

where P is the diameter of the parent conduit, C_1 and C_2 are those of the child conduits, and k is a constant, typically between 2.5 and 3.

These diameters affect the branch angle: wider tubes require a larger angle to fully separate the two branches. Examples of these angles include θ_1 and θ_2 in figure 3.12. These principles also restrict larger conduits (which typically carry more material at higher speeds) from feeding into branches with a large θ, since the abrupt change of direction creates turbulence, which reduces the system's throughput and stresses the tubing itself. As an analogy, compare the sharpness of turn angles found on small backroads (roads with fewer cars, typically traveling at lower speeds) to those on superhighways.

Hence, for hydrodynamic efficiency, larger conduits must branch at smaller angles. Similarly, in other systems involving competing but balanced forces, where all pull on a central point, the branch angles opposite the strongest forces are smaller. Imagine a three-way tug-of-war between a strong participant A and two weaker opponents, B and C. For the weaker two to have any chance of neutralizing A, they must reduce the angle between themselves to one that maximizes the resultant (in the direction of A) of their combined forces. Natural systems as diverse as water basins, tree roots, cardiovascular networks and brains all display this topological feature (Bejan and Zane, 2012), which emerges from the interplay of these and other forces during network development, a process easily characterized as a search for both target nodes and structural equilibrium.

Interestingly, these mechanical and hydrodynamic interactions bode well for volume-minimizing Steiner trees, as shown in figure 3.12. For the simple four-node networks, note that by moving point P closer to A, the branch angle (BPC) naturally decreases ($\theta_2 < \theta_1$) and total conduit volume decreases. Thus, Steiner trees in natural neural networks emerge from nothing more than the interactions of basic physical forces among growing branches. Development does not search for Steiner trees, but it finds them.

3.4 Biological Search and Problem Solving

The bio-inspired approach to solving problems differs quite dramatically from traditional AI. The difference is summarized in figure 3.13. Problem solving is often characterized as a search through solution space in which solutions/hypotheses are generated and then tested for feasibility, optimality, and so on. Traditionally, an efficient problem solver has been viewed as one that exploits as much intelligence as possible to generate reasonably good solutions and thereby avoids wasting time testing bad hypotheses. In fact, a popular measure of problem-solving improvement (i.e., learning) in knowledge-based systems was the degree

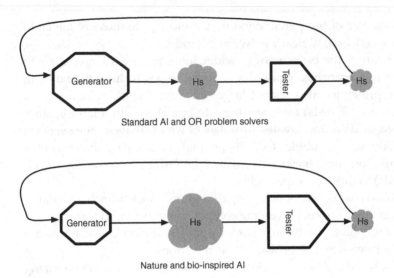

Figure 3.13
The generate-and-test view of problem solving with respect to bio-inspired and more traditional artificial intelligence and operations research approaches. The size of each octagon and pentagon indicates
the amount of problem-relevant knowledge that it employs, while the size of each Hs cloud denotes
the number of active hypotheses (full solutions) under consideration.

to which test knowledge or constraints could be re-expressed (or *operationalized*) in
the generator (Mostow, 1983).

The A* algorithm illustrates knowledge operationalism at a few different levels.
First, most nonexhaustive searches among partial-solution states (e.g., best-first
search) attempt to prioritize states for expansion, where A* favors those with the
lowest combination of current and projected future cost. A*'s heuristic function (h)
embodies knowledge of the problem domain operationalized as expected future
costs of partial solutions. It enables A* to give low priority to many nodes, effectively pruning them, and thereby preventing many (suboptimal) complete solutions from being generated. Thus, a good h function incorporates a good deal of
intelligence into the generator by carefully prioritizing nodes for expansion.

The node expansion process is another story, involving the application of operators. In naive A* applications the algorithm simply throws the whole gauntlet of
transformation operators at each state that is chosen for expansion, immediately
producing C new children, where C can be quite large. More sophisticated versions
of A* may only generate one or a few children at a time but keep the generator *on
hold*, ready to produce more as needed. The naive approach employs little context-
appropriate knowledge: it simply uses the set of operators originally designed for
the problem, whereas the delayed-generation model may produce only children

that have the most promise (based on knowledge of what operators work best on particular types of parent states), or it may simply produce the children randomly, but just one at a time.

In A-STP an important piece of compiled knowledge is that any Steiner point (P) will probably link up to three or more vertices, at least one of which should be part of the triple of preexisting points for which P was generated. Since each new Steiner point adds greater complexity to the search process—by moving a partial state farther from a solution (in terms of the number of remaining spanning tree edges to add)—it is imperative to rein in this complexity by adding edges whenever possible. Immediately after a Steiner point addition, some or most of the points to which it should connect already exist, so A-STP restricts the expansions following a Steiner point addition to those involving one or more edge additions. Hence, the operator set is context-sensitive, as an important piece of Steiner tree knowledge gets operationalized as a constraint on the legal subset of expansion operators for different classes of partial solutions. This helps reduce the number of partial and (thereafter) complete solutions generated.

Now contrast this with the approach of local search methods. These work with complete solutions, some of which are randomly generated, just thrown out there for evaluation; others are random variations on parent states. The number of ineffective (if not completely worthless) solutions produced and evaluated can be quite high, particularly for bio-inspired approaches such as EA-STP. However, these solutions admit concrete evaluation, unlike partial solutions, which demand evaluation of their *potential*, a process fraught with inaccuracy. As a simple illustration, the Steiner tree of figure 3.9 (bottom left) found by EA-STP requires the evaluation of 20 solutions in each of 200 generations (a total of 4,000 evaluations), whereas A-STP rarely finds a good solution in cases where it produces and evaluates 4,000–5,000 states. For problems such as STP, the evaluations provided by complete solutions may provide more useful information to the search engine than do assessments of partial states. Good heuristics (h) are often hard to produce, whereas objective functions like equation (3.6) give an accurate quantification of the entire spectrum from failure to success.

In general, most GOFAI systems can take a *hypothesis* (an example of a *full solution*, as discussed) and its test results (i.e., the basis for its evaluation) as inputs and produce a new hypothesis that is almost guaranteed to be an improvement. For example, if a logic-based system uses disjunctions of primitive terms to produce classifications of input examples, and if the test results indicate that a hypothesis is too specific (in that it classifies many positive examples as negatives), then the system will normally add an extra disjunction in order to generalize the hypothesis and thereby admit more positive examples into the set of acceptable cases. This intelligence that the generator contains is simply a knowledge of basic logic.

Alternatively, if the representational form is an artificial neural network, then the knowledge of the backpropagation algorithm (a standard learning procedure for neural networks) leads it to increase and decrease connection weights in ways that decrease the network's output error. In this case, the intelligence lies in a very basic understanding of the effects of change in a system of interlinked equations. In either case, the intelligence is representation-dependent: the backpropagation algorithm could not handle a logical disjunction written in standard propositional or predicate calculus, and a logical engine would be lost if given the weights and activation values from a neural network.

Bio-AI algorithms tend to have very little intelligence in the generator. In general, the greater the effort required to convert the basic syntactic representation (e.g., the preceding integer vector) to a meaningful semantic form (e.g., a Steiner tree), the less intelligent the algorithm will appear in terms of its ability to generate good hypotheses/solutions. This stems from the simple fact that solutions are created by operators working at the syntactic level, which in these algorithms have little awareness of the corresponding semantic translations. Bio-AI methods (particularly EAs) permit a large representational gap between syntax and semantics, thus allowing the generator to operate relatively independently of the problem-solving context. For example, the same EA can mutate and recombine bit string representations that encode everything from logical expressions to neural networks to Bayesian probability tables at the semantic level. In this sense, EAs can be extremely task-independent. But, again, the manipulations made to the syntax have no guarantee of producing improved solutions.

From a different perspective, standard AI problem solvers include a good deal of information about *how* a good solution should be created, and they may even possess meta-knowledge about *why* they perform certain hypothesis manipulations. This bias helps avoid the generation of bad hypotheses; and the *why* knowledge may even enable the system to explain its choices to a human user.

In contrast, an EA has little *how* or *why* information but a good deal of knowledge about *what* properties a good solution should have. It just tosses solutions out there and filters out the bad attempts. This is exactly how nature works: many random variations are tried, and the environment determines who survives based solely on *what* they can do. Neither creationists nor classical engineers are comfortable with this approach to problem solving, but as biologists and evolutionary computationalists can document, it works.

A major point of this book is that many of the other neural processes underlying intelligence also work that way: by the relatively random production of attempts followed by selection of the successes, by processes reflecting little more knowledge than trial and error, and by mechanisms that more closely resemble local search than A*. And again, it works quite well.

4   Representations for Search and Emergence

4.1   Knowledge Representations

The success of automated search hinges upon the manner in which problem states are expressed in computer code. This, along with operators that convert one state into another, formally defines a search problem. GOFAI has a long history of research into these computational formalisms, known as *knowledge representations* (KR), whereas Bio-AI and ALife have much more strained relations to KR, both in terms of the K and the R.

First, the knowledge of GOFAI systems is very explicit (often in formats that resemble natural language) and very high-level, capturing core nuggets of *expertise* in domains ranging from biochemistry to banking. Few would argue that the critical element of most GOFAI systems is indeed human knowledge, in a declarative, machine-understandable representation (that humans can easily interpret as well). In Bio-AI, however, the knowledge, (i.e., the information that allows the system to perform its task) is normally very low-level, since most Bio-AI systems attack less cognitively challenging tasks. The experts for Bio-AI might be ladybugs, spiders, or the scientists who study them; and the knowledge is not about how to assess credit risk but how to put one foot in front of the other—hardly the stuff of nonfiction bestsellers. This knowledge is also far from explicit and often tightly interlaced with the decision-making mechanisms. It manifests itself as a bias on the sensorimotor controller of a robot, for example, without having any independent status. In this sense, it is often purely procedural: an embodiment of *how to*, as opposed to *what*, *where*, *when*, or *why*.

This predominantly procedural form of much Bio-AI knowledge has led influential AI and robotics researchers like Rodney R. A. Brooks (1999) and AI philosophers like Hubert Dreyfus (2002) to argue against the notion of internal representations. Instead, organisms are believed to rely on their advanced sensing abilities to exploit a world that becomes *its own best model*. Combine that with an organism's ability to off-load the intermediate results of cognitively challenging tasks to the environment (Clark, 2004)—for instance, writing down products and sums on *memory extenders* like paper while doing multidigit multiplication—and indeed the role of mental models/representations does come into question.

So neither knowledge nor representation is a dominant theme in Bio-AI, though the view of intelligence without representation invokes considerable opposition within the field (Steels, 2003; Kirsh, 1991; Clark, 2001). Furthermore, representational choices can strongly affect the performance of classic Bio-AI tools such as artificial neural networks (ANNs) and evolutionary algorithms (EAs). However, it often feels like an exaggeration to call these knowledge representations, since the high-level significance of the information encoded in these representations normally pales in comparison to that of a typical GOFAI system: the former may encode tactile sensory inputs from a robot's bumper, while the latter captures complex buy-sell decision-making rules for a stock trader. Also, the knowledge in GOFAI systems is usually portable: it can be moved around the system in the same way that books are moved among shelves. In Bio-AI, knowledge is often so tightly tied to the underlying architecture of the system that it loses all meaning if superimposed upon another part of the neural network, EA genome, or agent phenotype.

One important aim of this book is to keep KR issues in focus by asking, Where is the knowledge and in what form? when examining a wide variety of biological and Bio-AI systems. Another is to provide a deeper and more flexible understanding of the concept of a representation that might inform diverse inquiries into intelligence, whether natural or artificial, GOFAI or Bio-AI.

4.2   Models of Intelligence

Formal characterizations of the real world, often called *models*, perform important functions in science while also serving as good starting points for general discussions of representations. Typically, models are either descriptive or predictive. Descriptive models help to structure existing data about a system in support of interpretation and classification, while predictive models may foretell the future behavior of the system. For example, a statistical descriptive model showing that the size distribution of traffic jams in a particular city exhibits a power law can help scientists to determine that the underlying system is complex, not chaotic. On the other hand, a predictive model for traffic flow would allow the highway department to determine the consequences of closing a particular bridge during rush hour. The current discussion focuses on predictive models, though descriptive models using information theory are discussed in chapter 13.

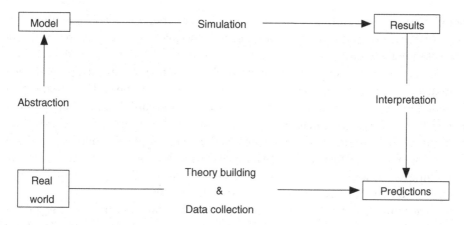

Figure 4.1
Two key paths to predicting real-world phenomena. (*Bottom*) Collect a lot of data about the world and infer. (*Top*) Build models based upon abstractions of real-world entities and processes, simulate the model, and interpret the results in terms of their real-world parallels.

As shown in figure 4.1, the process of making predictions in science can follow a few different paths. Traditionally, science works by formulating theories and gathering data, with or without overlap, and in varying order: in bottom-up approaches, the collected data motivates theory formation, while in top-down science, theories shape the pursuit of data. Alternatively, scientists can form operational abstractions of their domain knowledge, i.e., models, whose simulation provides results in this abstract world. If models capture the essence of reality and are not based on faulty assumptions, then the abstract results of simulation can, under careful interpretation, produce useful, concrete predictions for the real world.

Though some will argue that theories also function as models—they often abstract from reality and then provide results that require careful interpretation before supporting fully grounded predictions—the distinction here is between a predominantly human chain of reasoning from data to predictions and one where a computer makes many of the causal inferences.

Common examples are weather and climate modeling. Scientists codify their collective meteorological knowledge into extremely complex models, often composed of hundreds or thousands of differential equations that manifest what scientists believe are the most critical factors. When run on powerful computers, these produce weather or climate scenarios that scientists can then analyze and interpret with respect to their complete human set of knowledge and reasoning capabilities, i.e., using some of the information that was not incorporated into the model, possibly for reasons of expressibility (or lack thereof) of the modeling sub-strate (e.g., differential equations) or efficiency: too many factors require too many computational resources.

Contrast this with using a computer to solve a mathematical puzzle, such as a sudoku. Once again, the human must formalize a model of the sudoku puzzle and the standard logical reasoning techniques used to fill blank boxes. Then, given a puzzle, the model can be run to produce a solution. If the model includes enough of the typical cell-filling logic used by humans, then it can find solutions to puzzles of varying degrees of difficulty. And as long as the model does not include operators employing faulty logic, the machine-produced solutions will require no significant interpretation: humans should be able to *trust* the result.

What accounts for the varying levels of trust (or varying amounts of interpretive effort) in climate versus sudoku modeling? Though the sheer complexity of the meteorological domain accounts for some of the difference, a similar argument holds for chess simulators and automated solvers of large traveling salesman and Steiner tree problems: the results require minimal interpretation.

In these domains all the critical factors of the situation are digital and thus completely independent of the substrate. When we add two numbers, we can use paper and pencil, an abacus, our fingers, or a cluster of supercomputers, but all the results have equivalent significance. No 173245 is more meaningful than another 173245, and all the initial information and information-producing steps of the process are just as *real* on a computer, our fingers, or paper, or (maybe) even in our brains. If Joe beats Jake in an online chess tournament, there is no reason to believe that he cannot beat him in an indoor, face-to-face confrontation, with a real wooden board and ivory pieces, or with large human-sized plastic pieces in a Russian park. The game means the same regardless of the substrate. If a contractor needs to minimize total cable length in hooking up a dozen buildings on a flat college campus, the results of a two-dimensional Steiner tree search based on the relative locations of the 12 buildings should give a very reliable indication of the expected cable length.

On the other hand, weather and climate (or the movement of balls on a billiard table) have an acute dependence upon the substrate. Computer models may include many aspects of the situation, but they invariably fail to capture the complete essence of the domain. And the more salient features they lack, the more effort and general skepticism are associated with the interpretation of the results.

It seems clear that formal, computational representations of digital domains have the potential to provide more reliable information than do those in substrate-dependent (often called *analog*) domains.

The problem with intelligence (of the extreme, human-level variety) is that nobody knows if it's digital or analog. AI has produced impressive examples of digital intelligence while giving some indication of the limits of a pure-digital assumption, while neuroscience continues to uncover more evidence of the mind-boggling detail required of analog models (should they be necessary). If substrate is vital, then all forms of digital intelligence will fall far short of the mark, regardless of the number of rules in an expert system, probabilities in a Bayesian network, or artificial neurons in a neural network. If glial cells, the intricate branching patterns of dendritic trees, gap junctions, and the processes of neuro- and synapto-genesis all make *necessary* contributions to intelligence, then it will be many years before a complete working set of digital approximations will even get AI in the ballpark. However, there may be proper digital abstractions of neural substrate that can carry AI research a long way toward humanlike intelligence. This book is based on a belief in these digital representations of neural reality.

Keep in mind that DNA is a quaternary digital code. Though the chemistry of nucleic acids and their interactions are complex, the abstraction to the level of the four-element code retains all the essential information needed to re-create (re-present) an organism when placed in the species-preferred, prenatal environment. Since this digital code interacts with a chemical and physical environment of decidedly nondigital nature, the end result (i.e., body and brain) need have no allegiance to the digital world. Still, it is important to remember that a significant contribution to life is digital.

4.3   Representation and Reasoning

The representations of GOFAI are akin to climate models: abstractions of a real-world system of interest whose results will hopefully require minimal interpretation to produce useful predictions. The claim, though several decades outdated, is that collections of logical axioms and some initial knowledge, along with a logic engine to derive more knowledge, are sufficient (if not necessary) to explain human reasoning and intelligence.

Most GOFAI workers have given up on models of natural intelligence in favor of souped-up engines of superhuman inference (Kaufmann et al., 2011; Zhang and Malik, 2002; Moskewicz et al., 2001)—a great feather in the cap of AI—but then the question of what the system actually represents has little meaning. It's like asking what a toaster or a jet engine represents. They are tools for performing important jobs but representations of nothing. Of course, within the system, there are representations of the real world that the axiom cruncher manipulates to produce knowledge that solves important problems for its users. But the mechanisms by which all this is done have no relevance for those wishing to understand natural intelligence.

The general framework for intelligence in GOFAI systems involves representations of knowledge and methods for reasoning with that knowledge: methods for deriving more knowledge from that with which the system begins. Pure and simple, a good representation supports the derivation of lots of *relevant* knowledge, and relevance is very domain-specific. A good GOFAI engineer can bias the representation toward producing this salient knowledge while generating minimal superfluous information, whereas a good psychologist might criticize that same representation for being much too specific and not expressive of human thought.

Early on, AI students are taught that this derivation/reasoning process is just another form of search, typically among partial solutions. The relation to best-first search is straightforward: each state represents the sum total of all knowledge either given or derived so far, while the search operators are the basic rules of formal logical (or computer-friendly versions of them), which are applied to portions of the current knowledge state to derive new pieces of knowledge. Typically, the goal of a reasoning search is the answer to a particular query, such as the existence of a student with a double major in religion and evolutionary biology who also captains the swim team, or the most likely cause of a sputtering automobile engine. The problem-specific details vary, as do the types of formal logic and reasoning task, e.g., Boolean logic versus first-order predicate logic (FOPL), or deduction versus abduction. But in many such systems the search metaphor applies either indirectly or as an integral component of the actual computations.

In these approaches a trade-off between expressibility and efficiency often arises when formulating a representation. For example, consider the following collection of logical axioms composing a representation for family relationships (all of which are based on traditional families of the Norman Rockwell era):

$\forall$ X, Y: sibling(X,Y) and female(Y) $\leftrightarrow$ has-sister(X,Y)

$\forall$ X, Y: sibling(X,Y) and male(Y) $\leftrightarrow$ has-brother(X,Y)

$\forall$ X, Y, Z: mother(X,Y) and mother(X,Z) $\rightarrow$ sibling(Y,Z)

$\forall$ X, Y, Z: mother(X,Z) and married(X,Y) $\rightarrow$ father(Y,Z)

$\forall$ X, Y: parent(X,Y) and female(X) $\leftrightarrow$ mother(X,Y)

Using this representation/model of families, one can begin with a few given pieces of information/knowledge:

parent(Carol, Sandra)

parent(Carol, Bobby)

female(Carol)

female(Sandra)

male(Bobby)

married(Carol, Luke)

From this, we can derive all of the following:

mother(Carol,Sandra), mother(Carol,Bobby), sibling(Sandra,Bobby), sibling(Bobby,Sandra), has-sister(Bobby,Sandra), has-brother(Sandra,Bobby), father(Luke,Sandra), father(Luke, Bobby)

In other words, the system could easily answer questions such as, Does Bobby have a sister? Is Luke the father to anyone?

Some questions require more inference (the use of more logical axioms) than others. For instance, determining has-brother(Sandra,Bobby) requires four applications of axioms, whereas mother(Carol,Sandra) needs just one. Thus, one can claim that when given only parental, marital, and gender information, the model is more efficient at inferring motherhood than brotherhood. This is an inherent bias of the model. A stronger bias is that it cannot (by any combination of axioms) answer the simple query, How many children do Luke and Carol have? That answer would require an interpretive phase wherein a human (or some other algorithm) would go through the complete collection of given and derived information and count pairs of occurrences of father and mother relationships involving Carol and Luke and the same child. This is an easy procedure but not one handled by the five given axioms. In fact, counting (while avoiding double counting and an infinite accumulation) is a nontrivial process within the framework of FOPL: the system needs to mark each father and mother relationship as it tallies them using rules such as

$\forall$ W, X, Y, Z: num-kids(W,Y) and parent(W,X) and not(marked(X)) and successor(Y,Z) $\rightarrow$ num-kids(W,Z) and marked(X)

But even this simple fix adds problems by changing the state of marked(X) from false to true, which is forbidden in pure FOPL systems. The bottom line is that logics and all other representations are biased, extremely so, in favor of the types of reasoning tasks that they will perform. Representations have varying degrees of expressivity and efficiency, but they can never represent everything. They can efficiently derive the subset of information that they are designed to handle, but all other types of information are either inefficient or impossible to produce. Despite the claims of their inventors and investors, truly general-purpose knowledge representations do not exist; in the words of Ken Forbus (1997), a renowned AI researcher, "Representation without reasoning is an idle exercise."

The trick is to design representations that support the task and leave room for easy extensions to handle others. One does not want to have to redesign an entire representation every time a new type of query is introduced. Preferably, the addition of a new axiom or two will suffice. A good representation can, to some degree, adapt.

Unlike an engineer, nature has almost no freedom to overhaul its fundamental representation, DNA. The code has taken millions of years to fine-tune, but all future changes must be incremental (unless human intervention eventually introduces a whole new line of DNA and organisms, Life 2.0), since all modifications must properly interact with the *old codes*, lest the organism and its novel new codes die out. Nature the designer faces an overwhelming constraint of backward compatibility that human engineers rarely experience, but over the millennia, it has progressed toward a set of biases that provide rich support for levels of adaptivity that dwarf those of artificial systems. The details of these biases are gradually being discovered, and Bio-AI researchers are listening.

4.4   Bio-Inspired Representations for Local Search

Biology provides useful insights into representations for local search in general and evolutionary algorithms in particular. Some local search methods employ a primitive data structure residing at what GOFAI researchers often call the *syntactic* level. There, the components of the data structure have little intrinsic meaning. For example, they may be bits or integers or floating-point numbers, but at the syntactic level, that is the extent of their meaning.

The syntactic representation achieves meaning via translation within the context of the problem-solving situation. Then the bits and numbers become flags (for a robot's gripper state or for the decision whether or not to buy a stock) or velocities (for a robot's wheels or the rise/fall of a stock price), respectively. This translation process converts a syntactic to a *semantic* representation, one that has immediate meaning within the problem-solving context.

In biology, the syntactic data structure corresponds to genetic code, and biologists speak of the *genotype level*, while the semantic level mirrors that of the phenotype, the outwardly observable properties of an organism; and development is the process that converts genotypes into phenotypes. Bio-AI models often exploit the genotype-phenotype distinction, which fully embodies Deacon's third level of emergence: the genotype, when placed in the proper environment, produces a copy of the organism that the genotype represents.

Many Bio-AI supporters advocate *dropping a level* to consider representations at the genotypic level in order to create advanced, digital intelligence—not necessarily because an understanding of evolution is prerequisite to understanding the brain (though it helps) but because digital models of evolution and the genotype-phenotype distinction may be the best option for designing the degree of adaptivity required by sophisticated intelligence. This is best illustrated by the plight of a seventeenth-century Swedish warship, the *Vasa*.

From 1626 to 1628 the Swedish king Gustavus Adolphus oversaw the building of a large ship, one that was badly needed in the Thiry Years' War (involving most of Europe). Historical accounts vary, but in one of the more popular, the ship was originally designed to have two decks, the uppermost of which would house cannons. Then, late in the building process, after the timbers had been cut for the hull, the king demanded the addition of a second cannon deck. This was done without any redimensioning of the rest of the boat. On its maiden voyage the *Vasa* sank on the way out of port in very light winds; it was much too top-heavy. The boat was salvaged in 1961 and is now one of Sweden's most popular tourist attractions, although possibly a bit of an embarrassment for the Swedish military (Hocker, 2011).

The *Vasa*'s relevance to genotypes, phenotypes, and adaptivity is reasonably clear: when significant changes are needed to the final product (e.g., boat, airplane, or phenotype), it is often impractical, if not impossible, to make them *in isolation*. Rather, the entire design plan (e.g., blueprint or genotype) should be modified and a new product/phenotype generated from it. In the case of the *Vasa*, an extra deck should have entailed a wider hull, which, of course, would have affected nearly every aspect of the overall design. Some changes, such as the addition of an extra window or a few more rungs on a ladder, would not have sent the king and his architects back to the drawing board, but many others should have.

The same holds for adaptive computer systems. Small surprises from the environment (of system deployment) can often be handled by the system via a few small changes to a component or two, without human assistance. So, to some degree, the system can *learn* from environmental interactions and improve its performance over time. Larger changes, however, often require a careful consideration of a host of relations between components, many of which might need modification. This often requires human intervention either to alter many components of the current system in a nonconflicting manner or to return to the original system design, make modifications there, and then regenerate the system from the design plans. The goal of research into truly adaptive systems is to keep humans out of this loop; the system needs to make major changes on its own. Thus, the system should have a means of returning to the drawing board, modifying a low-level representation, and regenerating the higher-level, business end of itself. The alternative, making all changes directly at the high level, can be exceptionally difficult or impossible even for a human expert.

Computer programmers, especially those working close to a deadline, understand the frustration of making hacks to a system, often involving unnatural changes to a host of modules. Although these often get the job done, producing the necessary adaptation requested by the customer, a good programmer knows that when these unnatural tweaks accumulate, a return to the design plans is in order, and a complete rewrite of the system should be considered. Otherwise, the programmer can eventually tweak herself into a corner, followed by disasters similar to the sinking of the *Vasa*.

The genotype-phenotype distinction has another important advantage in nature: relatively small genotypes can generate very large and complex phenotypes. The genotype need not be a blueprint that directly describes each detailed aspect of the phenotype. Rather, it can serve as a recipe for developing the phenotype or, more accurately, a *critical part* of the recipe, a part that re-presents the organism's essence to the environment in third-order emergent fashion. Returning to the array-of-quads representation of Steiner trees for local search, note that the complete recipe for producing the spanning trees includes Kruskal's algorithm: the representation provides critical constraints on that algorithm but not the details of the edge-adding process itself.

Bio-AI attempts to harness these same types of third-order emergent, genotype-phenotype relations to design adaptive intelligent systems. In almost all such cases, a developmental algorithm is designed ahead of time. The genotype then provides critical inputs to that process. These Bio-AI systems do not *exhibit* third-order emergence, since the genotypic format and developmental routines are hand-coded a priori, but they *utilize* it to a much greater extent than other approaches to design search.

Another advantage of explicit syntactic and semantic levels, with clear separation, lies in the supported generality. For sufficiently general syntaxes, such as bit vectors or Boolean expressions, a basic set of operators exists for converting one syntactic form into another. For instance, bits can be flipped or swapped (between vector positions), or conjuncts can be added or removed, respectively. Hence, the movement of search is easily performed, and in the same manner for any problem that is amenable to the chosen syntax. The impetus for movement, the objective function, is problem-dependent, but the basic mechanisms of state-to-state transition are generic and need not be recoded for each new problem domain. Human problem solvers who forfeit this separation must design special-purpose operators for state transitions, while those who maintain it can take advantage of decades (or centuries) of research into representations with well-defined syntax and operators.

Similarly, all living organisms take full advantage of the genetic code, a syntactic substrate tested and tuned over hundreds of millions of years, and one supported by a small but powerful set of operators such as mutation, inversion, and crossover. This one code produces millions of viable phenotypes.

In this book the terms *genotype*, *phenotype*, and *development* get plenty of use, often in the context of an evolutionary algorithm. However, they can often be re-expressed as *syntax*, *semantics*, and *translation*, respectively, and be generalized to the entire suite of local search methods, all of which must deal with representational issues in order to most effectively navigate design space.

In systems that maintain a genotype-phenotype distinction, the genotype/syntax provides the basis for the phenotype/semantics, but the reverse effect, i.e., the degree to which semantics influences syntax, varies considerably. In biology (and most EAs) the genetic operators working on genotypes pay no attention to the corresponding phenotypes. For example, mutation involves random changes to DNA, not changes designed to enhance some aspect of the phenotype. Nature and EAs rely on selection to filter out the deleterious mutations. In contrast, many local search methods, for example, the vast majority of those specialized for finding Steiner trees, make syntactic changes based directly upon the overlying semantics. Steiner points are not placed between randomly generated vertices, only between those known to be neighbors.

Most AI and operations research workers would not attack the Steiner tree problem using the integer vector representation presented earlier. Instead, they might employ a data structure that strongly reflects the semantics, such as a cluster hierarchy of vertices (with nearby vertices in the same or closely linked clusters). Search operators would then act directly upon that structure, abiding by the constraints imposed by the problem-solving domain. This would decrease the odds of spurious modification: most changes would be meaningful, reflecting *intelligence* and *knowledge* of the problem.

On the other hand, bio-inspired techniques often rely upon random changes to generic syntactic representations, which must then be converted into semantic-level solutions for evaluation by an objective function. Only at that point can the wisdom of a modification be fully assessed, i.e., only in retrospect. Nature and Bio-AI are replete with this act-first, evaluate-later behavior.

4.5   A Representation Puzzle

A simple example shows the potential advantage of a significant genotype-phenotype difference in cases where interdependencies abound among components of the system under design. As shown in figure 4.2, the task is to place four rectangles on the background grid so that no rectangles overlap or extend outside the grid. The algorithm penalizes overlap more harshly than grid boundary violations when evaluating a solution. Although puzzles of this type are often

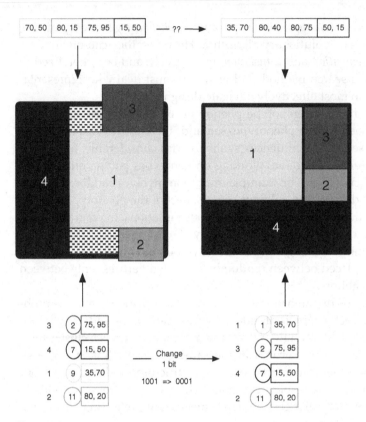

Figure 4.2
Comparison of a direct representation (*top*) and an indirect representation (*bottom*) for solving a puzzle via local search. The goal is to place all four rectangles on the background grid with no piece overlap or overrun of the grid border. The left-hand and right-hand representations encode the same puzzle solution, but the indirect ones require a complex translation process that involves an outwardly expanding search in grid space such that pieces *push* one another to locations that avoid overlap penalties.

successfully attacked by best-first search, a local search is now assumed in order to compare two different representations.

A local search algorithm will adjust whole solutions in search of improvement, with the choice of representation and state change operators having a strong influence on performance. First consider the representation at the top of figure 4.2. For each of the four pieces, a two-coordinate code specifies the midpoint of the rectangle on the 100 × 100 background grid. Pieces are presumably rotated 90 degrees if their default orientation overlaps another piece. The default orientations of all rectangles appear in the left-hand solution of the figure, which has no overlap but a few boundary violations.

Figure 4.2 (right) displays an optimal solution, which tightly packs the grid, with no overlap nor overrun. Note that the representations in the upper left and right vary considerably. Given the left-hand solution, the system might randomly (or methodically, intelligently) decide that piece 1, the largest rectangle, should move to the upper left grid corner. However, like adding an extra gun deck to the *Vasa*, this single change has repercussions throughout the design: it affects the (appropriate) position of every other piece. Changing only the position of piece 1 in the solution will produce a greatly inferior solution because of overlaps with pieces 3 and 4. However, even a methodical local search algorithm has trouble making many changes in one step; these constitute *long jumps* in search space that are often hard to predict. Any attempt to move only piece 1 or piece 4 will lead to overlap, whereas pieces 2 and 3 have some degree of horizontal freedom. But in no case can an isolated change bring an increased evaluation, so local search can easily bog down. The representation turns out to be too *direct* for the problem, more similar to a blueprint than a recipe.

Now consider another, indirect, representation, with a more complex translation routine. Shown along the bottom of figure 4.2, this coding format has a triple of integers for each piece, with each integer encoded as a binary vector. A triple's first integer (encircled) serves as a priority, while the other two comprise a *preferred location*. Pieces are ordered by ascending priority codes, with earlier pieces being the first placed on the grid. A piece added later may be moved from its preferred location (L) via a localized search that spirals outward from L until a spot without overlap (under either orientation) is found. In short, the translation routine uses search to resolve conflicts in the same way that a crowded population of cells could jostle around to fill a cavity.

At the bottom of figure 4.2, the representation in this indirect format is drawn vertically to show the priorities. On the bottom left, the encircled codes dictate that piece 3 is placed first, then piece 4, then 1, and finally 2. This produces the same solution as does the direct representation on the top left. However, with the indirect representation, a single bit flip suffices to reorganize the entire puzzle, producing the optimal solution: when the priority code bits for piece 1 change from 1001 (a 9) to 0001 (a 1), this single change bumps piece 1 to the top of the priority queue. It now receives its desired location without the need for a spiral search. Instead, pieces 3, 4, and 2 must deviate from their desired locations, with 3 and 2 finding nearby locations, 4 needing to spiral a bit to find a suitable center, and both 3 and 4 reorienting by 90 degrees.

One simple change to a portion of the *recipe* does the job because, with recipes, some minor tweaks can have large and global consequences. (If *some* becomes *most*, local search can encounter all sorts of problems, but more on that later.) The main point for now is not that jigsaw puzzles should be solved by local search instead of A*, or that this indirect representation is a highly desirable approach to spatial brain teasers, but that a sizable gap between a genotypic and phenotypic representation can facilitate efficient search when that gap is being filled by a translation/developmental routine that can account for some interactions (and enforce some constraints) between the components of a solution. In these cases, going back to the drawing board means making modifications to the genotype in hopes of achieving improvements in the phenotype, instead of looking at the latter directly and making adjustments there, because, as the example illustrates, one local change to a phenotype often has widespread effects that require many additional changes that can be hard to coordinate. It's often easier to return to the genotype, make changes there, and let the translation process iron out the conflicts to produce a viable phenotype.

4.6   A Common Framework for Emergent AI

The genotype-phenotype distinction lies at the heart of many bio-inspired attempts to build adaptive intelligence. These systems often follow the same basic scheme:

1. Individual solutions to the problem (of building an intelligent system) are represented in full (not partially). Each solution can be called an *agent*.

2. Each agent has a genotype and a phenotype.

3. The phenotype arises from the genotype via a predefined developmental process.

4. Phenotypes are exposed to an *environment* of some sort, and their performance within it is assessed.

5. Phenotypes have adaptive abilities that come into play during their *lifetime* within the environment. This may improve their evaluation, particularly in environments that change frequently, throwing many surprises at the phenotypes.

6. Phenotypic evaluations affect the probabilities that their agents become focal points (or forgotten designs) of the overall search process.

7. New agents are created by combining and modifying the genotypes of existing agents, then producing phenotypes from the new genotypes.

This framework employs several different representations. First, in many instances, the phenotype will be considered an abstract model of a natural system (e.g., the brain) or process (e.g., logical reasoning) and thus have legitimate representational status, such as that given to GOFAI systems designed to test psychological theories. To be adaptive, this phenotype must tackle environmental surprises via internal changes, the concerted combination of which may embody parallel search leading to emergent patterns. Second, the genotype constitutes an alternative encoding/representation of the phenotype, filling the role of a representation in Deacon's third level of emergence. Adaptivity at this level implies an amenability to genetic operators (such as mutation, inversion, and recombination) such that the encodings of novel phenotypes are produced, and most new genotypes develop into functional (though not necessarily optimal) phenotypes. Finally, the phenotype will possess internal states that correspond to aspects of the environment and the agent's experiences within it. These representations of the agent and world are generally of a more indirect nature than many of GOFAI's classic KR formalisms such as logical axioms, semantic nets, scripts, and frames (Rich, 1983).

These are the three most important representations in this book's approach to Bio-AI, in the sense that the *quality* really matters with respect to each mapping: (1) from phenotype to real-world system, (2) from genotype to phenotype, and (3) from phenotypic structure and dynamics to the agent's environment.

Other Bio-AI researchers may disagree, pointing to additional representations in this general scheme whose quality (i.e., trueness to the system or process being modeled) matters. For example, the developmental process captures considerable interest among Bio-AI researchers. Many feel that nature's ability to produce complex phenotypes from a single cell that replicates and differentiates (with each new cell holding its own copy of the original genome) is a vital prerequisite to the evolutionary design of complex systems. Hence, the quality of an artificial model of development vis-à-vis nature's approach has great significance for these scientists. Similarly, many feel that nature's evolutionary process, in all its detail, is critical, so any artificial models of it must remain true to neo-Darwinism. Others feel that the environments into which AI systems are deployed should have a strong resemblance to the real physical world, i.e., they should model it accurately, while still others feel that no model is good enough: the agents need to be robots that sense and act in our world.

The *inspiration* in bio-inspired AI has many degrees, however, and this book's is best characterized as mild. The genotype-phenotype distinction is important because, as discussed, (1) problems of scale can be combated with small genotypes that recursively produce large phenotypes, (2) well-established syntactic representations can be reused as the basis for many semantic representations, and perhaps most important, (3) an intelligent system often needs to return to the drawing board to adapt.

By sending a modified genotype through a developmental process, the necessary changes to several components can be automatically coordinated by the translation algorithm, which typically houses important constraints that forbid the production of unviable phenotypes. However, the particular process by which genotypes become phenotypes matters less, under the view that natural development is a critical *artifact* of the constraints of the real world, of the fact that reproduction, prenatal development, and childbirth go most smoothly when things *start small*. Computational systems certainly have resource constraints, but none of these demand that the designs/recipes for intelligent phenotypes be tailored for self-copying modules. Similarly, the three basic characteristics of evolution (variation, inheritance, and selection) seem vitally important for the success of evolutionary algorithms, but many of the genetic details seem only peripherally relevant for the success of parallel local search.

4.7   Representations for Evolving Artificial Neural Networks

The general scheme described earlier has many instantiations. One of the most popular, and the focus of this book, is the simulated evolution of ANNs (artificial neural networks) or EANNs (evolving artificial neural networks). Figure 4.3 depicts the overall process wherein a population of genotypes (whose syntax is often very simple, such as a vector of bits) is translated into neural network phenotypes, which are then subjected to some form of environment to assess their fitness. These evaluations then bias the choice of genotypes used by the genetic operators to create a new generation of genotypes. The phenotypic ANNs can be any of a wide variety of models found in the literature (Arbib, 2003; Callan, 1999; Haykin, 1999), and the concept of environment generalizes to anything from a data set (for a typical machine-learning classification task) to a full-blown simulation of a three-dimensional world to the world itself (with the ANN as the controller of a robot).

Here, representations involving the genotype-phenotype mapping and the ANN-brain mapping deserve special attention. In each, the general support of search is critical. Genotypes must facilitate the evolutionary search for successful phenotypes, while the structure and dynamics of ANNs support the exploratory, trial-and-error processes that underlie lifetime adaptation.

As discussed more thoroughly in subsequent chapters, Bio-AI researchers have extensively explored both types of representation. Literally thousands of different ANN models exist, many of which are designed to mimic some aspect of the mammalian brain and perform at least one type of online modification reminiscent of lifetime learning.

A comparable number of genotype-phenotype mappings have been attempted in the EA community. These rarely disappoint as examples of profound creativity, just perfect for demonstrating the joys of crafting search-friendly, low-level

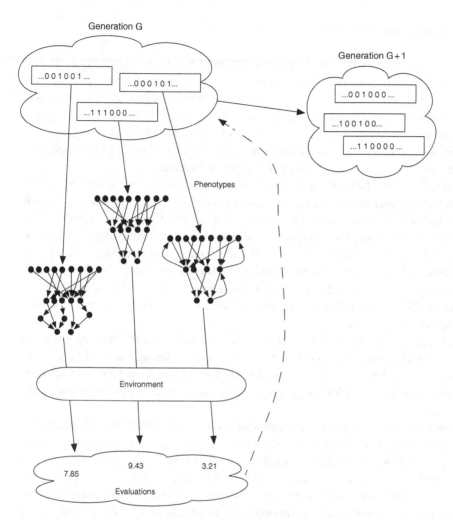

Figure 4.3
Overview of evolving artificial neural networks (EANNs). Genotypes (bit strings) of generation G are converted into neural network phenotypes, which are then tested in an environment. The resulting evaluations feed back on the genotypes by affecting their frequency of appearance in generation G + 1, which results from the application of genetic operators (such as mutation and crossover) to the genotypes of generation G.

encodings of high-level designs. Whereas biologists can only rarely and painstakingly show vertical slices of the transition from particular DNA strings to concrete phenotypes, EA programmers do it all the time, thus illustrating the power of genotype-phenotype interplay to life scientists and engineers alike.

4.8   Representational Trade-offs

The preceding discussion points to three different representational types of particular relevance for emergent intelligence. First, ANNs and EANNs can serve as models/representations of brains and the evolution of brains, respectively. Second, the patterns of synaptic strengths and neural firings within neural networks represent knowledge of the agent's history and environment. Third, the genotypic encodings of ANNs (and agents) function as representations of a phenotypic self, the pinnacle of emergence in Deacon's three-tiered scheme.

For these representations, modeling choices strongly influence a search process that, in turn, affects the emerging results, in much the same way that classic KR design decisions bias automated reasoning. The main difference is emergence. In classic KR the result of reasoning is typically a symbol or simple logical combination of symbols that already existed in the knowledge base but was not a priori recognized as part of the inferential closure. In Bio-AI the results tend to be complex, more unexpected combinations of the original primitives. There is legitimate surprise in the results. What Bio-AI representations lack in logical formality, they make up for in creativity.

Indeed, many logical formalisms are expressive, allowing the representation of many forms of knowledge. But the rules of logical entailment severely constrain reasoning to produce only sound inferences. Bio-AI has looser bounds on construction, allowing a wider range of products, some ridiculous but some quite impressive.

The move to Bio-AI requires an appreciation and embrace of this uncertainty, a willingness to trust emulations of natural processes over safer, mathematically verifiable procedures designed by humans to model rational thought. For cognitive science and its goal of understanding intelligence, the choice is between hoping that logic-inspired representations and their formal inference mechanisms suffice to explain intelligence, in all its rational and irrational splendor, or banking on relatively unbridled creative forces to do the same. Which is harder, producing irrational from rational, or vice versa? This book offers no definitive answers but does shamelessly promote self-organization and its ability to produce structures of undeniable cognitive significance.

5 Evolutionary Algorithms

5.1 Darwinian Evolution

In 1789, Thomas Malthus wrote his *Essay on the Principle of Population*, in which he recognized that population growth rate is a function of population size, and therefore, if left unchecked, a population will grow exponentially. However, since environments have only finite resources, a growing population will eventually reach a point, the *Malthusian crunch*, at which organisms will have to compete for those resources and will produce more young than the environment can support.

Crunch time was Darwin's (1859) groundbreaking entry point into the discussion:

> As many more individuals of each species are born than can possibly survive; and as, consequently, there is a frequently recurring Struggle for Existence, it follows that any being, if it vary however slightly in any manner profitable to itself, under the complex and sometimes varying conditions of life, will have a better chance of surviving, and thus be naturally selected. (5)

In short, the combination of resource competition and heritable fitness variation leads to evolution by natural selection. When the Malthusian crunch comes, if there is any variation in the population that is significant in the sense that some individuals are better equipped for survival and reproduction than others, and if those essential advantages can be passed on to offspring, then the population as a whole will gradually become better adapted to its environment as more individuals are born with the desirable traits. In short, the population will *evolve*. Of course, this assumes that populations can change much faster than geographic factors.

In addition to the pressure to favor certain traits over others (known as *selection pressure*) and the heritability of desired traits that enables the features of well-adapted individuals to spread throughout and eventually dominate the population, the concept of variation is also essential to evolution, not merely as a precondition to population takeover by a dominant set of traits but as a perpetual process insuring that the population never completely stagnates lest it fall out of step with an environment that inevitably changes.

Thus, the three essential ingredients for an evolutionary process are

- *Selection*   Some environmental factors must favor certain traits over others.
- *Variation*   Individuals must consistently arise that are significantly (although not necessarily dramatically) different from their ancestors.
- *Heritability*   Children must, on average, inherit a good many traits from their parents to insure that selected traits survive generational turnover.

These three factors are implicit in the basic evolutionary cycle, depicted in figure 5.1. As shown at the lower left of the figure, a collection of genetic blueprints (more accurately, recipes for growth) known as genotypes are present in an environment. These might be (fertilized) fish eggs on the bottom of a pond, or the collection of all one-day-old embryos in the wombs of a gazelle population. As one moves upward in the diagram, one sees that each genotype directs a developmental process that produces a juvenile organism, a young phenotype. At the phenotypic level, traits (e.g., leg length, coloration patterns, hair thickness) encoded by genotypes become explicit in the organism.

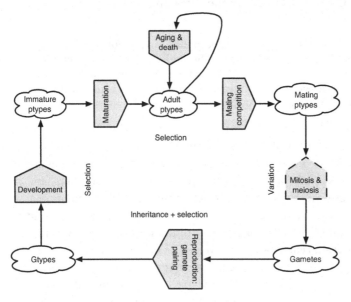

Figure 5.1
Basic cycle of evolution. Clouds represent populations of genotypes (gtypes) or phenotypes (ptypes); solid-line pentagons, processes that filter the pool of individuals; and the dashed-line pentagon, an increase (in this case, of genetic material).

In figure 5.1 selection pressure is present in all processes denoted by solid-line pentagons. These represent the metaphorical *sieve of selection* through which populations (of genotypes and phenotypes) must pass. Already during development, this sieve may filter out genotypes that encode fatal growth plans, e.g., those that may lead to miscarriages in mammals. The sieve is relentless, however, so even genotypes that encode plans for healthy juveniles have little guarantee of proliferation; selection pressure persists. Juveniles must survive the torrents and turmoils of life while growing to adulthood, only to enter a new arena of competition, at the highest level, for the right to produce offspring. Thus, by the time organisms reach the mating phenotypes (upper right) the genotypic pool has been narrowed considerably. In many species, individuals need not perish for lack of a mate; some can be recycled and try again next mating season, as depicted in figure 5.1 by the aging/death filter above the adult phenotype collection.

From the upper right corner of figure 5.1, one returns to the genotypic level via the production of gametes. This occurs within each organism via mitosis (copying) and meiosis (crossover recombination), resulting in many half-genotypes (in diploid organisms) that normally embody minor variations of the parent's genes. During mating, these half-genotypes pair up to produce a complete genotype, a new plan for development that is normally very similar to but slightly different from that of each parent. There is normally an overproduction of gametes; only a chosen few become part of the new genotypes. Hence, even reproduction (figure 5.1, bottom) is a sieve.

Although pinpointing the exact locations of variation and heritability is difficult, since genetic mutations can occur to germ cells (i.e., future gametes) at any time during life, it seems fair to say that the major sources of genetic variation are the (imperfect) copying and recombination processes during mitosis and meiosis, respectively. Inheritance is then located along the path from gametes to genotypes, since this is the point at which the parents' DNA (and the traits it encodes) officially makes it into the next generation.

Via repeated cycling through this loop, a population gradually adapts to a (relatively static) environment. In engineering terms, we can say that the population evolves such that its individuals become better *designs* or better *solutions* to the challenges that the environment poses.

5.2 Artificial Evolution in a Computer

The field of evolutionary computation (EC), detailed in a variety of excellent textbooks (M. Mitchell, 1996; Goldberg, 1989; Banzhaf et al., 1998; De Jong, 2006; Eiben and Smith, 2007), capitalizes on the metaphor of evolution as a problem-solving and design process. By developing evolutionary algorithms (EAs) that incorporate bits and pieces of the kernel evolutionary process, EC practitioners solve complicated problems that are often beyond the reach of more traditional, more formal optimization and design tools. This is not without its price, however, since evolution often finds rather strange solutions/designs that are hard to analyze and explain. But in many cases, it is precisely this off the board thinking that is required for creative, groundbreaking progress. EC has had its share of such successes, in areas as diverse as antenna design, music, art, movie animation, circuit design, and proteomics.

Figure 5.2 illustrates the basic cycle of an evolutionary algorithm (see appendix A for general implementation issues). Different types of EAs include different subsets of the components shown, but all incorporate mechanisms for selection, inheritance, and variation.

At the lower left of figure 5.2, one sees that some EAs use distinct genotypic representations, such as bit strings, that must be translated into phenotypes, which usually represent explicit designs or problem solutions. In other cases, the genotype and phenotype are, for all intents and purposes, the same. The conversion

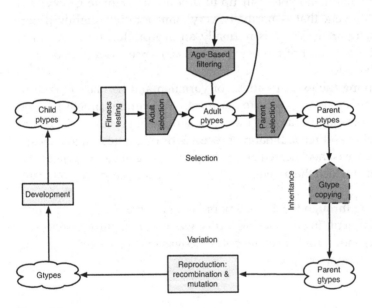

Figure 5.2
Cycle of an evolutionary algorithm. Clouds represent populations of genotypes (gtypes) or phenotypes (ptypes); shaded polygons, processes; solid-line pentagons, processes that filter the pool of individuals; the dashed-line pentagon, an increase (in this case, of genotypes); and rectangles, processes that produce no net change in the genotype and phenotype pools.

from genotype to phenotype, termed *development* in the diagram, can be everything from a simple copy operation to a detailed growth process. The relation between genotype and phenotype determines the form of the recombination and mutation operations (see chapter 8).

In population biology *fitness* generally refers to an individuals ability to produce offspring (Roughgarden, 1996). In evolutionary computation such a definition would appear circular, since EAs use fitness values to *determine* reproductive success. EAs, unlike biologists, cannot watch a population reproduce and then afterward assign fitness to the productive individuals. EAs must be more proactive by stepping in and restricting access to the next generation. The fitness evaluation is the first (and main) step in that process.

The fitness function is applied to each phenotype directly after it is generated from the genotype. These functions vary with the problem domain, but in general, they quantify each phenotype's success on a particular problem. For example, an EA for solving a traveling salesman problem would give higher fitness to shorter paths, while one for attempting circuit design might reward energy-efficient solutions with higher values. The most important aspect of the fitness function is that it should give *graded* evaluations such that good, but not necessarily optimal, solutions receive some partial credit. Without it, evolutionary search degrades to a hopeless, random, needle-in-a-haystack hunt.

Generally speaking, an EA has little knowledge about *how* a good solution/hypothesis should be designed but solid information about *what* properties it should have. Typically, the fitness function operationalizes a lot of this *what* knowledge. Whereas the phenotypic representation and the genotype-to-phenotype mapping embody knowledge of how to generate a *legal* individual, it is the fitness function that assesses the individual's *quality* with respect to the problem. So the extent of *how-to-construct* knowledge in the EA is normally restricted to legal phenotypes, not superior ones. Genetic operators generate hypotheses, and fitness values are instrumental in filtering them (see appendix A for details on fitness functions).

As soon as a new individual is created (during initialization, by cloning or by mating), its fitness value is computed and retained for all selection operations that follow: adult and parent selection. As shown in figure 5.3, the newly created children may simply take over the adult pool, a process known as *full generational replacement*. Alternatively, in *generational mixing* the children and previous generation of adults compete for these K pool spots; most EAs simply chose the K highest-fitness individuals from the the union of the two groups. In either case, *overproduction* of children may occur: the EA may generate more than K children during each round of reproduction.

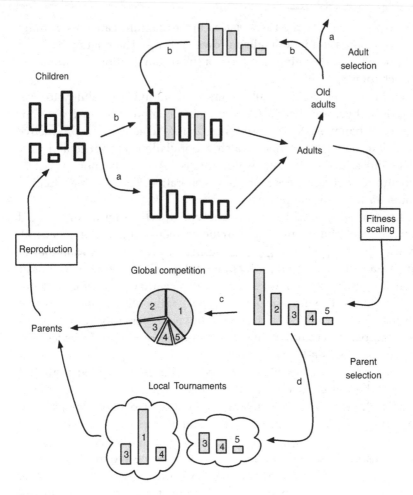

Figure 5.3
(*Top*) Adult selection: either newly generated individuals (children) automatically fill the adult pool and previous (old) adults are discarded (option a), or the children and old adults compete for a limited number of spots in the adult pool (option b). (*Bottom*) Parent selection: after their fitness values have been scaled, the adults compete either globally (against every other adult) or locally (within randomly selected tournament groups) to participate in one or more matings. Sizes of rectangles denote relative fitness values of individuals.

The adult fitness values are scaled to realize an important aspect of EA selection pressure. These scaled values are then used during parent selection, when adults compete to produce children either sexually or by asexual cloning. In global selection the EA repeatedly makes stochastic choices from the complete set of parents, a process akin to spinning a roulette wheel, where the size of an individual's slice

of the wheel is proportional to its scaled fitness. In local selection randomly chosen subsets of adults compete to participate in each mating (or cloning) event (see
appendix A for details on fitness scaling and selection strategies).

In living organisms mutation and recombination normally occur prior to gamete
union; essentially, children recombine their parents genes in preparation for producing their parent's grandchildren. EAs typically employ a simpler scheme: they
select two parents, copy their genotypes, and then mutate and recombine those
copies to produce one or two new genotypes. Hence, in EAs that use crossover,
children *receive* recombined genomes from their parents.

In EA terminology, mutation and crossover are *genetic operators*. They are performed on the genotypes, not the phenotypes, although constraints from the phenotypic level may be used to bias the genomic manipulations. Mutation involves
single genotypes, while crossover involves groups of two (or occasionally more).
In mutation, a small component of an individual is randomly changed. If the genotype is a bit vector, then one of the bits is simply flipped. For more complex genotypes, such as a permutation, two integers of the permutation are swapped.

In general, crossover involves taking parts from two or more parent genotypes
and combining them into one (or several) new child genotypes. With bit vector
genotypes, crossover swaps pairs of bit segments, as shown in figure 5.4. With

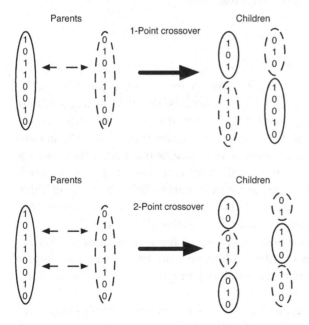

Figure 5.4
Single- and double-point crossover operations on bit vectors

more complicated genotypes, the design of a crossover operator requires careful thought, since genotypic segments may be less modular and hence more difficult to mix and match under the standard constraint that the resulting child genotypes should translate into legal phenotypes.

The dashed-line pentagon in figure 5.2 indicates that some of the mating adults may produce several copies of their genotypes, portions of which will then appear in newly formed genotypes. The Reproduction rectangle indicates that, for the most part, all portions of the copied parent genotypes will appear in a new child genotype.

As figure 5.2 shows, the three critical evolutionary components of selection, inheritance, and variation are cleanly modularized in EAs. Adult and mate selection are the only points where the population is filtered; chosen parents are guaranteed to pass on their genotypes via simple copying; and mutation and recombination operators create the genotypic variations so critical to evolutionary progress.

After many trips around the EA cycle, the average fitness of phenotypes increases as better and better solutions/designs arise. Eventually, optimal solutions may be found, although there is no guarantee of optimality with EAs, in contrast to exhaustive (and often computationally exhausting) brute force methods such as breadth-first search and dynamic programming.

5.3   Spy versus Spy

As a simple example of an EA in action, consider the *short taps puzzle* (Shasha, 2003), in which a CIA official (spy 1) attempts to send a series of N messages within a given time limit, T. The messages have different lengths, and their sending intervals can overlap with no limit on the number of parallel broadcasts. The enemy (spy 2) is known to be tapping radio signals, but since he must enter the sender's line of sight to steal signals, spy 2 must limit tapping to a single K-minute interval. To tap a message, spy 2 must intercept it from start to finish. Spy 1 (still bitter after a recent salary dispute) has dropped his vigilance level a bit and is therefore willing to have some messages tapped, but not more than M of them.

Spy 1's task is to schedule the sending times of the N messages so that no K-minute interval between times 0 and $T - 1$ fully contains more than M messages. Clearly, many versions of the problem exist, depending upon the parameters T, N, M, K, and the messages' durations.

A straightforward EA to solve this puzzle uses a genotype with N genes, one for each message. As shown in figure 5.5 (left), each gene simply encodes the start time for its message. Each phenotype is a complete schedule for the messages: a list

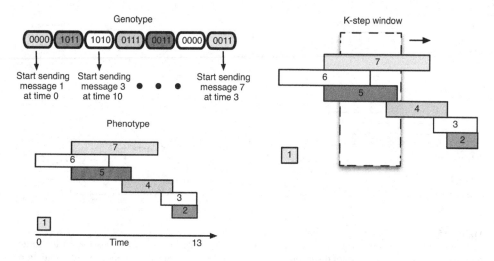

Figure 5.5
(*Left*) Conversion from bit string genotype to phenotypic schedule for a seven-message puzzle. (*Right*) Use of a sliding window to test the phenotype's vulnerability.

of sending intervals. These intervals are formed by the pairs $(s, s + d - 1)$ for each gene, where s is the start time and d is the message duration. For example, $(3, 7)$ is a five-timestep message starting at the beginning of timestep 3 and persisting until the end of timestep 7.

To assess fitness, calculate the *vulnerability*, V, of a phenotype by moving a K-unit time window along the time axis, checking each interval $(0, K - 1)$, $(1, K)$, ..., $(T - K, T - 1)$, as shown on the right of figure 5.5. At each location calculate H (*hits*), the number of messages that would be fully intercepted if the spy were to perform the tap during this exact interval. If $H > M$, then increment V by the amount $H - M$.

After all tap windows have been tested, V should give a good indication of the sending schedule's total vulnerability, which should be inversely related to fitness. A simple fitness function is thus

$$\text{Fitness} = \frac{1}{1 + V}. \tag{5.1}$$

One version of the puzzle involves $N = 7$ messages of durations 2, 3, 4, ..., 8 time units, with $T = 15$, $K = 10$, and $M = 3$. An EA with a population size of only five individuals can find a solution (the phenotype in figure 5.5) in less than 50 generations. A more complicated (ten-message) version of the puzzle supplements the original seven messages with three more, all of duration 4, and increases T to 20.

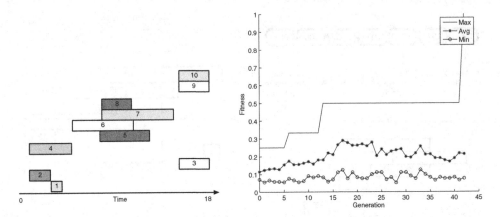

Figure 5.6
(*Left*) Solution for a ten-message spy versus spy puzzle. (*Right*) Evolutionary progression for the same ten-message puzzle, with plots of the minimum, maximum, and average fitness of the adult population for each generation.

Figure 5.6 shows a solution for this ten-message puzzle, with population size of 20, along with the evolutionary progression. In the figure (right) notice that the evolutionary progression is not one of continuous improvement but rather a series of *punctuated equilibria*, just as in nature.

5.4   The Main Classes of Evolutionary Algorithms

Evolutionary computation revolves around the belief that random variations and selection combine to produce incredible complexity of structure and behavior. However, the field is divided into several camps whose adherents share similar fundamental views but insist on minor (at least to the outsider) differences. This has led to some animosity.

The four main EA types are evolutionary strategies (ES), evolutionary programming (EP), genetic algorithms (GA), and genetic programming (GP). Each shares the basic EA design philosophy of (relatively) domain-independent hypothesis formation followed by domain-specific selection. Thus, each follows the basic scheme of figure 5.2. However, they differ in primary representations, the relative importance of mutation versus crossover, and preferred selection strategies.

5.4.1   Evolutionary Strategies
In the 1960s, Rechenberg and Schwefel invented evolutionary strategies (Schwefel, 1995) in an attempt to solve search problems in hydrodynamics by randomly

mutating existing solutions, assessing fitness, and applying selection. Their genotypes were vectors of real values that were mutated by making small changes to
the reals, with the magnitude of change drawn from a Gaussian distribution centered at 0. Hence, small mutations were much more common than large, and evolution in ES was normally very gradual, especially given the fact that, in the earliest
attempts, populations were of size 1.

Although ES initially ignored recombination, later work introduced several variations, including both simple vector segment swapping and the combining of corresponding vector values (e.g., via averaging) from the two parents. For example,
if parents A and B had values 6 and 10, respectively, for the same gene, then the
child would receive an 8 for that gene.

ES traditionally uses the overproduction of children followed by either full generational replacement or generational mixing. The latter provides a measure of
what EA researchers call *elitism*, since good parents can survive many generations
by outcompeting their offspring, and thus the best solutions (the elites) never disappear from the population.

ES researchers also introduced the concept of genetically determined variation
parameters, wherein factors such as mutation and crossover rates and Gaussian
variances are encoded in the genome and thus open to evolutionary control. For
example, in many simulations an optimal evolutionary search employs high mutation rates early on (for greater exploration) and lower rates (for more exploitation)
as the population converges on an optimal region.

ES is a strongly engineering-driven approach to evolutionary computation.
Thus, those who use it are most interested in finding optimal solutions to technical problems, not in testing theories of evolution or intelligence.

5.4.2 Evolutionary Programming

Lawrence Fogel (1966) had both engineering and artificial intelligence in mind
when he invented evolutionary programming in the 1960s. He viewed simulated
evolution as a potent tool for solving engineering problems and achieving human-
level machine intelligence, even consciousness.

The philosophical difference between EP and the other EAs lies in the view of
each phenotype as representing an entire species. Since, by definition, members
of one species cannot mate with members of another, the concept of genotype
recombination has no place in EP. Hence, it is the only EA in which crossover is
forbidden.

This philosophy also implies that each parent/species should produce some offspring, so early EP systems used *uniform mate/parent selection* in which each parent
produced the same number of offspring (originally just 1). Adult selection was
typically generational mixing.

EP was somewhat of a precursor to genetic programming in that the original applications were the design of finite-state automata (FSAs) for sequential prediction tasks. FSAs consist of states, rules for interstate transitions, and criteria for producing output. In theory, FSAs can be souped up to be Turing-equivalent, but the basic versions are not. Regardless, FSAs provide only rather cumbersome solutions to problems outside their primary domain of sequence generation and recognition, so their generality as a programming substrate is far inferior to even the simplified LISP-like syntax used in GP.

Typical EP genomes are integer vectors that encode the states, output conditions, and transitions for an FSA. Gaussian mutations similar to ES are frequently employed; and like ES, EP has a form of self-adaptation. In this case, Gaussian variance is directly proportional to the distance from the current population to the known optimal value. So once again, as the population approaches an optimal solution, mutation decreases (in magnitude although not necessarily in frequency).

One of EP's most celebrated achievements comes from David Fogel, son of Lawrence Fogel. He and Kumar Chellapilla used the basic EP framework to evolve checkers-playing neural networks. As detailed in Fogel's book (2002), their networks improved by playing against one another and against humans on a popular checkers website (zone.com). Their virtual player, Blondie24, climbed the site's rankings to the elite 99.5 percentile; on one occasion it even beat Chinook, the world checkers champion (a human-written computer program). This served as at least partial confirmation that evolution could indeed be used to achieve impressive artificial intelligence.

5.4.3 Genetic Algorithms

Also in the 1960s, John Holland recognized the potential of evolutionary mechanisms for artificial adaptive systems (Holland, 1992). His classifier systems combined learning and evolution using the same bit string representation. If there is one name that most often comes to mind in connection with EAs, it is John Holland's. Many, if not an overwhelming majority, of today's leading EA researchers are either students or "grandstudents" of Holland. Adjectives such as *monumental* and *ground-breaking* only begin to describe his contributions to EA and complex adaptive systems (CAS).

Today, the main difference between GAs and other EAs is the bit vector representation, which continues to be a staple in the field. In this respect, GAs are the only EAs that maintain any significant syntactic distance between genotypes and phenotypes, one key exception being developmental genetic programs (see chapters 8 and 12). Also, whereas EP forbids crossover and ES uses it sparingly, GA adherents usually consider it the most significant source of variation.

Holland's seminal theoretical contribution comes from the *schema theorem*, which helps predict the frequency changes of useful gene combinations, or *building blocks*, over the course of evolution. It takes both fitness and the sequence-disrupting effects of mutation and crossover into account to derive the minimum speed at which a useful building block will dominate a population and an inferior one will disappear. Not incidentally, the theorem also justifies the use of binary genotypes as opposed to closer-to-phenotype representations.

Holland's original GA employed binary genomes, full generational replacement for adult selection, and fitness-proportionate (global) parent selection. However, contemporary GA researchers seem open to just about any type of representation and selection strategy, although there still appear to be strong ideological (if not purely political) divides between the GA and GP communities.

5.4.4 Genetic Programming

Invented by John Koza in the late 1980s, genetic programming (Koza, 1992; Banzhaf et al., 1998) is clearly the most radical of the EAs. Few serious computer scientists would pursue the wild idea of randomly combining fully functioning computer programs to generate new, better ones. Koza (also a Holland student) was the clever exception, and his perseverance has yielded a gold mine of results that push EAs well beyond the level of interesting novelty to potent problem-solving and creative design tool. GP applications run the gauntlet from musical composition and artistic design (Bentley and Corne, 2001) to state-of-the-art (in some cases patentable or patent-infringing) inventions of electrical circuits, antennas, and factory controllers (Koza et al., 1999; Koza, 2003). Myriad successful applications are summarized by Banzhaf et al. (1998), who also give an excellent (and relatively concise) overview of GP, clearly showing its relation to other EA approaches.

Additions to the original simplified-LISP version of GP include looping constructs, memory, and strong typing. Perhaps the most important enhancements were subroutines, which greatly improved the efficiency of GP search, as thoroughly documented by Koza (1994). The ease of linearizing GP trees permitted a shift from LISP to C as the basis for large GP systems, although GP genotypes are still typically drawn as trees.

The discovery of cellular encoding (CE) (Gruau, 1994) (see chapter 12) set the stage for a long line of *human-competitive* GP results. Originally designed to evolve artificial neural networks (ANNs) with GP, CE proved effective in evolving graph topologies for a broad range of domains, including biochemistry, electronics, and control theory (Koza et al., 1999). CE also widened the (normally thin) gap between GP genotypes and phenotypes by using the genotype as a procedure for developing the phenotype.

5.4.5   Convergence of EA Types

Evolutionary biologists often speak of *convergence* in reference to similar traits that evolve independently. Eyes are an excellent example, having evolved on many separate occasions.

The general field of evolutionary computation has displayed this form of convergence in that ES, EP, and GA researchers all (relatively independently) realized the huge potential of the evolutionary metaphor in computational problem solving.

Now, another form of convergence, the mixing and homogenization of ideas, is clearly evident in all EA subfields, as each approach becomes less and less distinct from the others. Only GP remains somewhat segregated, since the evolution of programs is quite a bit different from that of parameter lists.

Table 5.1 summarizes the features of the traditional versions of each EA. Of course, the partial integration of the four subfields has broken down many of the barriers, but, for example, EP folks still eschew crossover and ES researchers focus on engineering problems best represented by vectors of reals. In the GA and GP communities fitness-proportionate selection has given up some ground to tournament selection, but almost all selection mechanisms can be found in a random sampling of GA and GP applications.

To the outside observer, with no academic bloodlines to the founding fathers of EC, the differences between the approaches seem vanishingly petite and the disputes petty. For long-term success with EAs, the wise researcher will adopt (or design) a general system such as ECJ (Luke, 2013) and experiment with different representations, fitness functions, and selection mechanisms with cautious disregard for which of the four EA types the eventual configuration most resembles.

5.5   Let Evolution Figure It Out

If ever a biological theory deserved to be carved on a golden tablet, it would probably be Darwin's theory of evolution by natural selection. So simple, so elegant,

Table 5.1
Four EA types and their traditional characteristics

Evol Alg.	Gtype Level	Mutate	Crossover	Adult Selection	Parent Selection
ES	Real vector	Yes	Yes/No	OPC, GM	Uniform
EP	Int vector	Yes	No	OPC, GM	Uniform
GA	Bit vector	Yes	Yes	FGR	FitPro
GP	Code tree	Yes	Yes	FGR	FitPro

FGR = full generational replacement; GM = generational mixing; OPC = overproduction of children; FitPro = fitness-proportionate.

and so extremely powerful, it goes a long way toward explaining the history of life on earth. Its elegance stems from the three foundational mechanisms: variation, inheritance, and selection. That's it!

EAs inherit these three concepts from Darwin, and they solve a wide spectrum of problems whose diversity is rivaled only by that of the species on our planet. For the most part, if one can devise a reasonable genotype-phenotype representation, genetic operators to achieve variance and inheritance, and a fitness function tuned to the task, the EA will take care of the rest. One need not have many deep insights into *how* to solve the problem by hand (or by machine), just some idea of *what* a good solution does. Though it may seem like a cop-out at times, there are plenty of problem-solving situations where we lack these deep insights but can just *let evolution figure it out*. This is the beauty of Darwinism that EAs embody: complex functionality emerges from only minor preparation (and a little computing power).

6   Artificial Neural Networks

6.1   The Mental Matrix

When it comes to emergent phenomena, particularly those directly involved in intelligent behavior, the inner workings of the brain are unsurpassed. From the interactions of 100 billion neurons and 100 trillion synapses, cognition arises, and in a manner that fascinates and can still baffle neuroscientists. What does seem clear is that central controllers such as the legendary homunculus are no longer legitimate parts of the puzzle. Considerable evidence (Fuster, 2003; Kandel et al., 2000; Granger, 2006) points to brain regions such as the prefrontal cortex and thalamocortical loop as important facilitators of high-level cognition, but no one area runs the whole show. The basis of intelligence is fully distributed throughout the brain, across a matrix of relatively simple processing elements, the neurons.

Artificial neural networks (ANNs) are based on this abstract characterization of the brain as a network of simple distributed processors. These networks generally house many modifiable connections to produce an extremely adaptable substrate for knowledge representation, reasoning, and sensorimotor control. The list of scientific and engineering applications is voluminous (Arbib, 2003).

Roughly speaking, the behavior of neural networks, whether natural or artificial, involves two primary processes: information transmission and learning. In the former, firing patterns promote other firing patterns in long sequences of repeatable state changes. These embody everything from our ability to recall the words of a song to our skills in playing instruments to movement sequences in sport.

Learning involves alterations to the firing sequences such that pattern P1 gradually becomes an instigator of P2 while simultaneously inhibiting patterns P3 and P4, for example. These modifications are often driven by correlations in the activity patterns of pairs of neurons, such that frequently co-active neuron pairs become more likely to have simultaneous activity in the future.

Complex chemical transfers and interactions underlie both transmission and learning, and many of the primitive processes are now codified by neuroscientists.

Fortunately, one need not understand all the biochemical pathways to get a good feel for the chemical essence of information flow and adaptation in the brain. The following sections touch on some of these chemical details and then abstract them into equations and computer models that form the basis for many modern AI systems.

6.2   The Physiology of Information Transmission

Information flows through brains in the form of action potentials (APs) and synaptic potentials (SPs), which are both traveling waves of electrochemical potential energy. They arise from the flow of ions into and out of neurons, both their cell bodies (soma) and their extensions, i.e., their axons and dendrites, known collectively as *processes*.

   As shown in figure 6.1, SPs travel along dendrites and eventually converge upon the soma, where the cell *integrates* the SPs (of which there are often thousands). If the cumulative charge exceeds a particular threshold, the soma produces an AP that travels outward along its main axon and its branches, eventually arriving at interfaces with the next level of dendrites. This AP production by the soma is often referred to as the *firing* of a neuron, with the AP termed a *spike*.

   APs and SPs differ in a few key respects (Kandel et al., 2000). An SP is a continuous signal that degrades rather quickly with distance, and it can rarely travel more than a few millimeters. Its initial potential varies positively (negatively) by the amount of excitatory (inhibitory) influence stemming from the neurotransmitter-receptor binding at the synapse. An SP can be excitatory or inhibitory and may last anywhere from milliseconds to minutes, and its amplitude varies between 0.1 and 10 millivolts. APs, on the other hand, have a higher amplitude (70–110 mV), are always excitatory, have short (1–10 ms) durations, and manifest discrete on/off signals that are continually regenerated along the axon, enabling them to travel centimeters or even meters without attenuation. The difference in amplitude between SPs and APs explains why the production of an AP often requires the integration of many SPs.

   APs travel as sharp waves because of internal potential changes from the highly negative, *polarized* state (resting potential about −65 mV) through a depolarized state (about 0 mV) to a brief spike at a positively charged state (near +40 mV), as shown in figure 6.1. SPs involve similar ion flows but exhibit less dramatic voltage swings and no spiking.

   These potential changes are driven by the flow of sodium (Na^+) and potassium (K^+) ions, while calcium (Ca^{++}) flows become more important at synapses. At rest, in the polarized state, a neuron contains more K^+ and less Na^+ internally than externally. As shown in figure 6.1, depolarization begins with the influx of

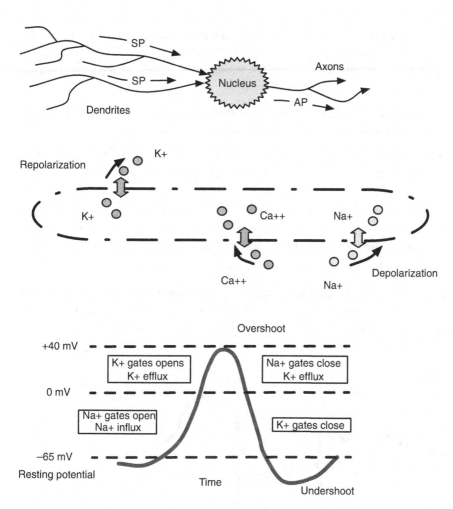

Figure 6.1
(*Top*) Transmission of synaptic potentials (SPs) toward the nucleus, which may produce an action potential (AP) on the axon leading away from the nucleus. (*Middle*) The main three ion channels in axons and dendrites. Larger chemical symbols denote larger concentrations. (*Bottom*) Changes in membrane potential as a function of time-varying Na^+ and K^+ flux that produce APs.

Na^+ ions. This drives internal charge upward. However, soon afterward, K^+ ions begin flowing outward, thus reducing internal charge back toward the negative, polarized state.

This *vertical* ion flow pattern then travels horizontally along the axon until it reaches a *synapse*, which is a microscopic gap between the axons of an upstream neuron and the dendrites of a downstream neighbor. Here, the signal is *ferried* across the gap by purely chemical means: by the neurotransmitter. Figure 6.2 portrays a synapse, where an action potential arrives at the presynaptic terminal and sets off a chain of events that effectively transfer the signal to the postsynaptic terminal and on to its corresponding soma:

1. Calcium ion (Ca^{++}) influx to the presynaptic terminal causes vesicles (holding neurotransmitter) to migrate to the edge of the synapse.
2. Upon reaching the edge, the vesicles release neurotransmitter (NT).
3. NT diffuses across the gap, attaching to the NT-gated ion channels of the postsynaptic terminal.
4. NT-bound channels open, allowing positive ions (such as Na^+ and Ca^{++}) to enter the postsynaptic terminal, causing it to depolarize, thus creating an excitatory postsynaptic potential (EPSP) (i.e., an excitatory SP).
5. The EPSP travels toward the soma of the receiving terminal.

Nothing guarantees the transfer of an AP across this gap. If the synapse has a low concentration of vesicles or too few receptors, then not enough ion channels will open on the postsynaptic side to achieve a robust depolarization.

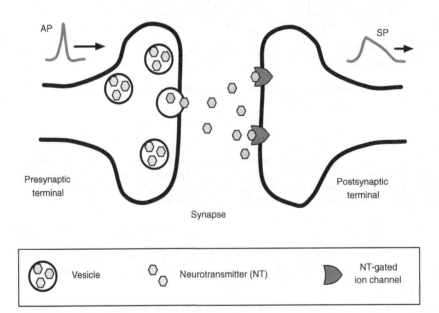

Figure 6.2
Basic structure of a synapse

Furthermore, many neurons in the brain have inhibitory effects upon their downstream neighbors. They accomplish this by secreting neurotransmitters of types such as GABA that bind to (and open) Cl^- channels, thus allowing the influx of negatively charged chlorine ions and making depolarization all the more difficult. According to Kandel et al. (2000), the postsynaptic receptors actually determine whether an SP is excitatory or inhibitory, since different receptors can have qualitatively different responses to the same neurotransmitter. Thus, in theory, a presynaptic neuron should be capable of having both positive and negative influences upon downstream neighbors, depending upon the synaptic composition. In practice, this seems to be a rarity: neurons can generally be classified as promotors or inhibitors, not both. ANNs often lack this characteristic, with one neuron providing stimulation to some neighbors and inhibition to others.

The strength of excitation or inhibition depends not only upon the physiochemical status of the synapse but also its proximity to the soma. Because of the attenuation of SPs, a more proximal synapse tends to have a much stronger influence on AP generation. Figure 6.3 shows the four general cases of distal versus proximal promotion and inhibition. In the brain proximal inhibition (with synapses right on the soma) is quite common, so a single inhibitory neuron can counteract many attenuated excitatory signals from distal dendrites. By analyzing the topology of interconnections between two neurons, or two layers of neurons, a neuroscientist gets hints as to the strength of interaction between the two.

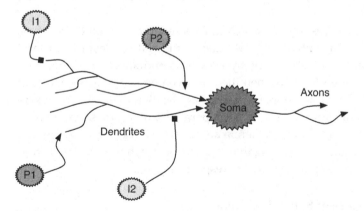

Figure 6.3
Four general types of neurons in terms of their effect and synaptic location: Distal inhibitors (I1) open Cl^- channels far out along the dendrite, while proximal inhibitors (I2) open Cl^- channels as well, but they synapse close to the soma and thus have a stronger effect. Distal promoters (P1) open Na^+ and Ca^{++} channels far from the soma, while proximal promoters (P2) stimulate SP formation near the soma.

Neuron processes may also interact by direct ion transfer across *gap junctions*. These are prevalent in the brain, but this type of connection has little room for adaptation: the network has no systematic means of altering the transfer efficiency across these junctions. However, the ability to modify the efficacy of synapses enables learning, a critical property of all animal brains.

6.3   The Physiological Basis of Learning

In 1949, Donald Hebb (1949) predicted,

> When an axon of cell A is near enough to excite a cell B and repeatedly or persistently takes part in firing it, some growth process or metabolic change takes place in one or both cells, such that A's efficiency as one of the cells firing B, is increased.

In short, correlated activity leads to fortified connectivity. Neurophysiological research has revealed the underlying *metabolic changes* (to which Hebb alludes) and largely confirmed Hebb's hypothesis (Kandel et al., 2000). This mechanism is central to associative memory formation in both real brains and artificial neural networks (Hopfield, 1982).

This basic principle of *fire together, wire together* is instrumental in learning a wide range of associations, such as those between segments of an image, words in a song, multisensory perceptions, sensory inputs and motor outputs, and the primitive movements of a complex action sequence. Associative learning constitutes a pivotal aspect of intelligence, and one that the brain has evolved complex chemical pathways for achieving.

Hebbian learning involves the detection and strengthening of correlations between firing neurons. It embodies *cooperation* at the cellular level in at least two respects. First, and most obviously, by (nearly) simultaneously firing, neurons A and B work together to strengthen their intervening synapse. Second, and often overlooked in Hebb's quotation, is the clear implication that A is one of several neurons that actually stimulate B. This cooperation among presynaptic neurons underlies classical conditioning, wherein an initial C-B association essentially bootstraps an A-B association, assuming that A, B, and C are the conditioned stimulus, response, and unconditioned stimulus, respectively.

6.3.1   Basic Biochemistry of Synaptic Change

Hebb's writings were largely based on speculation, since neuroscientists of the 1940s lacked today's advanced neural monitoring equipment. However, in the final decades of the 1900s, they began to unravel the mysteries of learning at the synaptic level. Although some mechanisms remain opaque, a fairly clear and comprehensive account now exists. The description that follows is a summary of much more detailed explanations found in two seminal neuroscience textbooks (Kandel et al., 2000; Bear et al., 2001).

Synaptic modification occurs at both presynaptic and postsynaptic terminals, albeit by different means. However, both mechanisms rely on *coincidence detection* wherein at least two different signals must coincide to stimulate lasting change. The key modifications affect

- the propensity to release neurotransmitter from the presynaptic terminal, and
- the ease with which the postsynaptic terminal depolarizes in the presence of neurotransmitters.

Thus, a change that strengthens a synapse would entail an increase in the amount of neurotransmitter released when an AP arrives at the presynaptic terminal, or an increase in the rate of positive ion influx at the postsynaptic terminal in response to neurotransmitter. Either change increases the likelihood that an AP arriving at the presynaptic terminal can generate a corresponding SP at the postsynaptic terminal. Conversely, a weakened synapse shows reduced neurotransmitter release or less sensitivity to it on the postsynaptic side.

At the presynaptic terminal the key coincidence-detecting chemical is adenyl cyclase (AC). When it binds to both calcium ions (Ca^{++}) and serotonin, it triggers a chemical sequence that increases neurotransmitter release. While Ca^{++} is another ion that enters the cell during depolarization, with particular prominence near synapses, serotonin is a *neuromodulator* secreted by Raphe nuclei of the brain stem. Unlike neurotransmitters, which are local signals between two neurons, neuromodulators are broadcast to whole regions of the brain, affecting behavior throughout. Serotonin is known to have a major influence upon many physical and mental factors, including body temperature, anger, mood, appetite, and sleep.

The details of AC's effect upon neurotransmitter release are visualized in figure 6.4 and described here (vertical arrows denote increases and decreases, while horizontal arrows represent causality):

- ⇑ serotonin and Ca^{++} concentrations →
- ⇑ AC activity →
- ⇑ cAMP (cyclic AMP) →
- ⇑ activity of protein kinase A (PKA) →
- ⇑ protein phosphorylation →
- ⇓ efficiency of K^+ channels →
- ⇓ K^+ efflux during repolarization →

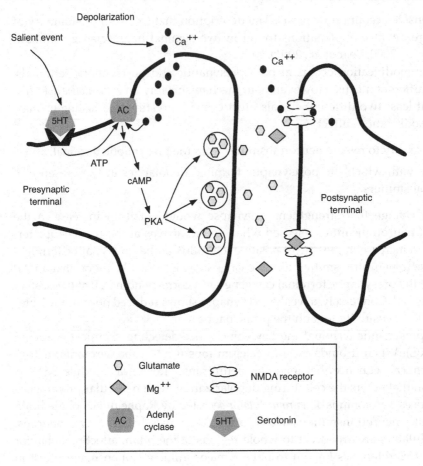

Figure 6.4
Key processes involved in coincidence detection by adenyl cyclase (AC) and the ensuing increase in propensity for neurotransmitter release in the presynaptic terminal. The biochemical coincidences (serotonin and Ca^{++}) are results of two other correlated activities: the detection of a salient event and the depolarization of the presynaptic terminal.

- ⇑ depolarization phase →
- ⇑ duration of APs →
- ⇑ duration of Ca^{++} influx →
- ⇑ migration of vesicles to the synaptic cleft →
- ⇑ neurotransmitter release

Since changes to the K$^+$ channels can be long-lasting, this process insures prolonged presynaptic action potentials and thus greater neurotransmitter release in the future.

Neuromodulators tend to be secreted when the body experiences *salient conditions* of one form or another, whether painful, pleasurable, or surprising. Hence, those presynaptic terminals that are active (i.e., depolarized) during significant events will experience an enhancement of their effect upon downstream neurons.

At the postsynaptic terminal, the pivotal coincidence detectors are the NMDA receptors, which serve as gates on Ca^{++} channels. They open, allowing calcium ions to enter the postsynaptic terminal, only when

- glutamate, a very common neurotransmitter, binds to them, and
- the postsynaptic terminal depolarizes.

Note that this second condition means that the dendrite depolarizes as a result of APs transmitted over *other* synapses than the one in question. Essentially, depolarization at nearby synapses causes SP transmission in all directions along the dendrite, both toward the soma and away from it along alternate branches of the dendritic tree.

As summarized in figure 6.5, the basic chemical activity at an NMDA receptor involves the expulsion of a magnesium ion (Mg^{++}) and binding of a glutamate molecule to open the gate for Ca^{++} influx. When the terminal is in a polarized state (with negative electropotential), the net negative charge attracts the positive Mg^{++} ion, which gets *stuck* in the NMDA receptor, effectively blocking it. However, when the terminal depolarizes, becoming positively charged, this expels the Mg^{++}. If a glutamate molecule attaches to the receptor during the same time period, it opens the Ca^{++} channel.

Ca^{++} is an important *second messenger* in the nervous system. Its presence in significantly high concentrations stimulates the production of CaMKII and protein kinase C. These then appear to

- alter K^+ channels, thereby prolonging action potentials;
- increase the synthesis of NMDA receptors on the postsynaptic terminal; and
- stimulate retrograde signals back to the presynaptic terminal, which further increase the production of neurotransmitters (such as glutamate).

Thus, the (slightly anthropomorphized) neurochemical explanation of *fire together, wire together* is the following:

- *Fire together* When the pre- and postsynaptic terminals of a synapse depolarize at about the same time, the NMDA channels on the postsynaptic side notice the coincidence and open, thus allowing Ca^{++} to flow into the postsynaptic terminal.
- *Wire together* Ca^{++} (via CaMKII and protein kinase C) promotes post- and presynaptic changes that enhance the efficiency of future AP transmission.

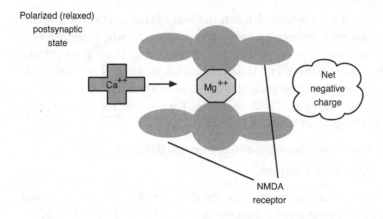

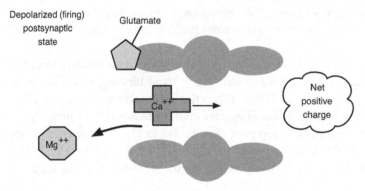

Figure 6.5
Two states of an NMDA receptor. (*Top*) When the postsynaptic cell is polarized (i.e., has a net negative internal charge), the magnesium ion is pulled toward the cell and sits fast in the receptor, blocking entrance to calcium. (*Bottom*) During depolarization the net positive internal charge repels the magnesium ion. When a glutamate molecule simultaneously binds to the receptor, calcium can flow through.

The discovery of NMDA receptors and their functionality as coincidence detectors was the big breakthrough in uncovering the mystery of Hebbian learning. Although the complete account of synaptic Ca^{++} effects has yet to be written, Kandel et al. (2000) provides a wealth of information on the subject in their classic textbook.

6.3.2   Long-Term Potentiation
Long-term potentiation (LTP) refers to synaptic changes that persist well beyond the duration of the training period. LTP manifests long-term learning in that the modifications affect future behavior for hours, days, months, or years after training. There are two distinct phases to LTP:

- In the early stage chemical changes to pre- and postsynaptic terminals, due to AC and NMDA activity, respectively, increase the probability (and efficiency) of AP transmission for minutes to hours after training.

- In late-phase LTP structural changes occur to the link between the upstream and downstream neurons. This often involves increases in the numbers of axons and dendrites linking the two and seems to be driven by chemical processes triggered by high concentrations of Ca^{++} in the postsynaptic soma.

In short, learning not only involves the strengthening of individual synapses but a general increase in synapses as well. On the other hand, synaptic weakening and reduction is also a vital element of learning. So even though few new neurons are created in mature brains—although there is increasing evidence of neuron production into adulthood in a few isolated regions of avian and mammalian brains (Shors, 2009)—the connections between existing neurons are extremely dynamic.

6.4   Abstractions and Models

The basic mechanisms of neural activity lend themselves to computational models of benefit to both science and engineering. Scientifically, these models serve as representations of biological reality that give brain researchers virtual test environments for their theories, sometimes saving millions of wet lab dollars. In engineering and other practical disciplines, computational systems grounded in the essence of biological neural networks solve untold problems in pattern recognition, data clustering and classification, forecasting, control, and so on.

These computer programs are only simplifying abstractions of reality, and it is not always clear which of the plethora of neuroscientific details can safely be ignored. Abstraction can occur spatially, temporally, or functionally: we can ignore details concerning the physical components and layout of a neural network, or the time course of events, or the mechanisms underlying a phenomenon. The different approaches fill thick books on connectionism and computational neuroscience, so the following examples serve as merely a patchy overview.

In general, AI workers ignore ion channels and have no interest in waiting for traveling pulses to emerge from the physicochemical dynamics. Instead, they simply assume that signals sent by an upstream neuron's soma will reach all downstream neighboring soma, albeit in different states of attenuation or amplification. When the upstream neuron is an inhibitor, a *negative signal* is transmitted. The only preserved component along the path from one soma to the next is the synapse, which most ANN designers abstract into a simple real number, known as the *weight* of the connection between the two soma. They also drop distinctions between axons and dendrites by viewing the neural network as a matrix of nodes, where any two nodes that are connected have a single weight that modifies the signal sent from the upstream to the downstream node. It is common to retain the unidirectionality of these links; so in those cases where two nodes are bidirectionally connected, the weights in each direction can differ.

In short, AI researchers disregard considerable biology in their pursuit of efficient engineering solutions. They seek to exploit the basic neural paradigm of numerous simple processors that integrate inputs from several neighboring nodes before determining their own output, and modifiable internode connections. Beyond that, the details may derive more inspiration from statistical mechanics or machine learning, for example, than from neuroscience.

6.4.1   Spatial Abstractions

Probably the most obvious sources of simplification in the mapping from real brains to ANNs are physical components and the spatial relations among them. Whereas most ANNs require simple interconnected processors (i.e., nodes with weighted links), few use nodes as anything more than integrators of signals and links as conduits with a universal, one-timestep delay. All other structural and topological details of biological neural networks add unnecessary complexity to most practical ANN applications. Thus, in figure 6.6, only the topmost diagram pertains to the vast majority of ANNs. However, models of relevance to many neuroscientists employ compartments (middle of figure) and their ion channels.

Microscopic images of natural neural networks tend to resemble spaghetti, someone's bad-hair day, or a European city's street map ravaged by graffiti. Experts recognize the essential underlying patterns, but laymen see only chaos. Experts see right through the noisy graffiti to the salient neuroanatomy and its functional significance. These patterns of connectivity (along with the manner in which nodes are grouped into layers) constitute a network's topology, which is a common starting point when designing ANNs with particular desired capabilities, such as memory, pattern completion, or context detection.

For example, the vast majority of ANN applications in engineering, banking, business, and other fields employ feedforward arrangements in which each layer sends signals only to the next layer in the input-to-output sequence. These networks facilitate use of the backpropagation learning algorithm for supervised learning but can also be trained by other means, such as evolution. Figure 6.7 (top) shows three different feedforward topologies that vary only in the number of hidden (i.e., noninput, nonoutput) layers.

Figure 6.7D illustrates another common topology, found in competitive networks, that involves a layer of neurons that have predominantly inhibitory interaction with one another. When one activates, it tries to shut down the others. This

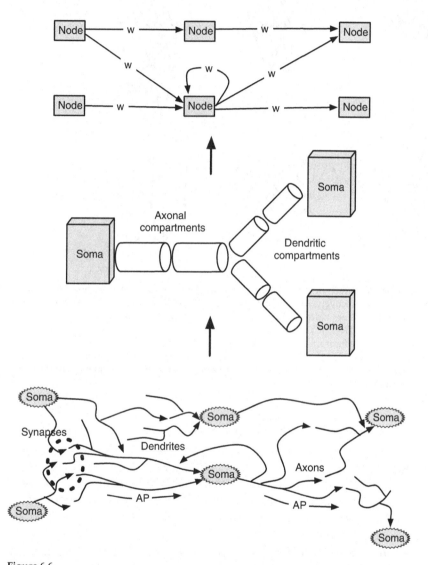

Figure 6.6
Spatial abstraction in neural networks typically begins with a complex compartmental model of the biological network (*middle*), while the highest level (and most common) ignores all but nodes, their basic connectivity, and the strengths of each link.

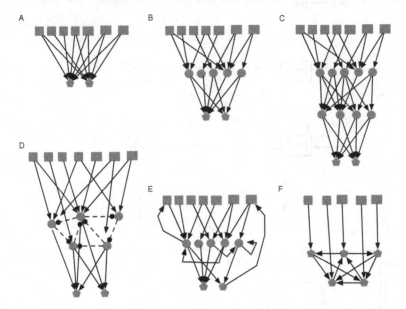

Figure 6.7
Collection of standard ANN topologies, where input, hidden, and output nodes are denoted by squares, circles, and pentagons, respectively. All solid connections can be excitatory or inhibitory, while dashed lines with balled endpoints are strictly inhibitory. (A) Simple feedforward network. (B) Feedforward network with one hidden layer. (C) Feedforward network with multiple hidden layers. (D) Hidden nodes inhibit one another as basis for a competitive network. (E) General recurrent network where neurons can signal themselves or other neurons in the same or upstream layers. (F) A pattern-completing network in which neurons excite or inhibit others in the same layer, thus completing a partial pattern arriving from the input layer.

competition often manifests a *context-detecting* layer, where each neuron becomes specialized to fire on a special type of activation pattern in the immediate upstream (afferent) layer.

Recurrent connections are a common element of ANNs with memory for their previous state. By feeding their outputs backward or sideways (and not merely forward), an ANN insures that outputs from time t will still be available (though modified by a synaptic weight) at time $t + 1$. As shown in figure 6.7E, recurrence can take on many forms: intralayer connections, intraneuron links, and feedback across one or many layers. Recurrent topologies frequent the ALife and evolutionary robotics literature, where the controllers of real or simulated agents often have a strong need for historical information when performing cognitively challenging tasks when simple sense-and-react behavior does not suffice.

Finally, figure 6.7F depicts a pattern-completing network in which the output nodes—a hidden layer could also do the job—have a high degree of intralayer recurrence. This enables partial patterns (mapped from the input layer) to be completed and returned by the output layer. Hopfield networks (Hopfield, 1982) for unsupervised learning (see chapter 10), are prime examples of this network category, and, in general, of emergence in neural networks.

6.4.2 Functional Abstractions

These typically involve a spatial and temporal aspect, as many mechanisms (with time courses of behavior and spatial extent) are simplified, ignored, or aggregated into higher-level models. Naive computer programmers who pick up a neuroscience textbook are quickly overwhelmed by the brain's myriad physicochemical processes. Learning when to just say no to an eager neuroscience colleague's request to include just one more *essential* biological reality is paramount to achieving modeling success. Without this ability to digitally cut and cauterize, the modeler spends more time writing grants for supercomputers than writing code.

However, for building biologically authentic models, a standard starting point is the relatively low level of ion channels, familiar territory for Hodgkin and Huxley (1952), whose pioneering model of axonal membrane potential (V_M), based on giant squid experiments, appears in equation (6.1):

$$\tau_m \frac{dV_M}{dt} = -g_K(V_M - E_K) - g_{Na}(V_M - E_{Na}) - g_L(V_M - E_L). \tag{6.1}$$

Here, the g terms denote conductances for potassium, sodium, and general leak channels, and the E terms represent Nernst equilibrium values for each flow. Behind this simple version lie detailed calculations for the conductances (which change constantly as a function of V_M), all of which are well documented (Hodgkin and Huxley, 1952; Dayan and Abbott, 2001). The middle of figure 6.8 illustrates the basic abstraction of the Hodgkin-Huxley model, which represents neural components as electrical circuits.

Note that for models of this high spatiotemporal resolution, the AP/spike becomes an emergent phenomenon; and when multiple axonal compartments are modeled, a *moving* AP arises (Dayan and Abbott, 2001). Although AP movement is inherently bidirectional, spreading in all possible directions, an important local constraint comes into play: when a sodium channel closes, it cannot reopen immediately. This prevents an AP from traveling backward and insures the emergence of unidirectional flow (with some exceptions).

Though a good deal of computational neuroscience involves this (and more) physiological detail, AI researchers typically work at a higher abstraction level. One of the first steps upward comes from ignoring ion channels and dynamic conductances, thus yielding a standard *leaky integrate-and-fire* neuron model, such

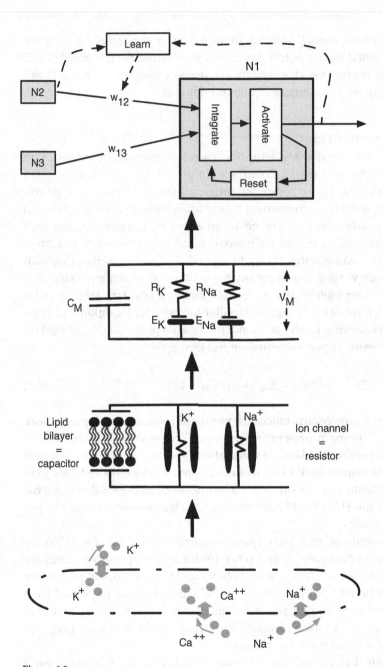

Figure 6.8
Hierarchy of functional abstractions beginning with the brain's ion channels (*bottom*), which can be modeled as electrical circuits for neuroscience purposes (*middle*) but which are typically abstracted even more for AI (*top*). *w* denotes a weight on the link.

as equation (6.2):

$$\tau_m \frac{dV_i}{dt} = c_L(E_L - V_i) + c_I \sum_{j=1}^{N} x_j w_{ij},$$

(6.2)

where V_i is the membrane potential for neuron i, and E_L is the (normally constant) extracellular potential, x_j is the output (current) from neuron j, and w_{ij} denotes the weight on the connection from neuron j to neuron i.[1] Also, c_L and c_I are the leak and integration constants, respectively. Finally, τ_m represents the membrane time constant, which is inversely proportional to the rate of change in V_i; high time constants entail slow change.

By assuming that $E_L = 0$ and $c_L = c_I = 1.0$, we take another step up to the simpler leaky integrate-and-fire model of equation (6.3):

$$\tau_m \frac{dV_i}{dt} = -V_i + \sum_{j=1}^{N} x_j w_{ij}.$$

(6.3)

From here, ignoring leak current $-V_i$ yields an *integrate-and-fire* model. Alone, this modification can be problematic, since neurons that accumulate charge but never leak tend to saturate, thus producing the same output on every timestep. So abstracting away current leak often coincides with another abstraction: dissipating all potential after each timestep. This yields a neuron without a lasting internal state (i.e., memory), as in equation (6.4), where V_i at each timestep depends only upon the current sum of weighted inputs, not upon its own previous value:

$$V_i = \sum_{j=1}^{N} x_j w_{ij}.$$

(6.4)

This is equivalent to equation (6.3) with $\tau_m = 1$.

In most AI and ALife research, ANNs operate at these higher levels, many abstractions away from what neurophysiologists consider essential, though this level is less problematic for systems neuroscientists investigating the emergence of intelligence from widescale cerebral interactions (Rolls and Treves, 1998). Equation (6.3) aptly generalizes over many such AI and ALife models, so it forms the basis of further discussion. Figure 6.8 (top) summarizes this standard functional abstraction, which consists of

- a leaky integration function, which normally sums the weighted outputs from all upstream neighbors (the weights come from the connections) and then adds that sum to (and subtracts any leak from) an internal state variable, V;

- an activation function, which transforms V into an activation level for the neuron, which then uses it as an output value, to be weighted and summed by downstream neighbors;
- a reset function, which determines whether V should retain its current state or be set to a predefined boundary value, such as zero;
- a learning function, which uses the activation values of connected nodes (plus, possibly, global signals akin to neuromodulators) to compute weight changes to the arc(s) that connect them.

Figure 6.9 summarizes several of the options for activation functions and resetting.

Leaky Integration   Equation (6.5) updates the neuron's internal state by adding the sum of weighted inputs and extracting leakage, with both contributions scaled by the time constant. In many models, $\tau_m = 1$, and thus the neuron has no enduring state.

$$\Delta V_i = \frac{-V_i + \sum_{j=1}^{N} x_j w_{ij}}{\tau_m}.$$ (6.5)

The vast majority of ANNs follow this protocol for leak and integration but tend to differ with respect to activation, reset, and learning.

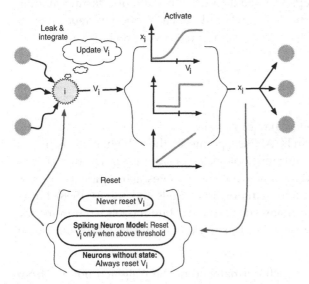

Figure 6.9
Alternatives for leaky integrate-and-fire neural networks.

Activation A wide variety of activation functions are used for ANNs, with the logistic equation (6.6) being one of the more popular:

$$x_i = \frac{1}{1 + e^{-V_i}}. \tag{6.6}$$

Figure 6.10 plots the logistic function (a sigmoidal curve) along with several others. All except the identity function exhibit a thresholding effect, which nicely mirrors the behavior of real neurons. The logistic function has the advantage of

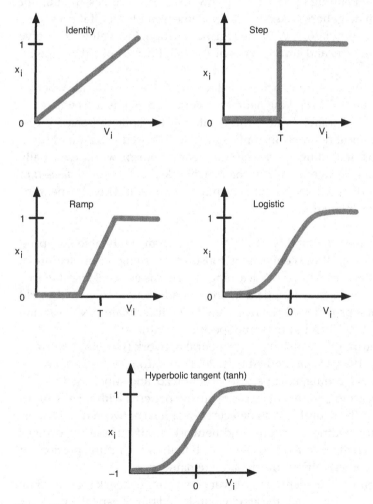

Figure 6.10
Activation functions commonly used for artificial neural networks. Of these, only the logistic and hyperbolic tangent equations have the (often desirable) combination of thresholding and continuity.

being continuous and thus differentiable at all points in its domain. Differentiability is a critical property for ANNs used in backpropagation learning.

$$x_i = \frac{e^{2V_i} - 1}{e^{2V_i} + 1}. \tag{6.7}$$

The hyperbolic tangent, equation (6.7), gives a similar continuous sigmoidal curve, but one in which x_i ranges from -1 to 1, which is a useful property for networks in which some neurons need to have powerful influences whose sign can quickly toggle, depending upon V_i: if it is low, the node becomes an inhibitor of those downstream neighbors linked to with positive weights, but if it goes high, the effect switches to stimulatory. With the logistic function, low values of V_i give a near-zero activation level and curtailed downstream influence (of either type).

Reset   After calculating a neuron's activation level, an ANN must decide whether to reset the internal state, V. In spiking neural network models V is reset whenever it exceeds a particular threshold value, thus imitating the spiking behavior displayed by most biological neurons. In contrast, the ANNs used in standard backpropgation scenarios, including much classical connectionism work, essentially reset V to 0 after every timestep.[2] Finally, many networks, such as *Continuous-time recurrent neural networks* (CTRNNs), have state but never reset it. Thus, the neurons have memory but never spike.

Learning   The extreme adaptability of ANNs stems from modifiable synapses, and processes that change these connections constitute learning, a core feature of most ANNs. A huge repertoire of ANN learning algorithms exists. Some stay true to neuroscience, and some stray far afield in order to adequately solve real-world problems. This section provides a brief overview of the three main ANN learning paradigms (see chapters 10 and 11 for some specific algorithms).

In the machine learning (T. Mitchell, 1997), neural network (Haykin, 1999), and neuroscience (Doya, 1999; Dayan and Abbott, 2001) literature, authors often distinguish three types of learning: unsupervised, reinforced, and supervised.

In unsupervised learning, the agent learns recurring patterns without any tutoring input. Essentially, the neural system detects correlations between neuronal firing patterns and between those patterns and network inputs. These correlations are strengthened by changes to ANN weights such that, in the future, portions of a pattern suffice to predict/retrieve much of the remainder.

In supervised learning, the agent receives very frequent feedback that not only signals good/bad but also indicates the action that should have been taken in cases where the agent makes a mistake. This is the classic form of learning handled by neural networks in many practical applications, with gradient-descent methods, such as the classic backpropagation algorithm, used to modify weights so as to reduce error.

In reinforcement learning, the agent receives an occasional reward or punishment that essentially indicates that the net result of many activities was good or bad. The trick, known formally as the *credit-assignment problem*, is to figure out, from this general global signal of good or bad, how to modify the individual weights so as to improve performance in the future.

The three learning paradigms are summarized in figure 6.11, with an example of a maze-following agent and a sketch of typical ANN controllers.

Clearly, the supervised approach, when feasible, can yield much faster learning of useful situation-action pairs than the other two methods. Unfortunately, omniscient tutors are not available for the brunt of biological learning, and their assistance, when available, normally comes closer to reinforced than supervised. In nonhuman animals, the level of detailed supervision is even less. Yet, all animals are capable of learning useful behavioral information in a relatively short period. But if there is no element of supervision in this process, then what provides the feedback?

One reliable feedback source is prediction. Essentially, an agent can be its own teacher when doing predictive tasks. At time t, it predicts the future state of its body and immediate surroundings at $t + 1$. Then, at time $t + 1$, it learns the correct answer and can use that knowledge to adjust its own mapping from current to future states. This can be done almost continuously as the agent moves about the world, as often seen in evolutionary robotics (Nolfi and Floreano, 2000).

For each of the three primary learning methods in natural neural systems, Hebb's rule frequently holds: neurons that are active within the same time window (of about 50 ms) often initiate chemical mechanisms that strengthen the synapse between them. The computational details and implications of Hebb's rule are further explored in chapter 10, but for now, suffice it to say that the locality and simplicity of this fundamental concept motivate many interpretations of learning as a highly emergent phenomenon.

6.4.3   Temporal Abstractions

The preceding spatial and functional simplifications tend to have accompanying temporal abstractions. For example, disregarding axonal length often means ignoring signal transit time, while glossing over ion channel dynamics enables simulations to run at coarser temporal resolutions. However, one key phenomenon constitutes a common reference point for temporal abstraction: the action potential (AP), or spike.

As shown in figure 6.12 (bottom), some models capture the actual buildup of membrane potential, spike formation, and refractory period. This requires a

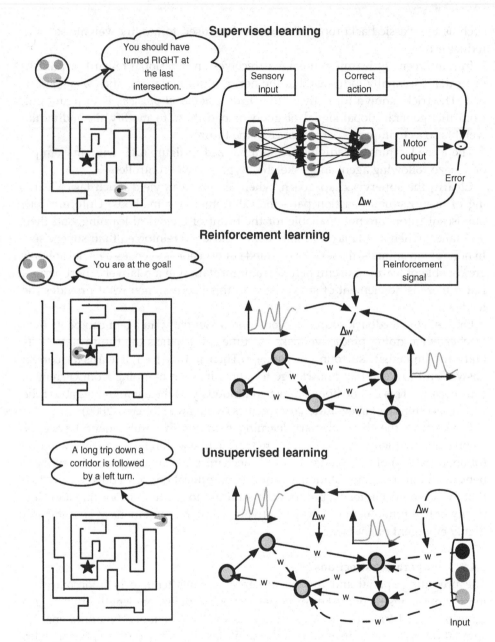

Figure 6.11
Three basic neural network learning paradigms: supervised, reinforcement, and unsupervised. (*Left*) A maze-following task and an idea of the feedback that the network would receive from a tutor. In unsupervised learning, there is no feedback, so the network learns associations based solely on correlations among neural firing patterns and between those patterns and sensory inputs or motor outputs.

millisecond or smaller timestep. However, other models (middle of figure) deal only with the spikes themselves as the atomic unit: each neuron keeps track of the time points at which it spikes, and these spikes are driven by the spike times of other, upstream neurons rather than a simulation of ion channels. A $V_i(t)$ term still comes into play but largely as an accumulator of afferent spikes that produces an output spike when it exceeds a threshold. These are traditionally called *spiking neuron models* (Rieke et al., 1999).

As discussed in chapter 7, a point of contention among neuroscientists involves the significance of spikes as carriers of information. Some insist that actual spike times matter, while others view a neuron's firing rate (i.e., spikes per second) as the essential signal. Thus, the former group prefers models like those diagrammed in the middle of figure 6.12, whereas the latter group considers models as shown in the top diagram sufficient. Most ANN applications to practical problems employ the model in the top diagram. Bio-AI benefits from both types, although it has traditionally been difficult to achieve the learning of high-level concepts with spiking ANNs.

An important aspect of temporal abstraction involves the distinction between the potential, V_i, and activation value (often labeled x_i), for neuron i. This can easily lead to confusion, since x_i mirrors no precise biological phenomenon. In brains, V_i simply rises and falls, with sharp rises constituting spikes, or action potentials, which travel along axons and dendrites. In ANNs, V_i typically represents the general state of a neuron, a state that reflects the recent density of incoming signals and often accumulates (and leaks) over time to form a decaying history. This state is then converted by the activation function into an output signal, x_i, that travels to downstream neurons. The biological counterpart to x_i can be any of the following:

- The membrane potential, i.e., a simple copy of V_i
- A binary spike/no-spike based on V_i's relation to a firing threshold
- The absolute firing rate (in spikes per second) of the neuron
- The firing rate scaled to a positive range, such as $[0, 1]$
- The difference between the current firing rate and an average firing rate
- The preceding difference scaled to a range about 0, such as $[-1, 1]$

In an ANN that explicitly models spiking behavior, x_i is either absent or just a copy of V_i, which periodically gets reset to a low value after it exceeds a spiking threshold. Another alternative is that when V_i spikes, x_i becomes a simple signal value, such as 1, and gets sent downstream. Otherwise, $x_i = 0$.

However, in models that ignore spiking behavior and use the general activity level of neurons as the currency of information exchange, the transition from V_i to

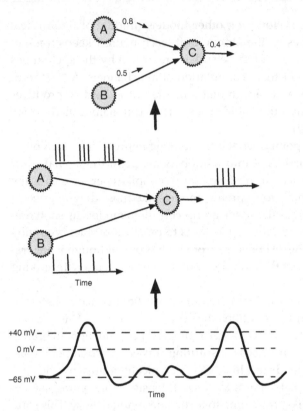

Figure 6.12
Common hierarchy of temporal abstractions in ANNs, from simulations of the rise and fall of membrane potentials (*bottom*) to the recording of only their spike times (*middle*) to the transfer of normalized firing rate values between nodes (*top*)

x_i is somewhat precarious biologically though very practical from a computational standpoint. It captures the intuition that a highly stimulated node should output a high value, under the basic assumption that a neuron with many active excitatory afferents should have a high firing rate. However, absolute firing rates (in, for example, Hertz) merely add unnecessary complexity to most AI applications, so functions such as the sigmoid and hyperbolic tangent scale activity to standard ranges while also mimicking the firing rate saturation exhibited by biological neurons.

6.4.4   The Spike Response Model (SRM)

The ANN and computational neuroscience literature is full of diverse spike-timing models (Rieke et al., 1999; Dayan and Abbott, 2001), including the SRM (Gerstner and Kistler, 2002), which captures several of the essential concepts. It is expressed

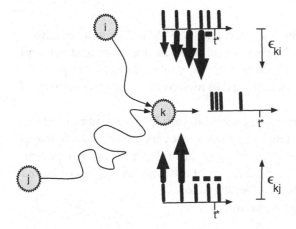

Figure 6.13
Basic influences of afferent spike trains from neurons i and j upon neuron k in the spike response model. The arrows above or below each spike train indicate the strength of its effect upon $V_k(t*)$ for a particular time point $t*$. Since i is closer to k than j is, its spikes experience less delay in their effects upon k. At time $t*$, neuron k has not spiked for awhile and thus has negligible refractory inhibition.

in equation (6.8) as an update rule for the membrane potential:

$$V_i(t) = \kappa(I_{ext}) + \eta(t - \hat{t}_i) + \sum_{j=1}^{N} w_{ij} \sum_{h=1}^{H} \epsilon_{ij}(t - \hat{t}_i, t - t_j^h), \tag{6.8}$$

where the first tem, $\kappa(I_{ext})$, represents a complex function of the history of all external input to neuron i. The second term, $\eta(t - \hat{t}_i)$, models the refractory period of the neuron, where $\hat{t}_i$ is the time of i's most recent spike. To capture the fact that a neuron cannot build up potential immediately after firing, $\eta(\Delta t_i)$ is an increasing function with $\eta(0) < 0$. Finally, the third term examines all N of i's afferent neurons and the last H spike times of each; the function $\epsilon_{ij}(\Delta t_i, \Delta t_j^h)$ models the strength of the afferent signal from neuron j based on (1) j's hth most recent spike, (2) any delays along the transmission line from j to i, and (3) i's receptiveness to afferents at time t (because of the refractory period). Figure 6.13 illustrates the relative effects of different upstream spikes upon a downstream neuron as captured by $\epsilon_{ij}(\Delta t_i, \Delta t_j^h)$.

Note that this model requires the use of (potentially long) spike train histories, along with the definition of different versions of ϵ_{ij} for each $i - j$ neuron pair. In short, SRM incorporates many details of neuron dynamics and network topology, e.g., connection lengths and the transmission delays that they incur.

6.4.5   The Izhikevich Model

A much simpler, yet extremely effective, alternative to SRM is the model of Izhikevich (2003), which seems to have the best of both worlds: it is more abstract, and
thus computationally cheaper, than SRM and its many relatives, yet it can produce
dynamics that closely match those of many brain regions via the proper settings of
a few critical parameters.

The two key variables for any neuron i of the model are V_i (the membrane potential that doubles as the activation level), and U_i (a recovery factor), which acts to
draw down V_i. Both of their values depend upon each other and four key parameters, a, b, c, and d, as shown by the calculations of the time derivatives of V_i and
U_i, and the threshold-driven resets, all of which follow.

The two key variable update equations are

$$\tau_m \frac{dV_i}{dt} - 0.04V_i^2 + 5v + 140 - U_i + I_i, \quad \text{and} \tag{6.9}$$

$$\tau_m \frac{dU_i}{dt} = a(bV_i - U_i), \tag{6.10}$$

where τ_m is the effective time constant, I_i is the sum of all input signals, and coeffi-
cients 5 and 140 are based on empirical data from cortical neurons. Computationally, I_i is updated whenever an afferent neuron emits a spike, thus obviating the
need for maintaining spike histories. When V_i exceeds a spiking threshold (typically 35 mV), the following resets occur:

$$V_i \leftarrow c. \tag{6.11}$$

$$U_i \leftarrow U_i + d. \tag{6.12}$$

Izhikevich likens the biological plausibility of this model to that of Hodgkin
and Huxley's, yet at a fraction of the computational cost. Figure 6.14 provides a
small sampling of the many spike trains produced by the model. The generation
of complexity and diversity from simplicity is the source of the model's popularity.

6.4.6   Continuous-Time Recurrent Neural Networks

The two previous models explicitly model spikes and, as a result, must run with a
small timestep; Izhikevich uses 0.1 milliseconds. Randall Beer's *Continuous-time
recurrent neural networks* (CTRNNs) (Beer and Gallagher, 1992) exhibit complex
internal dynamics by allowing each neuron to have its own time constant (and gain
term) yet abstracting away spikes completely such that outputs represent normalized firing rates. These have become very popular ANNs for AI and ALife research
into the evolution of minimally cognitive behaviors (Beer, 1995; 1996; 2003).

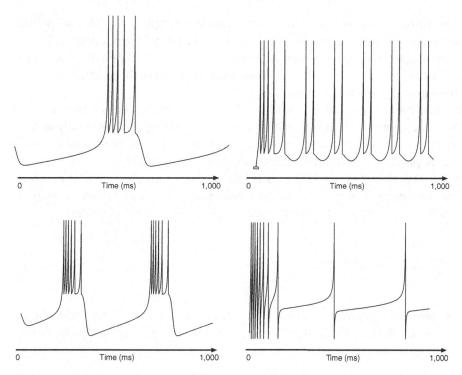

Figure 6.14
Spike patterns generated by Izhikevich's model, with each produced by a different setting of parameters a–d. For a larger collection of spike trains and their correspondence to neurophysiological data, see Izhikevich (2003).

Beer often describes the CTRNN neural model in a dynamic systems format, while the following reformulation adheres more closely to the standard leaky integrate-and-fire framework.

First, integrating the inputs from all upstream neighbors involves the standard equation (6.13):

$$s_i = \sum_{j=1}^{N} x_j w_{ij} + I_i. \tag{6.13}$$

As before, x_j is the activation level of neuron j, while w_{ij} is the weight on the arc from neuron j to neuron i. I_i is the sum of all external inputs to neuron i. A typical external input is the value of the single sensor associated with an input neuron. Noninput neurons typically have no external inputs.

Once again, V_i denotes the internal state of neuron i. Equation (6.14) shows that the derivative of V_i is a combination of s_i and a *leak term*, wherein a portion ($\frac{V_i}{\tau_i}$) of the internal state from the previous timestep drains out. In addition, Beer incorporates a neuron-specific *bias* term, θ_i. This is easily modeled as the fixed weight on a link from a *bias neuron* (that emits a 1 on each timestep).

The time constant, τ_i, determines how fast the neuron changes internal state. A low value entails fast change with the new state being predominantly determined by current conditions, whereas a high τ_i produces more *memory* in the neuron for its previous state(s) (because of less leak and less influence of s_i), which are reflections of earlier conditions. It is this memory, and the ability of neurons to vary in their levels of it (via different τ values) that enables CTRNNs to exhibit extremely rich dynamics, and hence convincingly sophisticated cognition.

$$\frac{dV_i}{dt} = \frac{1}{\tau_i}[-V_i + s_i + \theta_i]. \tag{6.14}$$

As shown in equation (6.15), Beer typically employs a logistic activation function to convert the internal state to an output, x_i. However, V_i is multiplied by the neuron-specific gain term, g_i.

$$x_i = \frac{1}{1 + e^{-g_i V_i}} \tag{6.15}$$

CTRNNs appear in all of Beer's minimally cognitive simulations as well as those of many other Bio-AI researchers (Floreano and Mattiussi, 2008; Nolfi and Floreano, 2000). Figure 6.15 shows the topology for one of Beer's video game agents. Each application involves different ranges of key parameters (gains, biases, weights, and time constants), depending upon the task demands. CTRNN topologies typically consist of an input layer—with each neuron attached to one sensor and simply outputting the value of that sensor—a hidden layer, and an output (motor) layer, which controls the agent's lateral movement. The *recurrence* in CTRNNs normally stems from (1) connections among the neurons of the hidden layer, (2) backward links from the motor neurons to the hidden layer, (3) connections between the two motor neurons, or (4) connections from individual hidden (or motor) neurons to themselves. Recurrence adds a second form of memory to the CTRNNs, since the states of neurons are fed back into the system at the next timestep. CTRNN weights are almost exclusively trained by evolution, not learning.

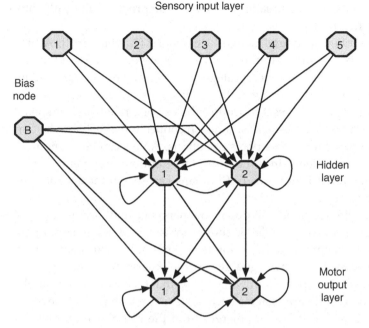

Sensory input layer

Bias node

Hidden layer

Motor output layer

Figure 6.15
CTRNN topology for one of Beer's many video game agents. All connections are shown (arrows). The looping connections imply that a node feeds its own output value back to itself on the next timestep. Each connection in the diagram has an evolvable weight; none are hardwired by the user.

6.5 Net Picking

Chapter 5 stated that the key aspects of evolution captured by almost all evolutionary algorithms are variation, selection, and inheritance. These are cornerstones of neoDarwinism and prerequisites for success in evolutionary search, whether natural or artificial.

The brain, on the other hand, admits no simple subdivision into core elements from which all ANNs should be built. For scientific purposes, ion channels and traveling action potentials are foundational, whereas engineering applications often need little more than distributed processing elements coupled by weighted links. In fact, some models even abstract the nodes away.

In between lie a whole host of alternatives, only some of which are biological or even bio-inspired. Wading through these options in search of the appropriate essence for a particular problem can be the hardest part of ANN research and deployment. Researchers like Hodgkin, Huxley, Beer, Izhikevich, and Hopfield have devised abstractions of widespread utility, but not all problems fit neatly into one of these predesigned frameworks.

As thoroughly described in neural network textbooks (Haykin, 1999; Arbib, 2003), the choices of mix-and-match building blocks can appear endless, although a good many ANNs fall under the general categories described or diagrammed in this chapter. The key decisions in many ANN applications are the numbers and sizes of layers, the activation and reset functions, the use of a rate-based or spike-timing code, learning rules, and basic constraints on connectivity, such as whether to include recurrent links. From these pools of options, ANNs can be configured to tackle many types of problems. In some cases, these options are opened up to evolutionary control, so that EAs instead of humans can search for appropriate combinations (see chapter 12).

In the end, a lot of the decisions boil down to emergence and how many levels of it a problem seems to require. Classic connectionism, with its reliance upon feedforward nets and backpropagation, exhibits no significant self-organization due to widespread instructive global signaling. At the other extreme are systems that evolve developmental recipes for growing ANNs that then learn using purely Hebbian means. In the middle are systems such as Beer's, which typically evolve parameters for CTRNNs, and behavior emerges from the complex local interactions among neurons; but no development nor learning occurs. Each level of emergence entails greater computational complexity, less user control, and thus greater unpredictability.

The true beauty of ANNs is their ability to conform to all these adaptive options. Few AI representation and reasoning substrates admit this level of flexibility, and thus few seem as well suited for computational explorations of emergent intelligence.

7 Knowledge Representation in Neural Networks

7.1 Activity, Connectivity, and Correlation

Recall the earlier definition of intelligence as "doing the right thing at the right time, as judged by an outside human observer." As a key facilitator of intelligence, knowledge can then be defined as "background information or general understanding (of a variety of domains) that enhances the ability to act intelligently," where domains include the natural, social, and even virtual worlds as well as mathematics, music, art, and so on. A *knowledge representation* is an encoding of this information or understanding in a particular substrate, such as a set of if-then rules, a semantic network, conditional probability tables, a Venn diagram, a mind map, or the axioms of formal logic. Thus, patterns (i.e., relations among primitive elements) within the substrate correspond to patterns in the target domain. The relations between the patterns (e.g., how each promotes or inhibits others) are key determinants of intelligence.

Neural networks (both natural and artificial) are the focal substrate of this chapter, with the key question being how patterns of both neural connectivity and activity encode knowledge. Many substrates have such strong roots in natural language that their knowledge content is completely transparent. For example, we can easily translate a mind map or a set of logical axioms into an explicit natural language description of knowledge content, such as "the combination of air masses from Canada, the desert Southwest, and the Gulf of Mexico facilitate tornado formation in the midwestern United States."

Unfortunately, deciphering the knowledge content of a neural network requires much more work; in many cases, the salient information does not map nicely to natural language, or when it does, the apparent concepts of import lack the crisp definitions preferred by linguists, philosophers, and GOFAI knowledge engineers. Instead, a neural pattern might encode a large (and vague) set of preconditions that embodies a complex definition with many exceptions. The interactions between these patterns produce highly intelligent behavior, but reduction of that behavior to the primitive patterns fails to produce the satisfaction that one usually gets from, for example, decomposing the description of a play into its characters, plot, conflicts, and climax.

A good deal of philosophical quicksand surrounds the concept of knowledge, with distinctions often drawn between a simple piece of data, such as a random ten-digit number, and data that *connects* to other information, such as a ten-digit number that happens to be your brother's phone number or the population size of China. The data acquires significance, or *meaning*, via these connections, and only via these links can the data contribute to reproducible behavior. Hence knowledge, as a key contributor to "doing the right thing," needs these ties to distinguish it from untethered (and essentially meaningless) data. Furthermore, *doing* may comprise either actual physical or subconscious mental activity, in which case the knowledge is often termed *procedural*, or explicit cognitive events, whereupon the knowledge is *declarative*.

In a physical system (e.g., animal, robot, smart room) controlled by a neural network, patterns of neural activity acquire meaning (semantics) via connections to the external world, the system's physical body, or other neural activity patterns. These connections, manifested as temporal correlations that the brain becomes wired to sustain, facilitate re-creation or re-presentation, since the occurrence of one pattern helps invoke another. Causality (in terms of causes preceding effects) plays no specific role. A sensory situation may precede the neural pattern that represents it, just as a neural pattern may precede a body's motor sequence. But alternatively, a neural pattern may predict an impending sensory experience, or the pattern, experience, and motor sequence may co-form as part of the looping brain-body-world interaction. Correlation, not purported causality, is crucial, and patterns lacking it may have short-term behavioral significance but little representational import. This correlational view of representation paves the way for formal tools such as information theory (see chapter 13) to help clarify network functionality.

Intertwined with this level of representation is another: synaptic networks and their ability to stimulate neural interactions. The complex matrix of synaptic strengths forms attractors in the landscape of neural activity patterns such that some patterns (and temporal sequences of patterns) become much more probable than others. Thus, the synapses represent the firing patterns, which represent knowledge. Since neural activity affects synaptic strength via learning, the relations quickly become circular, thus precluding a straightforward linear interpretation of this representational hierarchy. However, the review of each factor in (relative) isolation has explanatory merit, so this chapter treats each of these representation forms separately.

7.2   Representations in Firing Patterns

Any thorough discussion of firing patterns as representations begins with the atomic components of a pattern, which, like so many other aspects of intelligence, remain controversial. At the level of an individual neuron, there is considerable debate as to the fundamental carrier of information, with the two main candidates being the firing time of a neuron versus its overall firing rate, as explained by Rieke et al. (1999).

Firing rate coding implies that the information encoded in a neuron is due to the average number of spikes that it emits per time period, typically per second. Spike-timing code does not refer to the absolute time point when a neuron spikes but rather to a relative temporal delay: the phase (of some ambient clocklike oscillation) at which a neuron spikes (or possibly begins to emit spikes). These background oscillations stem from a large population of neurons that simultaneously fire, thus exciting or inhibiting downstream neighbors. More generally, a spike-timing code looks at a spike train starting at some reference time, 0, (e.g., the peak of a particular background oscillation) and attributes information to the exact times (after 0) at which spikes occur, whereas a rate code only counts the total number of spikes and divides by the duration of the spike train.

Instead of delving deeply into this debate, it suffices for our current purposes to view the two codes as more or less equivalent, using a slightly simplified version of an argument given by Mehta et al. (2002). Consider the situation in figure 7.1,

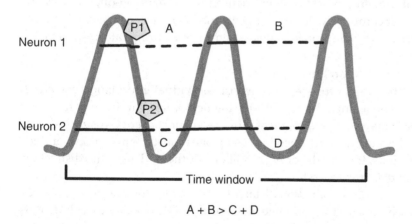

$$A + B > C + D$$

Figure 7.1
Two neurons that overcome inhibition and begin firing (dashed lines) at different phases (P1 and P2) of the inhibitory cycle (sine curve)

wherein two neurons are affected by the same inhibitory oscillation, but because of different levels of excitation, neuron 1 is able to overcome inhibition and begin emitting spikes at a phase (P1) when inhibition is higher than during phase P2, where neuron 2 overcomes inhibition. A similar but inverse argument holds for an excitatory oscillating input.

The dashed lines indicate time windows during which the neuron's excitation dominates inhibition, and thus the neuron emits spikes. From the figure, it is clear that neuron 1 is emitting spikes over larger windows (A and B), than is neuron 2 (with windows C and D). Thus, if we average over the complete time window (here composed of three inhibitory cycles), neuron 1 would be spiking more often (A + B > C + D) and would thus have a higher calculated firing rate than neuron 2. This appears to be a relatively monotonic progression in that the higher up on the inhibitory curve a neuron begins to fire (i.e., the phase at which it fires), the higher will be its firing rate.

So in this fairly simple sense, the spike-timing code (expressed as the phase at which the neuron overcomes inhibition) correlates with the rate code. Thus, although spike timing appears to play a key role in synaptic tuning via STDP (see chapter 10), the useful information in a relative spike time seems no different (nor more significant) than that encoded by a firing rate. Though this explanation certainly does not resolve the debate (and surely irks those who champion the importance of spike timing), it justifies ignoring it for the time being and simplifying the analysis of neural representation to that of the collective firing rates of neurons. Furthermore, I follow a good deal of the neuroscience and connectionism literature by abstracting the state of any neuron to a binary, on-off, value. Since neurons are never completely dormant, the distinction between on and off is typically characterized by being above or below some *normal* firing rate, respectively.

7.2.1　The Syntax of Neural Information

Given a population of neurons, each of whose individual information content is now assumed to be a single bit (on or off), the next important issue is how this collection represents a particular piece of information or concept C (using the term very loosely). Does the network encode C using one or just a few neurons, or does it employ hundreds, thousands, or even millions of them? This is the distinction between *local* and *distributed* coding.

In its extreme form this distinction dictates that if a collection of n neurons stores information, then a local coding scheme would use exactly one of the n cells to represent a single concept, whereas the distributed code would use a sizable fraction of all n neurons (and the synapses between them) to represent *every* concept. Thus, in the latter case, each concept is distributed over many nodes, and an entire suite of concepts *shares space* in the neurons and synapses of a single network.

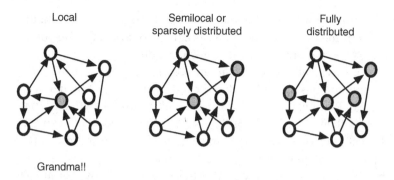

Figure 7.2
Three general coding schemes used in neural networks. Neurons are considered active (shaded circles) or inactive (open circles) when the concept is being processed.

A further distinction between sparsely and fully distributed codes is also commonly drawn. A sparse code uses only a few of the nodes per concept and is thus *semilocal* (another frequently used term), whereas a fully distributed scheme uses a large number of neurons per concept, with $\frac{n}{2}$ being the optimal number for representing the most concepts.

To understand this optimality, consider that the number of different concepts using exactly k nodes (i.e., k neurons are active for each concept) that can be represented in an n-node network is $\binom{n}{k}$, which has a maximum at $k = \frac{n}{2}$.

Figure 7.2 portrays these three representational forms: local, semilocal/sparsely distributed, and fully distributed. Cells that perform local coding are often refered to as *grandmother cells*, meaning that individual cells represent specific concepts such as bicycle, ice cream, or grandmother.

The number of distinct patterns that can (theoretically) be stored in an n-neuron network (for $n = 20$) is as follows:

- Local. $\binom{n}{1} = n$ patterns.
- Sparsely distributed. $\binom{n}{k} = \frac{n!}{(n-k)!k!}$ patterns; $n = 20, k= 3 \rightarrow 1{,}140$ patterns.
- Fully distributed. $\binom{n}{\frac{n}{2}} = \frac{n!}{(\frac{n}{2}!)^2}$ patterns; $n = 20 \rightarrow 18{,}4756$ patterns.

These capacities derive purely from the nature of binary codes. For distributed representations the theoretical values are several orders of magnitude greater than the practical storage limits of neural networks.

7.2.2   Pattern Completion in Distributed Memories
The mathematics of distributed representations always sounds impressive. In fact, one could load all 184,756 patterns into a 20-node network and get a unique

signature of on-off neurons for each. But most of them could never be unambiguously retrieved from the network based on partial information. So one could concurrently put all 184,756 patterns in but only get a few out.

Distributed memories typically operate in an *associative, content-addressable* manner, meaning that memories are indexed and cued by portions of themselves (or of related memories). Retrieval works by presenting the network with a partial pattern or cue, which is then completed via the exchange of signals between the nodes (as in the Hopfield networks described in chapter 10). Memory retrieval is simply distributed pattern completion.

So, indeed, with n bits, one can represent 2^n different patterns, but not at the same time or in the same place. In an n-node, fully connected, distributed-coded neural network, multiple patterns need to be stored across the $\frac{n(n-1)}{2}$ weights such that they can be retrieved and displayed on the same set of n neurons. The presence of a quadratic number of weights does help, but $\frac{n(n-1)}{2}$ is still much less than 2^n for n larger than 15 or 20, so the naive representational promise of distributed coding is completely unrealistic when the 2^n (or any significant fraction thereof) patterns must share space. To get a quantitative idea of the ubiquity of pattern interference in distributed memories, see the discussion of k-m predictability in appendix B.

7.2.3   Sequential Pattern Retrieval

In a single-pattern completion task a partial pattern cues the rest of the pattern. In sequential pattern retrieval one complete pattern cues the next pattern, which in turn cues the next pattern. In theory, the network can complete the entire sequence when given only the first pattern.

Sequence retrieval also illustrates the differences between local and distributed coding. In the former each sequence element corresponds to a single neuron, which provides an excitatory link to the neuron representing the next item in the sequence. The key advantage of this representation is that n linearly interlinked concepts are easily incorporated into an n-neuron network with only $n-1$ links. The key disadvantage is fault tolerance: if any neuron n^* in the sequence is damaged, all concepts represented by downstream neurons from n^* cannot be cued for recall and are thus inaccessible.

On the other hand, for a sequence of items stored in a distributed manner, a faulty neuron may have little adverse effect, since each concept is coded by a combination of many neurons. Unfortunately, to pack all those distributed codes into the same network can quickly lead to ambiguities during retrieval because of overlap (i.e., interference) between the patterns. Thus, a single concept could easily have several, not one unique, successor pattern.

7.2.4   Connecting Patterns: The Semantics of Neural Information

The most intuitive way to add meaning to neural patterns is to ground them
in reality by finding correlations between external events (such as the lightning
bolts of figure 7.3) and combinations of active neurons, termed *cell assemblies* by
Donald Hebb (1949). These neural activity patterns can later serve as stand-ins or
representations for the actual event. These correlations arise by learning and may
require several exposures to similar events before they consistently evoke the same
activity pattern.

   Sameness is an elusive relation in these situations; for mammals and their large
brains, two neural states at two different times are surely never identical, owing
to the ever-changing nature of complex natural neural networks. However, two
states can probably be very similar, and more similar to each other than to any
other states, for example. To further complicate the issue, the complete brain state,
which takes into account all the brain's neurons, reflects many more sensory and
proprioceptive factors than, for example, the sight of a lightning bolt. Whether
one sees lightning while sitting indoors versus running down the street will surely
impact the complete mental state. Even though the lightning concept could evoke
the same cell assembly in both cases, the complete brain states could differ significantly. For example, the neural effects of fear could be widespread in the running
scenario but not in the indoor setting. Regardless of the unavoidable philosophical quicksand related to mental states and their meanings, the general fact that
certain mental states do seem to correlate with particular sensory experiences provides some solid footing for explorations into the pattern-processing functions of
the brain.

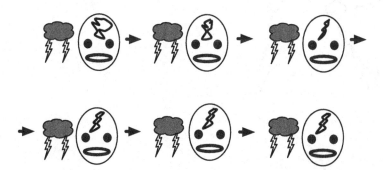

Figure 7.3
One of the brain's key functions is to form correlations between real-world events and brain states.
Repeated exposures to the event may be required to form a consistent correlation, as shown by the
sequence of event-brain-state pairs.

Once neural correlates of world events exist, they can become linked to one another via synaptic ties between the respective cell assemblies, as depicted in figure 7.4. This allows the presence of one event, such as lightning, to evoke the thought of another event, e.g., thunder, prior to or in the complete absence of that event.

When cell assemblies include many intragroup synapses among the neurons, pattern completion becomes possible. Thus, as shown in figure 7.5, when a partial sensory experience, such as an occluded view of a lightning cloud, stimulates part of the assembly, the internal synapses quickly activate the rest of the assembly, thus putting the brain into its normal lightning-viewing state, from which it can predict normal lightning successors, such as thunder, and plan normal lightning response actions, such as running for cover.

Although two events in the world are never truly identical, living organisms often need to treat them so. Otherwise, the appropriate responses learned from prior experience could never be reused in similar situations, since the organism would never recognize the similarity. To fully exploit the advantages of a learned association between a stimulus and a response, an organism must have the ability to generalize across stimuli. Thus, at some level of mental processing, two similar stimuli should evoke approximately the same cell assembly, which could then evoke the same predictions or actions associated with that general class of stimuli.

For example, if a small child learns that small, dark, short-haired dogs are dangerous (because of the unfortunate accident of stepping on one's leg), he may

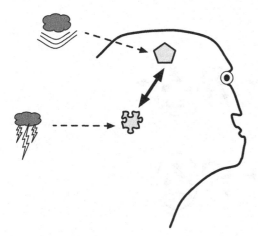

Figure 7.4
Another important brain function is to achieve correlations between brain states whose association has some utility for the organism. These states may or may not correlate with external events. Here, the brain states for lightning and thunder become correlated (double arrow).

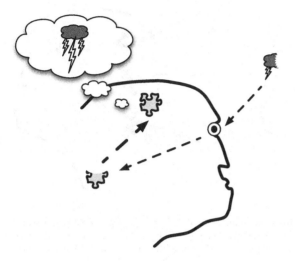

Figure 7.5
Another key function of brain is to complete partial patterns, thus putting the brain into a similar state to that achieved by a more complete sensory input, and allowing the brain to predict similar consequences and plan similar responses

also begin to watch his step around small, light, long-haired dogs. The child thus exhibits a general cautious behavior around all small dogs based on one experience. So in terms of the sensory preconditions for that behavior, the child simplifies considerably, ignoring most of the significant differences between dog species and mapping all small-dog experiences to a "tread lightly" behavior. In psychological or connectionist terms, the child has *categorized* these animals into a group that most adult observers would call "*small dogs*," and then the child chooses the same behavior in response to all members of that group. The categorization may be purely implicit, as only the child's actions reveal the underlying generalization; he cannot consciously contemplate or discuss it with anyone.

Figure 7.6 depicts this general mechanism, known as *sparsification*, which has the character of a reductive process in that many different events or neural patterns reduce to a single pattern, thus requiring a sparser population of neurons in the downstream assembly to capture a given concept.

In contrast, the process of orthogonalization, shown in figure 7.7, accentuates the differences between world events or upstream cell assemblies. Thus, even if two stimuli overlap significantly, the brain may map each to a different response. Much of the detailed knowledge of expertise would seem to involve orthogonalization, as small differences in problem contexts may invoke vastly diverse actions by an expert, who notices salient discrepencies that a novice might overlook. In the case

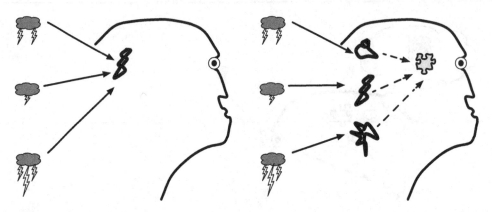

Figure 7.6
Neural process of sparsification, wherein the same cellular assembly is invoked by either different external events or different (prior) mental states. Differences are reduced by further mental processing. In
these diagrams, as in the brain, primitive visual processing begins at the back of the head, with higher-
level patterns forming toward the front.

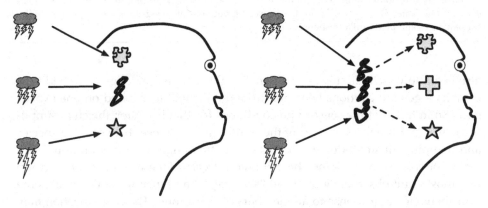

Figure 7.7
Neural process of orthogonalization, wherein markedly different cellular assemblies are invoked by
similar world events or similar (prior) mental states. Differences are accentuated by further mental
processing.

of the canine-wary child, orthogonalization means differentiating small dogs that
are playful from those that are angry.

Together, these pattern-processing activities facilitate intelligent behavior. Correlations between neural assemblies and world states enable organisms not only to
recognize familiar situations but also to think about those contexts when not actually in them. This ability to reason about situations other than those of the immediate sensory present constitutes somewhat of a breakthrough in mental evolution,

according to Deacon (1998); it supports complex faculties such as general symbolic reasoning and language use.

Pattern association and completion are often put forward as the hallmarks of human intelligence (Clark, 2001). They are tasks that we still do much better than computers. Additionally, a good deal of our intellectual prowess stems from sparsification and orthogonalization. The former allows us to generalize over similar situations, and the latter preserves our ability to tailor special behaviors for specific contexts. An interesting combination of the two is topological mapping (see figure 7.8) wherein similar situations elicit comparable (though not identical) responses via a correlation between one neural space (e.g., sensory) and another (e.g., motor). The structural backdrop for mammalian intelligence is a mixture of sparsifying, topological, and orthogonalizing circuits.

This characterization of neural representation in terms of pattern grouping and separation boils down to what the Nobel laureate Gerald Edelman and the renowned neuroscientist Giulio Tononi call *differences that make a difference*: distinct neural patterns that have distinct effects upon thought and action (Edelman and Tononi, 2000). They compare the human brain and visual system to a digital camera. Both can accept megapixel inputs spanning an astronomical pattern space, but only the brain has the ability to *differentially process* a huge number of them; the camera treats most of them identically. Thus, although both systems can receive and handle an extraordinary number of data points, the brain, via its elaborate

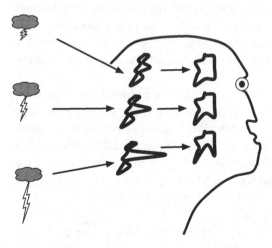

Figure 7.8
Neural process of topological mapping across three different spaces: one of external stimuli and two of neural patterns (in different areas of the brain). Similar stimuli evoke similar, but not identical, brain states.

and nuanced scheme for consistently linking those input patterns to internal states and actions, exhibits more sophisticated intelligence.

Clearly, the brain's circuitry is more extensive than a camera's, but this alone gives little ground for claiming that the brain has more knowledge. After all, newborn infants have more neural connections than adults but no more (arguably less) visual intelligence than today's cameras. It is the ability of the brain's synaptic network to create correlations among a huge number of patterns that produces intelligence. In this mature state of pattern matchmaker, the brain's neural networks embody a high level of knowledge that clearly separates us from the vast majority of our machines. That difference, however, is diminishing.

7.2.5   Hierarchical Cluster Analysis

The degree to which an ANN properly balances sparsification, orthogonalization, and topological mapping to best represent a domain (and most judiciously link differences to differences) is a very important characteristic—one that ANN researchers like to gauge either formally or informally. Quantitatively, one can measure the correlation between input and output patterns, or between inputs and the activation state(s) of any hidden neuron(s). This single correlation/covariance value provides a global assessment of the ANN's *understanding* of a domain, but a finer-grained assessment also has virtue, particularly in more complex networks that differentiate several categories and employ multiple responses.

The hierarchical cluster plot, a very popular connectionist tool, provides a very detailed but easily understandable overview of the implicit categories formed by an ANN. It displays each input case (C) in a graphic that shows the other input cases that the ANN treats most similarly to C. The behavioral basis for this similarity judgment can vary, but it often involves the patterns of activation of the hidden or output layers.

Consider the six cases in table 7.1, which indicates the hidden layer activation pattern produced by a hypothetical ANN upon receiving each case as input. Each data instance presumably includes several descriptive features that serve as inputs to the ANN. (Most of the detailed features are omitted from the table.) The hidden layer activation patterns that the ANN produces for each case form the basis for hierarchical clustering. Based solely on these activation patterns, the six cases can be hierarchically clustered to determine if, for example, the network treats dogs and cats in a distinctively different manner.

A wide array of hierarchical clustering algorithms exists. This section makes no attempt to summarize them, it merely employs the following basic approach:

1. Begin with N items, each of which includes a *tag*, which in this example is the hidden layer activation pattern that it evokes.

Table 7.1
Portions of a data set and its treatment by a hypothetical three-layered ANN with eight hidden nodes

Animal	Name	Hidden Layer Activation Pattern
Cat	Felix	11000011
Dog	Max	00111100
Cat	Samantha	10001011
Dog	Fido	00011101
Cat	Tabby	11011001
Dog	Bruno	10110101

2. Encapsulate each item in a *singleton cluster* and form the cluster set, C, consisting of all these clusters.

3. Repeat:

   a. Find the two clusters, c_1 and c_2, in C that are *closest*, using distance metric D.

   b. Form cluster c_3 as the union of c_1 and c_2; it becomes their parent on the hierarchical tree.

   c. Add c_3 to C.

   d. Remove c_1 and c_2 from C.

   until size(C)=1.

In this algorithm the distance metric D is simply the average hamming distance between the tags of any pair of elements in c_1 and c_2:

$$D(c_1, c_2) = \frac{1}{M_1 M_2} \sum_{x \in c_1} \sum_{y \in c_2} d_{\text{ham}}(\text{tag}(x), \text{tag}(y)), \tag{7.1}$$

where M_1 and M_2 are the sizes of c_1 and c_2, respectively. Applying this algorithm to the dog-and-cat data set yields the cluster hierarchy (also known as a *dendogram*) of figure 7.9. In the first few rounds of clustering, the tight similarities between the tags of, respectively, Fido and Max, and Samantha and Felix, trigger those early groupings. Bruno and Tabby are then found to be closer to those new clusters than to each other, so they link to their respective two-element clusters at the next level of the hierarchy. Finally, at the top, the dog and cat clusters link up.

The important information in this dendogram is that the ANN is treating dogs more similarly to each other than to cats, and vice versa. Thus, the ANN implicitly knows something about the difference between these two species, as reflected in the behavior of its hidden layer.

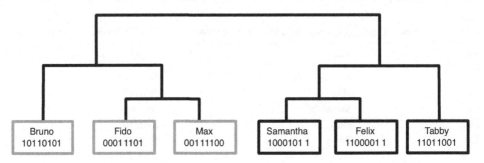

Figure 7.9
Dendogram for the dog-and-cat data set, using the hidden-layer activation pattern (displayed below
each animal's name) as the basis for similarity assessments, and the distance metric of equation (7.1)

In large ANN applications, where the training method may be supervised
unsupervised, or even reinforced, the dendogram can reveal intricate distinctions
in the complex *decision making* performed by an ANN. In many cases, the diagrams
indicate the learning algorithm's uncanny ability to form many-leveled hierarchical categorizations that closely (or exactly) parallel a human's approach to carving
up the world. They are proof that indeed the ANN has learned some very significant knowledge about the data set (and how to react to it) but without that
information being in any convenient human-readable form. One has to look hard
to see it, and cluster analysis with dendogram visualization is a fine tool for this
type of detective work.

7.3   Representations in Synaptic Matrices

From the activation patterns of neurons as representations of real-world events,
we now turn to the representations embodied in synapses via the neural activity that they induce. Synaptic networks house the potential to re-present neural
activation patterns, with particular combinations of strong and weak, excitatory
and inhibitory, immediate and delayed, connections providing a strong bias as to
which patterns form and which ones cling together in temporal sequences. Just as
genetic material, when placed in the proper environment, produces the organism
that it represents, a synaptic configuration, when properly stimulated (by external
inputs, an oscillating background, and so on) can produce the activity combinations that it represents.

The synaptic perspective supports the interpretation of neurons as detectors of
feature combinations. The concept embodied in these preferred features may map
nicely to standard mathematical or natural language expressions, such as "a man
wearing a bright shirt and dark pants" or, as is often the case, to very complex
concepts which can only be expressed by very long descriptions (i.e., not easily

compressible) or to concepts so diffuse that no expressions in natural language cover them.

When features are dimensions in a multidimensional space, and individual cases/examples are points in that space, then straightforward geometric analysis of its firing thresholds and input weights reveals the concept that a neuron detects. Consider a simple neuron, z, as depicted in figure 7.10, which accepts weighted

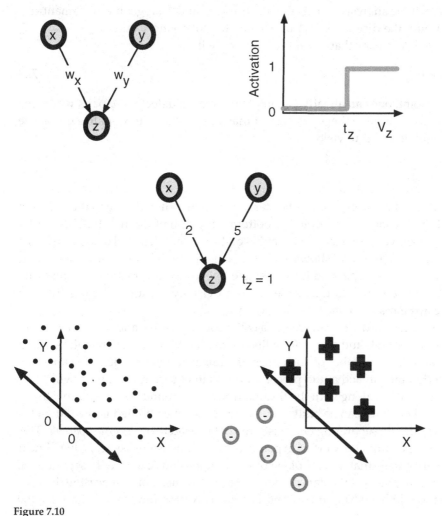

Figure 7.10
(*Top left*) Simple neural network with two inputs and one output node. (*Top right*) Step activation function for the output node, z. (*Middle*) An instantiation of this network with real values for the weights and threshold. (*Bottom left*) Border (line) and region (dotted area) represented by the network and expressed by equation (7.3). (*Bottom right*) Separation of positive (crosses) and negative (circles) instances of the concept detected by neuron z.

inputs from neurons x and y, with weights w_x and w_y, respectively. As shown, z has a step activation function with threshold t_z.

Since the sum of the weighted inputs will only exceed t_z, thus causing z to fire, under certain conditions (i.e., certain combinations of x and y values), neuron z essentially functions as a *detector* for those conditions. One very important step in understanding the knowledge content of ANNs is the determination of *what* various neurons have evolved or learned to detect. In the example of figure 7.10, a simple algebraic analysis reveals these conditions and their geometric semantics.

To simplify the discussion, let us assume the following values: $w_x = 2$, $w_y = 5$, and $t_z = 1$. This means that z will fire if and only if

$$2x + 5y \geq 1. \tag{7.2}$$

At this point, one can already get an idea of what z detects: x–y pairs where the sum of twice x and five times y is greater than or equal to 1. To visualize this, solve for y in equation (7.2) to yield

$$y \geq -\frac{2}{5}x + \frac{1}{5}. \tag{7.3}$$

Then draw the corresponding border and region denoted by equation (7.3) in the Cartesian plane, as shown in the bottom left graph of figure 7.10. This border separates positive from negative instances of the concept detected by neuron z. The bottom right of figure 7.10 shows several of these positive and negative instances, all of which can be separated by this single line, derived directly from the firing conditions of neuron z. In this sense, neuron z cleanly separates the positive and negative instances of a concept, with no error.

Working backward, one can receive a collection of positive and negative concept instances (a data set) and search for a line that separates them. There will be either none or an infinite number of such lines. If a line is found, its equation can easily be converted into an activation precondition, as in equation (7.2), from which the weights w_x and w_y, along with the threshold t_z, can be found. Thus, a simple three-node neural network can be built to detect the concept embodied in the data set.

In cases where a separating line exists, the data is said to be *linearly separable*. This concept extends from the Cartesian plane to any k-dimensional space, where linear separability entails that a hyperplane of $k - 1$ dimensions can cleanly separate the k-dimensional points of the data set. Of course, in k-dimensional space, the detector neuron z would have k input neurons, but the activation function would have the same form:

$$x_1 w_1 + x_2 w_2 + \cdots + x_k w_k \geq t_z. \tag{7.4}$$

In general, a data set is linearly separable if and only if (iff) a neural network with one output neuron and a single layer of input neurons (whose sum of weighted activation values are fed to the output neuron) can serve as a perfect detector for the positive instances of that data set.

Interestingly enough, a data set need not be large or exceptionally complex to fail at linear separability. For example, as shown in figure 7.11, simple Boolean functions of two variables can evade linear separability. Whereas AND and OR are linearly separable, XOR is not.

To build an ANN detector for XOR, one needs a network with three-layers: input, middle, and output, as shown at the top of figure 7.12. The middle layer consists of AND nodes that detect each of the two precise instances of XOR, while the output layer computes the OR of the middle layer outputs. Clearly, this is not a very satisfying example of a *general solution*, since each of the two instances requires special treatment: each demands a dedicated detector.

Building detectors, by hand, for more complex functions can be arduous, but if several straight lines can separate the data into positive and negative instances, then a straightforward procedure using multiple layers does the job (see appendix B). However, the tedious nature of this process indicates why all serious ANN users employ automated learning algorithms to train their networks, i.e., to find the proper connection weights.

In well-trained ANNs the salient concepts (embodied in detector neuron behavior), whether easily comprehensible to humans or not, work together with the concepts embodied by other detectors to achieve an effective overall functionality. Unfortunately, reductionist attempts to explain that functionality in terms of sequences of neural firings can be extremely difficult because the *meanings* of those firings (in terms of the concepts detected by the firing neurons) can be complex or diffuse.

There is no simple solution to this problem, but one can get a sense for a detector's preferred concept by examining the weights on all incoming connections. As shown in figure 7.13, if the input patterns are faces, then the weights for a hidden layer neuron can be projected back onto the plane of the input image (with pixel intensity corresponding to weight strength, relative to the other input weights to the same neuron). The resulting image gives a visual summary of the face that most strongly excites the detector neuron. This technique is commonly used in explaining the workings of the immediate downstream layer to the input layer of an ANN. When the input data is not of a visual form, such as acoustic or infrared sensor data, then the visualizations of preferred concepts are equally easy to produce in visual graphic form (using the same process as the preceding one) but typically harder for humans to interpret.

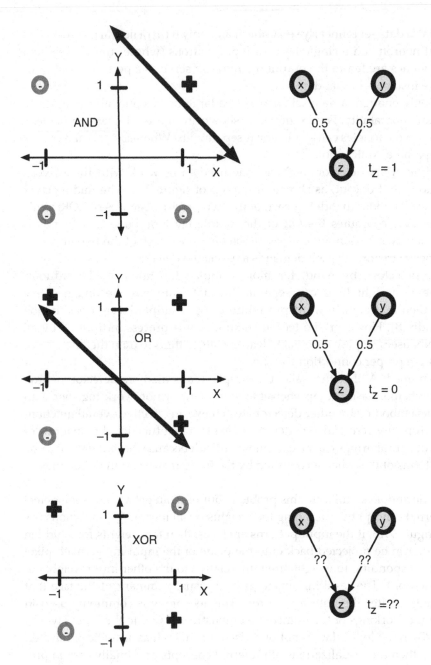

Figure 7.11
(*Left*) Cartesian plots of three standard Boolean functions, where True = +1 and False = −1. (*Right*)
Simple ANN detectors for four-element data sets for AND and OR. The borders (lines) corresponding
to the activation conditions for these ANNs are drawn as double-arrowed lines on the Cartesian plots.
Since XOR is not linearly separable, no simple ANN detector exists; a multilayer network is needed.

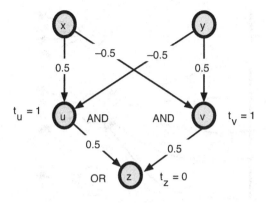

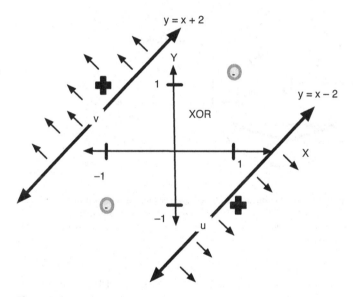

Figure 7.12
(*Top*) Three-layered ANN that realizes the exclusive-or (XOR) function. (*Bottom*) The borders (double-arrowed lines) and regions (small arrows) delineated by the activation conditions of the AND neurons *u* and *v*. Points in either of the two regions are positive instances (crosses) of XOR, while all other points are negative instances (circles).

Churchland (1999) describes a classic example of this decoding process for a face recognition ANN, where interesting, almost eerie, *facial holons* are the preferred stimuli of fully trained hidden layer neurons. These holons have no direct resemblance to any of the input faces but appear to contain a potpourri of features

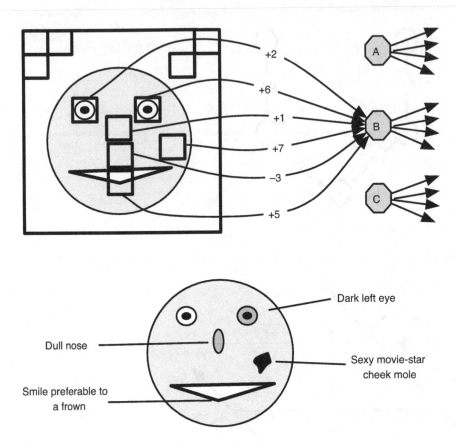

Figure 7.13
Deciphering the preferred input of a particular hidden node, B. (*Top*) Sketch of a neural network used for facial recognition. The weights on incoming connections to neuron B indicate those features of the image toward which B is most sensitive. (*Bottom*) Based on the incoming weights, a dream face for neuron B can be constructed, i.e., a face that would cause node B to fire strongly.

from several of the original faces. One look at these holons provides convincing evidence that the concepts embedded in automatically trained ANNs have few counterparts in natural language.

7.3.1   Principal Component Analysis

By properly tuning synaptic weights, a neural network can achieve an intelligent balance between orthogonalization and sparsification. In fact, Hebbian learning can produce networks that realize a simple version of principal component analysis: they differentiatiate (orthogonalize) along the high-variance component(s) and abstract (sparsify) along the rest. Thus, the network as a pattern detector/classifier separates input patterns based on the features that have the most variance in the sample population; and these are often considered the most significant features in a classification task. For example, when organizing a group of people to perform a complex task composed of many diverse subtasks—such as playing a team sport, giving a concert, or planning the economy of a large city—the manager normally overlooks low-variance features of individuals, such as the age of junior high band members, and focuses on high-variance properties, such as the particular instrument a teenager plays. These sources of diversity are normally the salient pieces in the organizational puzzle.

Formally, the principal components of a data set are vectors that capture the highest amounts of variance in that data. If the data set has two attributes, student age and instrument tone (for the junior high band members), then one expects little variance in age but considerable variance in tone. The principal component vector of this data will then have a large bias (reflected in a large vector component value) toward tone, with a much smaller value for age.

One very significant discovery of ANN research is the following:

> *If* (1) the values of a data set are scaled (to a common range for each feature such as [0, 1]) and normalized by subtracting the mean vector from each element; (2) the modified data values are fed into a single output neuron, z; and (3) the incoming weights to z are modified by general Hebbian means (wherein the correlation between input values and z's activation level have a direct influence upon weight change), *then* z's input weight vector will approach the principal component of the data set.

An important detail is the simple mathematical fact that *the border between regions carved out by a single output neuron is perpendicular to the weight vector.* This is easily shown by the following derivation:

$$xw_x + yw_y \geq t_z \leftrightarrow y \geq -\frac{w_x}{w_y}x + \frac{t_z}{w_y}. \tag{7.5}$$

This defines a region whose borderline has the slope $-\frac{w_x}{w_y}$.[1] Then, any vector with slope $+\frac{w_y}{w_x}$ is perpendicular to that border. Since neuron z's incoming weight vector is $\langle w_x, w_y \rangle$, it has slope $+\frac{w_y}{w_x}$ and is therefore perpendicular to the border line. This implies that the border will separate the data set based on those factors with highest variance.

Consider the general significance of this property in the context of another simple example. In figure 7.14 the instances of a hypothetical data set consist of two attributes: gray-scale color and size. Further, assume that the instances either involve mice or elephants. Since both mice and elephants are often gray in color, their gray-scale values will be similar, whereas their sizes will differ considerably, as the upper plot in the figure partially indicates—the actual difference is five or six orders of magnitude.

First, scale each feature value to lie within a small range, such as [0, 1]. Next, compute the average vector, which is the vector of the independent averages of each scaled feature (color and size) across the full data set; it is depicted as a diamond in figure 7.14. Next, normalize each scaled data point by subtracting this average from it. This yields scaled, normalized data points, shown at the bottom of the figure.

Now, assume that a simple output neuron (z) with two inputs (x and y), one coding for size and the other for gray-scale color, were trained on this data. Postponing (for a moment) the details, if this training involved Hebbian learning, then the input weight vector for z would become tuned to something akin to the nearly horizontal, dashed arrow at the bottom of figure 7.14. This vector contains a large x component and a small y component, not because elephant sizes are so large but because the *variance* in size between mice and elephants is so much larger than the variance in color. Remember that the average vector has already been subtracted out, so if all animals in the data set were large (i.e., elephant-sized), then the normalized x values would be small.

Since the border line vector (for the concept detected by z) is perpendicular to z's weight vector (as explained earlier), it resembles the nearly vertical double-arrow line at the bottom of figure 7.14. This border separates positive from negative examples of z's concept, and the groups that it most clearly differentiates are small animals (e.g., mice) and large ones (e.g., elephants). That is, the neuron will only fire on data points from one of those categories, not both. It detects the *most significant difference* in the data set: the *principal component*.

If the data set consisted solely of African elephants (the largest species of land animal), then the weight variance would be greatly reduced compared to the mice-elephants scenario. This would reduce the extreme horizontality of the weight vector, thus reducing the verticality of the border line, possibly to a level at which neuron z would appear to primarily differentiate between light and dark animals. If the variances in color and size were of similar magnitude, then the border line (perpendicular to the weight vector, determined by Hebbian learning) might only differentiate between concepts that, to a human observer, seem useless, such as "very large light-gray animals" versus "moderately large dark-gray animals".

To see how Hebbian learning nudges weight vectors in the direction of the principal component, consider a quantitative version of the mice-elephants scenario. First, all inputs to an ANN should be scaled to within the same range, such as [0, 1] or [−1, 1] to insure that each feature has an equal opportunity to affect the behavior of the network, both during and after learning.

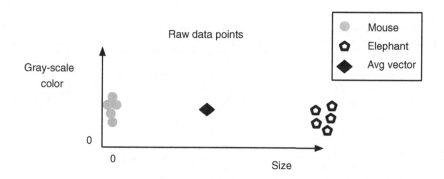

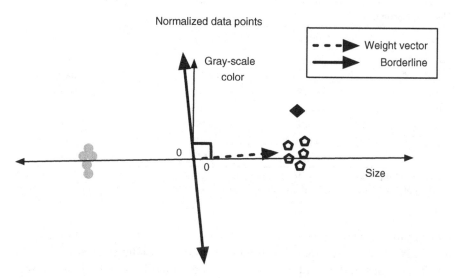

Figure 7.14

(*Top*) A sketch of hypothetical data points taken from mice and elephants, where each point consists of size and color attributes. No scale is specified, although it would have to be logarithmic to capture the huge discrepancy in size. The average data point/vector (diamond) appears in the middle. (*Bottom*) After scaling and then subtracting the average vector from each data point, the plots of these scaled, normalized data points straddle the x and y axes. When a simple neural network is trained (using Hebbian learning) on these normalized data points, the weight vector of the ANN's output node would show a strong bias toward size, since it has the highest variance. This produces a horizontal weight vector and thus a vertical border line, to separate the positive and negative instances of the concept detected by the output neuron.

Table 7.2
Hypothetical data set consisting of three mice and three elephants, with features being size (in kg) and gray-scale color (in the range 0 (black) to 100 (white))

Animal	Raw Data	Scaled Data	Normalized Data
Mouse	(0.05, 60)	(0, 0.6)	(−0.27, −0.04)
Mouse	(0.04, 62)	(0, 0.62)	(−0.27, −0.02)
Mouse	(0.06, 68)	(0, 0.68)	(−0.27, 0.04)
Elephant	(5,400, 61)	(0.54, 0.61)	(0.27, −0.03)
Elephant	(5,250, 66)	(0.53, 0.66)	(0.26, 0.03)
Elephant	(5,300, 69)	(0.53, 0.69)	(0.26, 0.05)

Note: The data are first scaled to the ranges (0, 10,000) for size and (0, 100) for color. They are then normalized by subtracting the average vector (0.2700, 0.64).

Table 7.3
Result of training a two-input, one-output ANN on the scaled, normalized data of table 7.2, using a simple Hebbian learning rule

Input (Size, Color)	Output	Δw_{size}	Δw_{color}
(−0.27, −0.04)	−0.031	+0.0017	+0.0002
(−0.27, −0.02)	−0.029	+0.0016	+0.0001
(−0.27, 0.04)	−0.023	+0.0012	−0.0002
(0.27, −0.03)	+0.024	+0.0013	−0.0001
(0.26, 0.03)	+0.029	+0.0015	+0.0002
(0.26, 0.05)	+0.031	+0.0016	+0.0003
Sum weight change:		+0.0089	+0.0005

Note: It is assumed that the training is done in batch mode such that none of the weight changes are applied to the network until each data point has been processed.

Assume an initial data set (size, gray-scale) consisting of the six raw data points shown in table 7.2. The data points are then scaled and normalized, as described in the table. The values in the rightmost column are then fed into an ANN with two inputs (one for size and one for color) and one output.

Assume that the initial weight on both connections is a small positive value, 0.1. Also assume that the activation function for the output node is straightforward linear scaling: the activation level directly equals the sum of weighted inputs. Finally, assume the following Hebbian learning rule:

$$\Delta w_i = \lambda x_i y, \tag{7.6}$$

where x_i is the activation level of the ith input node, y is the activation level of the output node, w_i is the weight on the arc from the ith input to the output node, and λ is the learning rate, assumed to be 0.2.

As shown in table 7.3, the changes to the weight between x_{size} and the output are much larger than those from x_{color}. Hence, after one round of batch learning there will be an increase in w_{size} that is nearly 20 times larger than the increase in w_{color}. After several rounds of learning w_{size} will completely dominate the detector's weight vector.

Why did this dominance occur? In looking at the data of table 7.3, note the order-of-magnitude difference in the absolute values of the scaled, normalized size values compared to those of the color values. Remember that this no longer reflects any bias stemming from the fact that elephants are so large; it arises solely from the large variance in size between mice and elephants compared to the small variance in color between the two animal species. This variance imbalance entails that the size value will dominate the color value during each of the six runs of the ANN. Hence, the output value will have the same sign and general magnitude as the size input. Since the Hebbian learning rule multiples these two values (along with λ), and since they have the same sign, the weight change will be positive.

The color input may or may not match the sign of the output, incurring positive and negative weight changes with nearly equal probability. Furthermore, the small magnitude of the color value will restrict the size of those weight changes. Hence, w_{color} will change in small increments, both positive and negative, whereas w_{size} will change in larger increments, all positive.

The general implications of this rather specific technical detail are very significant:

> If the detectors of a network modify their input weight vectors according to basic Hebbian principles, then, after training, the activation levels of detectors can be used to differentiate the input patterns *along the dimensions of highest variance*. Hence, those detectors will differentiate between objects (or situations) that are *most distinct* relative to the space of feature values observed in the training data.

As an example, if an ANN is trained on human and cat faces, then it would probably learn to fire (an output) on only one or the other species, not both—certainly a sensible way to partition the data set. But if the ANN were trained only on human faces, then the separation would occur along another dimension (or combination thereof) of high variance, such as male versus female, bearded versus clean-shaven, or maybe happy versus angry. The beauty lies in the fact that the ANN figures out the proper combination of discriminating factors, even if that combination evades concise formulation in human language.

7.4 Neuroarchitectures to Realize Effective Distributed Coding

As mentioned, the often-praised storage capacity advantages of distributed coding are frequently exaggerated vis-à-vis the demands of content-addressable, associative memories. However, organisms derive substantial benefits from distributed codes in terms of generalization (derived from pattern sparsification).

Figure 7.15 abstractly illustrates this fundamental advantage of a distributed code. Assume that an organism's flight response is driven by a single motor neuron, X, at the bottom of the figure. Tuned to trigger on the firing pattern 101110 of its upstream neighbors, X will probably also trigger on similar patterns, such as 101100 and 001110, which involve many of the same active neighbors. This makes great life-preserving sense. A gazelle that only flees from tigers with exactly seven stripes cannot expect to live very long. The ability to generalize/sparsify sensory patterns such that many map to the same successful behavior(s) is paramount to adaptation, intelligence, and ultimately, survival.

Distributed coding in networks of simple neurons provides this flexibility as a by-product. Via synaptic change, neurons become tuned to fire on particular combinations of afferent stimuli but not necessarily those *exact* combinations. As long as the net input to the neuron exceeds a threshold, it fires. Thus, as long as the network uses distributed coding, and the detector neuron maintains a threshold that prevents it from firing on all-too-sparse input patterns, the detector should easily generalize over many dense distributed codes and thus fire in many (similar) situations. This is a huge advantage of distributed memory systems composed of simple nodes joined by plastic links.

So the use of distributed codes involves trade-offs, and a key question is whether a neural system can somehow avoid their pattern-corrupting dangers while still taking advantage of their generalizing power? The brain appears to have found a way.

Note that the distributed afferent patterns on the dendrites of figure 7.15 do not necessarily originate from a pattern-associating layer. These afferent neurons may have very little influence upon one another. Although the advantages of

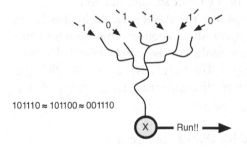

101110 ≈ 101100 ≈ 001110

Figure 7.15
Distributed code (and several similar codes) that stimulates a motor neuron

incorporating them in such a layer would include the ability to complete partial patterns, the disadvantages of interference and memory corruption could quickly override the positive effects, and neuron X may not require a completed afferent pattern to fire; anything reasonably close to its learned trigger pattern may work.

In short, the brain may exploit dense distributed codes at various sites without having to store (and thus retrieve from partial patterns) many of them within a common subpopulation of neurons. The brain may reserve associative memories for only relatively sparsely coded information.

But clearly humans (and other animals) have complex memories that involve thousands, if not millions, of active neurons and that are retrievable from partial instantiations of themselves. How might this work?

Three brain regions: cerebellum, basal ganglia, and hippocampus, all give some hints of a solution, as recognized by Marr (1969; 1971) and later elaborated by Rolls and Treves (1998). Each of these areas contains an entry-level layer of neurons that exhibit very competitive interactions: when active, they either directly or via interneurons inhibit their neighbors. This leads to very sparse coding at these entry layers, since only a relatively small subset of neurons can be active at the same time. However, fan-in to these layers can still be high (as it is in the basal ganglia and hippocampus). Thus, a dense distributed code in the space of afferents compresses to a sparse code within the competitive layer.

If this competitive layer then feeds into an associative network, the latter will only be expected to store and recall sparse codes, which can be achieved with little interference. This is precisely the case in the hippocampus (figure 7.16), where the dentate gyrus (DG) serves as a competitive layer; it sends axons to CA3, which, with its high degree of internal (recurrent) connections, appears optimally designed for associative pattern processing.

Figure 7.17 illustrates this basic combination of layers. Dense distributed codes are detected by neurons in the competitive layer, with ample flexibility to generalize across similar dense codes, just as the neurons of artificial competitive networks often fire on all vectors in the neighborhood of their prototype. Competition leads to sparse coding, since only a few neurons are active at any one time, and these sparse patterns are sent to the associative layer, which uses Hebbian learning to strengthen the synapses between co-active cells.

Thus, the sparsification of one layer by another normally involves an *expansion* in layer size, a concept known as *expansion recoding*. As detailed in appendix B, a sparse layer needs many orders of magnitude more neurons than an upstream dense-coding layer if the former is to faithfully detect and differentiate patterns in the latter. Figure 7.18 portrays this general topology, evident in both the hippocampus and the cerebellum (Rolls and Treves, 1998).

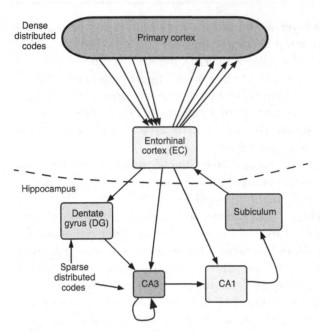

Figure 7.16
Overview of the hippocampus and its relation to the cortex, divided here into the primary cortex and the EC, which serves as the gateway to (and from) the hippocampus for the rest of the cortex. The sizes of each area roughly indicate relative sizes between the different brain regions. For example, CA3 contains fewer neurons than DG and CA1 and is much smaller than the primary cortex.

7.4.1   Pattern Processing with Dense Distributed Codes

Figure 7.17 shows how a sparsely coding competitive layer can serve as an interface between two layers, one densely coding and one sparsely coding. This allows the dense-coding layer to perform effective pattern processing despite the high interference incurred by dense codes. In this scheme the dense-coding layer can, with the help of competition and sparsification, transfer its partial patterns to a sparse layer, which performs relatively interference-free pattern completion. Then, recurrent links from the sparse to the dense layer can reinstate the complete dense pattern. This basic architecture is shown in figure 7.19, where dense coding occurs in a *weakly associative* layer, i.e., one containing a relatively low number of intralayer excitatory connections.[2]

Consider a dense pattern, P_d, presented to layer A_d of figure 7.19. Assume that P_d consists of two component patterns, p_1 (the four active (shaded) neurons on the left) and p_2 (the three active neurons on the right). Competitive layers often learn to detect recurring subpatterns, so if p_1 and p_2 are frequent components of dense patterns, then such dedicated detectors (the two shaded circles in C_s) can

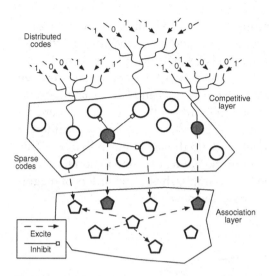

Figure 7.17
Using a competitive layer to sparsify dense distributed codes, which are then sent further to an associative layer

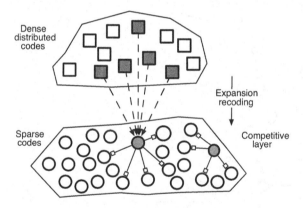

Figure 7.18
Expansion recoding between a densely coded layer and a sparsely coded, competitive layer with considerably more neurons that serve as detectors for many activation patterns in the upper layer.

easily emerge. These, in turn, send low-fan-out axons to layer A_s. In the figure low fan-out is represented as a one-to-one correspondence between the two layers.

In layer A_s of figure 7.19 the high number of internal recurrent connections facilitates the synaptic coupling of co-active neurons via standard Hebbian learning. Thus, the two active (shaded) nodes in A_s, which essentially represent p_1 and p_2, become linked such that either can stimulate the other. To complete this circuit, neurons of the strong association layer must link back to neurons representing the individual parts of p_1 and p_2. Thus, when either of these fire, they can activate all or most of p_1 or p_2, with the recurrent excitation of A_d stimulating the remainder.

With all these layers and links in place, the network can store and retrieve densely coded patterns in A_d, as detailed in figures 7.20 and 7.21. This relatively elegant solution appears too good to be true. After all, why should pattern-processing problems at one level be solved by a pattern processor at a different level? How can all the patterns in A_d be stored (and thus remembered) in A_s? This question is particularly difficult to answer in the brain, since the region that most convincingly exemplifies A_s, CA3 of the hippocampus, has many orders of magnitude fewer neurons than its corresponding A_d, the neocortex.

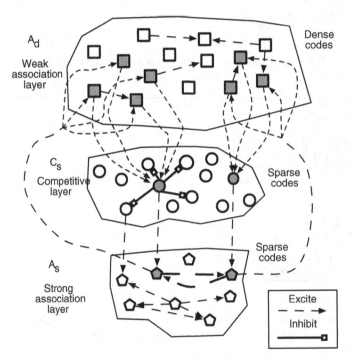

Figure 7.19
Interleaving associative and competitive layers to support pattern completion in a dense-coding distributed layer, A_d, which is only weakly associative. The lower layer, A_s uses sparse distributed codes and is strongly associative because of higher intralayer excitation. The middle layer, C_s, also codes sparsely but is competitive, because of prevalent internal inhibitory links. C_s detects subpatterns in A_d and sends sparse patterns down to A_s, which can complete partial patterns. High fan-out, recurrent links from A_s to A_d (dashed lines), enable single neurons in the former to activate neural assemblies in the latter.

Clearly, A_s cannot store all of A_d's possible patterns. However, if the patterns of A_d are biased such that they often consist of common subpatterns, then a competitive layer can detect these subpatterns, and a smaller but more densely interconnected associative layer can store links between these subpatterns. The key lies in the biased, strictly nonrandom nature of the pattern set; and living organisms inhabit a world full of these biasing regularities (invariants). Most humans are not capable of remembering a large corpus of random patterns, but gifted individuals or those with some training in memory techniques can easily remember large collections of *meaningful* patterns, in the sense that they consist of subpatterns that occur in everyday life.

Interestingly enough, patterns with common subpatterns are those most likely to cause interference in a distributed memory. Hence, the basic repetitive nature of our sensory world may have forced evolution to design network memory systems involving multiple layers of well-segregated competitive and associative functionalities.

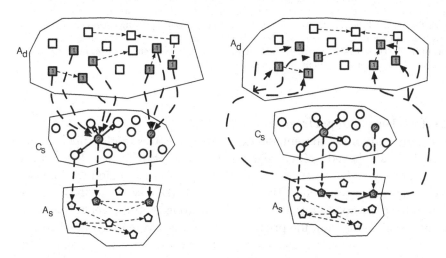

Figure 7.20
Storage (learning) of a dense distributed pattern (P_d) in A_d with the help of C_s and A_s, as defined in figure 7.19. Shaded shapes denote active neurons, and the numbers in each shaded neuron indicate the timestep at which they become active. (*Left*) After the complete pattern is entered into A_d, signals feed forward to C_s, where winner nodes have their active afferent synapses strengthened (thick dashed lines) by Hebbian learning. The winners send their signals to corresponding neurons in A_s. (*Right*) The synapses between co-active neurons in A_s become strengthened by Hebbian mechanisms (thick dashed lines). In addition, feedback collaterals from A_s to A_d (thick dashed lines) are strengthened when their presynaptic and postsynaptic neurons are active, thus binding P_s (the sparse representation of P_d) in A_s to P_d in A_d. In the right-hand diagram, some connections from the left are removed for ease of readability.

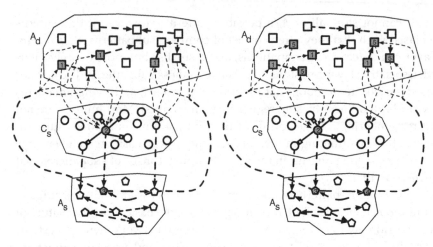

Figure 7.21
Pattern completion (recall) in A_d with the help of C_s and A_s, as defined in figure 7.19. Shaded shapes denote active neurons, and the numbers in each shaded neuron indicate the timestep at which they become active. (*Left*) The partial pattern (P_d*) enters A_d at time 1. This stimulates a detector for the left-most partial pattern in C_s at time 2. The competitive node then directly stimulates its counterpart in A_s at time 3. (*Right*) At time 4 the simple two-node pattern is completed in A_s owing to previously learned connections between these two neurons, which then send excitatory signals back to those neurons of A_d that constitute P_d. These feedback signals combined with the few recurrent synapses in A_d stimulate the remainder of P_d's neurons to activation at time 5.

7.4.2   Cortical-Hippocampal Interactions

In the mammalian brain the interaction between cortex and hippocampus (see figure 7.16) is believed to involve a *re-presentation* of densely coded cortical patterns as sparsely coded sequences in CA3, a hippocampal region with a very high level of intralayer recurrence. The dentate gyrus (DG) serves as the competitive layer entry point to the hippocampus. This, along with the general pattern of convergent collaterals from the primary cortical areas to the entorhinal cortex (EC), indicates that dense distributed codes in the neocortex are greatly sparsified within the hippocampus.

Return signals from CA3 to the cortex—through CA1, the subiculum, and EC—allow the completion of dense codes in the primary cortex after their corresponding sparse codes have been completed in CA3. After many instances of CA3-assisted retrieval, a cortical pattern may become self-completing as the weaker associative capabilities of the cortex eventually permit Hebbian learning between the dense pattern's component parts.

One often-cited theory (McClelland et al., 1994) posits that CA3 utilizes high recurrence and a high learning rate to quickly learn abstract (hence sparse) patterns, while the cortex must rely on a much slower learning rate and lower recurrence. Thus, CA3 learns a sparse pattern quickly and then gradually *off-loads* a denser version to the cortex via repeated recall, much of which is believed to occur during sleep and dreaming (Andersen et al., 2007). Since CA3 has relatively few neurons compared to the cortex, it cannot house many cortical representations at once, only those not yet fully imprinted upon the cortex.

7.4.3 Third-Order Emergence in the Brain

To review, the brain has evolved a topological structure, present in several areas, that performs expansion recoding. In general, this helps reduce interference between activation patterns. When the expansion recoding level pipes its outputs to a second, sparse-coding, associative layer, (A_s), and when that layer can feed back to the dense-coding entry level (A_d), the resulting circuit enables A_d to gradually tune its synapses to become an effective pattern-processing machine. For the patterns that it has learned, A_d no longer needs help carrying out the duties of a content-addressable memory: it can independently complete partial patterns.

Essentially, an associational network relies upon second-order emergence to learn and reproduce patterns, but doing so accurately becomes difficult when patterns begin to interfere with one another. One can argue that third-order emergence, as supported by a multileveled topology such as the cortico-hippocampal loop, solves the problem: the sparse patterns learned in CA3 have the power to re-create their corresponding dense patterns in the cortex. A key role of the hippocampus is producing these correlations between sparse CA3 patterns and denser cortical patterns. In other words, the hippocampus builds representations of cortical patterns, and these CA3 patterns generate their cortical counterparts by priming a second-order emergent process. During waking hours this re-creation is aided by sensory input, but during dreaming the sparse patterns manage quite well on their own, though their cortical correlates tend to be noisy.

The hippocampus directly nurtures third-order emergence by combining two second-order emergent layers into a complex circuit. Many brain regions build correlations between neural populations, but CA3 appears to be one of the most prolific pattern-learning areas. A layer with so many recurrent connections should produce very strong internal correlations, and these are a much better basis for high interlayer correlation. Consequently, the hippocampus is the queen of the neural matchmakers.

7.5 The Fallacy of Untethered Knowledge

Analogies between computers and brains frequent the science literature, although less so now than in AI's adolescence (see Boden, 1990; Haugeland, 1997 for collections of the classics). Some significant differences between minds and machines, such as the parallel versus serial nature of their computations, have blurred (with the ubiquity of parallel computing). But others cannot be ignored. As discussed in chapter 4, the hardware-software distinction appears to have no direct counterpart in the brain, since the *code* run by neurons is fully distributed across the brain and intimately tied to the synapses, which, as physical components of neural circuitry, would seem to mirror hardware, although their dynamic nature might earn them the label *softwire*. So one cannot just pull the code out of one brain and transfer it to another. But if not the code, what about the data? What about the knowledge, the representations?

As this chapter shows, the data is no less engrained in the neural hardware than the program is. A neural activation pattern make little sense outside of the brain, but it gets even worse. In a GOFAI system a chunk of information, such as a logical axiom, can be passed among different modules via an internal operation that typically involves copying its bit representation into different locations, and each module interprets those bits the same way. In a neural system transfer by copying is rare. Instead, one k-bit pattern in region R1 may correlate with an m-bit pattern in region R2 such that the two patterns essentially represent one another. The presence of one will, in some contexts, lead to the emergence of the other. However, any attempt at directly imposing the k-bit pattern into R2 and expecting it to mean the same thing (or anything) there seems as likely as tossing vials of DNA onto a floor and expecting them to square dance together. Even internally the meanings of activation patterns are entirely location-dependent; they require reinterpretation at every level. Neural networks don't store knowledge; they embody it.

Science programs (and science fiction movies) often tease the imagination with projections of future technologies that can copy all thoughts from a brain and load them into another brain, whether natural or artificial, thus producing a mental clone. Though the process of recoding a hundred trillion synapses is mind-boggling enough, it would not, in all likelihood, yield anything close to a cognitive copy. The patterns (and attractors in synaptic space that promote them) derive little meaning from their basic structure but rather from their relations to one another and (most relevant to cloning claims) to the body and world. Brain, body, and world are so tightly intertwined that isolated cerebral activation patterns reveal only an inkling of the actual knowledge and intelligence of the agent itself.

8   Search and Representation in Evolutionary Algorithms

8.1   Search as Resource Allocation

Imagine a (low-budget) mining company looking for rare minerals. They buy up tracts of land and then use a team of diggers (paid by the hour) to locate the mother lode, whose existence is usually promised by the seller of the property. Initially, the diggers are randomly distributed across the area, while a foreman moves about, checking their progress and occasionally relocating diggers.

The goal, of course, is to find the mother lode, but from the surface nobody can see it. Each digger serves as a probe into the terrain. If small traces of the mineral are found in a digger's hole, then that indicates the potential presence of the mother lode in the vicinity, but no probes are conclusive until shovels full of the mineral appear. The foreman makes educated guesses concerning the placement of diggers based on the mineral contents (or lack thereof) of previous holes. For example, if one digger finds 2 grams of the mineral after an hour of digging, and another digger, just to his left, finds 8 grams (in a similar-sized hole), then the foreman might assign a third digger even farther to the left based on these two pieces of partial information.

For the foreman directing this search process, the diggers (not the mineral) are his *resource*. He must allocate them in an intelligent manner to find the target site while incurring the lowest possible labor cost. Many computational search problems have a similar character. Given a problem, P, educated guesses are made about the solutions to attempt. After being constructed, each solution is evaluated with respect to its performance on P. In essence, the solution is analogous to the precise location of a digger, while the evaluation corresponds to the amount of mineral taken out of his hole.

Typical strategic choices of the foreman include the following:

- Move a digger from a location that does not appear promising.
- Command a digger to make a small move in a random direction.

- Deposit a digger in an area where other diggers have found some of the rare mineral.
- Deposit a digger at an intermediate location between two diggers whose locations both show signs of a mineral deposit.

The judicious, parallel application of these options should, over time, allow the crew to home in on the mother lode.

In his classic book on genetic algorithms Holland (1992) uses a two-armed bandit (a hypothetical type of slot machine) to illustrate and analyze search as resource allocation. In this problem you walk into a casino with a small bucket of quarters. In the corner stands a bright shining slot machine with two arms, the two-armed bandit. The sign above the bandit says that one of the arms has better odds of hitting winners, but which arm? Your task is to find out which arm is better while simultaneously maximizing total profit. Your only means of information gathering is pumping quarters into the machine, pulling one arm or the other, and recording the results.

Ideally, you can determine the higher-paying arm early in the process, using a small, but statistically significant number of quarters. After this *exploratory* information-gathering phase, you can commence an *exploitative* stage by investing all the remaining quarters in the heretofore better arm. The catch, of course, is that you are never completely certain which is actually the better arm. Thus, there are obvious trade-offs between exploring and exploiting: the former does allow you to gather more statistically significant data, but at the price of many quarters invested in the inferior arm, while the latter can maximize profit but carries a risk, due to the uncertain identity of the better arm. Holland's book provides a detailed mathematical analysis of this situation.

In the mining problem the foreman has similar exploration-exploitation trade-offs. To explore, he records the findings of each digger but does not move diggers to promising locations, only to new locations (in order to gather more information). To exploit, he uses the recorded findings to distribute diggers to promising spots. As with the two-armed bandit, the foreman hopes to gather convincing data as quickly as possible (such that exploration can give way to exploitation) and thus to minimize the total digging effort needed to pinpoint the mother lode.

8.2   Evolutionary Search

In evolutionary algorithms the individuals of the population are the resource, and one tries to allocate them efficiently such that desirable points in design space can be found as quickly as possible: the goal is to minimize the product of the population size and the number of generations. The EA's selection mechanisms and genetic operators have obvious analogies to the strategic choices of the mining company foreman.

The EA search process spans three interconnected spaces: genotype, phenotype, and fitness. Essentially, the EA searches in genotype space, but the trajectory of that search is strongly determined by the phenotypes and fitness values to which they are mapped.

Consider the example shown in figure 8.1. An EA is used to find a satisfactory assignment of M items of various weights to K containers (labeled 0 to $K-1$), each with a maximum weight limit of W, such that the final weights of each filled container are as close as possible to being equal (i.e., the variance in the weights is minimized).

Assume that each genotype is a bit string of length $M\lceil \log_2 K \rceil$: one gene of length $\lceil \log_2 K \rceil$ for each item, where the gene specifies the container in which to place the item.

A phenotype is then a simple list of container labels, of length M. For simplicity, assume that K is a power of 2, so each $\log_2 K$ substring of bits translates into a unique integer between 0 and $K-1$.

Many adequate fitness functions are possible, and their details are unimportant at this point; but in general, they will give higher scores to partitions in which none

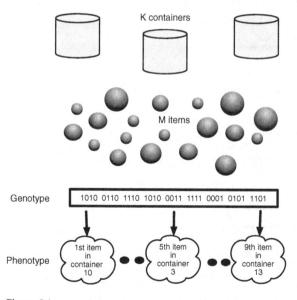

Figure 8.1
Simple search problem of assigning objects (balls) to containers (cylinders) so as to put all objects in containers without exceeding any container's weight limit while minimizing container weight variance

of the container weights exceeds W and the variance among the container weights is low.

In terms of the connections between the three spaces, the mapping between genotype and phenotype space will be bijective (i.e., each genotype will map to a unique phenotype, and all phenotypes will be mapped to by exactly one genotype), while that from phenotype to fitness space will be many-to-one, since several phenotypes may be awarded the same fitness.

Figure 8.2 shows the three spaces and their connectivity for a container problem with $K = 4$. The nth pair of bits on the genome encodes the destination container for the nth item. The developmental process is a simple conversion of bit pairs to integers (between 0 and 3), as shown at the phenotypic level. The phenotypes then map to fitness values in the upper landscape.

Various features of the EA determine the nature of, and interactions among, these three spaces and consequently the difficulty of the evolutionary search. The fitness landscape is probably the most critical of the three. If it is relatively smooth,

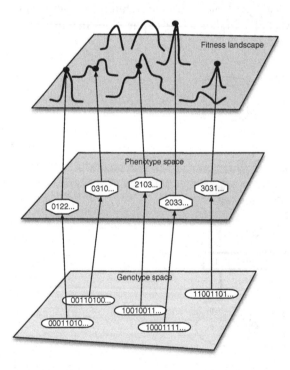

Figure 8.2
Mappings between genotype, phenotype, and fitness spaces

with gently rolling hills and gradual inclines leading to a summit corresponding to the optimal phenotype, then many search algorithms, not just EAs, will stand a good chance of finding that summit. However, if the summit sits atop a very steep peak that rises straight up from the zero plane, then search becomes a needle-in-the-haystack endeavor. Similarly, if there are many local (but not global) maxima on the landscape, then the search process is easily detoured toward, and bogged down at, these deceptive sites.

The fitness landscape is merely a visualization of the fitness function, so a landscape with one Mount Everest rising from the plains corresponds to a fitness function that gives no partial credit to any of the phenotypes directly *under* the plains. Hence, when the EA generates such phenotypes, it literally does not know which way is up and can only guess by searching out in random directions from the zero-fitness phenotypes. Hence, the mining scenario is a more appropriate real-life metaphor for evolutionary search than, say, wilderness navigation. Normally, hikers can see more than their immediate vicinity, and global peaks are often the most visible of all landmarks from almost any viewpoint. Diggers, on the other hand, have holes as their only windows onto the big picture, just as search algorithms only have the evaluations of previously generated individuals as their patchwork vision of the world. However, as an illustration of a search space and the current status of an algorithm grappling with it, the mountainous fitness landscape is both popular and appropriate.

A landscape with many peaks reflects a fitness function that gives a lot of partial credit to phenotypes. Unfortunately, the generous allotment of credit is not sufficient to insure a successful search. The basis for smooth landscapes with gradual ascents to promising summits is the *correlation* between phenotypes and fitness. Given a phenotype, P, and its fitness $f(P)$, if most phenotypes in the neighborhood of P also have a fitness close to $f(P)$, then the two spaces are well correlated. And clearly, if search moves a small distance from P, the movement in the fitness landscape will only be a small ascent or descent, if not a neutral movement along a local plateau.

Unfortunately, devising fitness functions that insure this correlation is nearly impossible in many problem domains. Even in the container example areas of the space are almost guaranteed to have low correlation. For example, assume that the items are ordered from heaviest to lightest. Then, small changes to the last container assignment (i.e., that for the lightest item) will probably not have a great effect upon fitness, since switching the container of the lightest item will not change total weights very much. Hence, in regions of phenotype space where all container assignments are the same (across all phenotypes in the region) except the last assignment, we can expect a rather high correlation between phenotypes and fitness.

However, in regions where only the *first* container assignment varies among neighbors, we can imagine very little correlation, since the phenotypes vary with respect to their assignment of the *heaviest* item, and there are surely many situations in which switching the container of the heaviest item can cause a transition from a reasonably good and completely legal solution to one in which a container has a total weight exceeding W. Since most fitness functions would punish such load violations severely, fitness would drop precipitously between neighboring phenotypes.

Assuming that the fitness and phenotype spaces are well correlated, the EA still has no guarantee of success. The mapping between genotypes and phenotypes must also show a smooth correspondence. Although genotypes and phenotypes are equivalent in some types of EAs, they often are not, and a nontrivial developmental process is required to convert genes into traits. Now, genetic operators such as mutation and crossover operate on genotypes, not phenotypes, so when a high-fitness parent genotype is mutated slightly (by, say, flipping a single bit), then if genotype and phenotype space are well correlated, the child should have similar traits to the parent and have a comparably high fitness. Thus, using mutation as a vehicle of change, search can move smoothly about the fitness landscape, without frequent abrupt dips and hops. Crossover normally incurs major changes to genotypes, so even in a well-correlated landscape, its does not produce smooth search.

In EAs mutation is critical for zeroing in on nearby maxima (whose general neighborhood may have been discovered by crossover), but it requires good genotype-phenotype and phenotype-fitness correlation in order to succeed. Once again, superior correlations are hard to guarantee. As the developmental process becomes more complex, the genotype-phenotype correlation often deteriorates dramatically.

However, even trivial developmental links can cause poor correspondence. For example, if bit strings are translated into integers in the straightforward manner of a base-2 to base-10 conversion, then the correlation always breaks down for the higher-order bits. To wit, the genotypes 0001 and 1001 are neighbors, since they differ by a single bit, but their phenotypes, 1 and 9, are quite distant. If these phenotypes represent settings for a thermostat, for example, then 1 and 9 would probably give much more divergent outcomes (and thus a larger fitness difference) than 1 and 2. So in this case, the lack of genotype-phenotype correlation could easily cause a poor genotype-fitness correlation and make evolutionary search difficult.

Conversely, in a $K = 10$ container problem, 1 and 9 would just represent container labels, so the difference between 1 and 9 versus 1 and 2 would not necessarily mean anything in terms of the actual problem solution and its fitness: moving an item between containers 1 and 9 need not be significantly better or worse than moving it between 1 and 2.

In general, the correlations between these spaces are rarely perfect, and when they are, the problem is usually either trivial or better solved using a traditional search method. EAs are designed to handle the tough cases: those with fitness landscapes containing scattered local maxima and steep peaks jutting out of flat planes, and those with nontrivial mappings between the syntax (i.e., genotype) of a problem solution and its semantics (i.e., the phenotype). When applied to a particular problem, this topological analysis gives tangible criteria for assessing approximately *how hard* these inevitably difficult searches will be, and for designing fitness functions and genotypic and phenotypic representations (along with their mapping) so as to make evolutionary search as computationally tractable as possible.

8.3 Exploration versus Exploitation in Evolutionary Search

As mentioned, the fitness landscape is a visualization of the fitness function. Unfortunately, for complex phenotype spaces, a complete visualization of the fitness values for each phenotype is computationally impossible. Hence, we cannot simply look at the fitness landscape (in a multidimensional space) and discern the global maxima. Also, we cannot simply solve the fitness function (the way we solve $x^2 - 3x - 18 = 0$ for x) using symbolic techniques to find the optimal phenotype.

Instead, we must perform search, meaning that partial or complete solutions must be generated, tested, and modified, with changes taken in directions that appear to lead to improved solutions, i.e., directions that seem to head toward a maximal summit of the fitness landscape.

Whereas classic AI search techniques such as A* deal with partial solutions, evolutionary algorithms work with complete solutions/designs. Also, while A* tends to process these partial solutions serially and independently, EAs work in parallel with many different solutions and often combine them to produce hybrid children.

Search is a partly blind process, since the search module can never see the entire fitness landscape; it only knows the terrain near its attempted solutions. Each such attempt is therefore a probe point into the fitness landscape, and the intelligence in search comes from the judicious choice of these probes such that global optima can be discovered using as few probes as possible.

A good search algorithm strikes a proper balance between exploration and exploitation, typically by beginning in an exploratory mode and then gradually becoming more exploitative. In biological terms, search procedures begin by accentuating variation before gradually moving to a more inheritance-centered strategy wherein apples never fall far from the tree.

In general, mutation embodies an exploitative change, whereas recombination via crossover is more explorative. Since EAs often do both, the *frequency* of each genetic operator is often a measure of the degrees of exploration and exploitation.

From a very general perspective, all genetic operators can be considered explorative, since they do produce variation, whereas selection strategies embody exploitation by giving priority to the best individuals. But again, this is a very broad-brush approximation, since both genetic operators and selection strategies have tuning mechanisms to adjust their balance of exploration and exploitation. Figure 8.3 illustrates both the broad and more-detailed views of these interactions.

Essentially, selection pressure mirrors exploitation. When pressure is high, the EA strongly shifts focus to the high-fitness solutions at the expense of the lesser solutions. This often means that one or a few highly fit individuals produce a large fraction of the next generation. Hence, population diversity drops, which effectively reduces exploration.

Figure 8.4 illustrates exploration versus exploitation for evolutionary search. At time T, the population is reasonably well spread across genotype space and

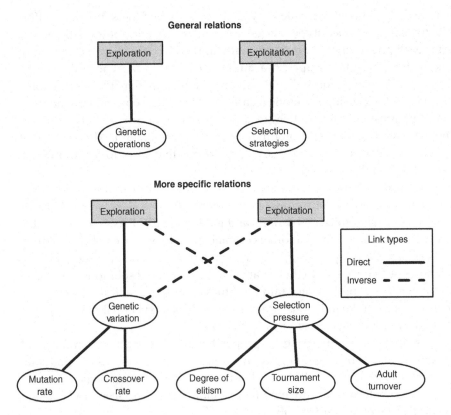

Figure 8.3
Broad and more detailed views of the relations between exploration, exploitation, genetic operators, and selection

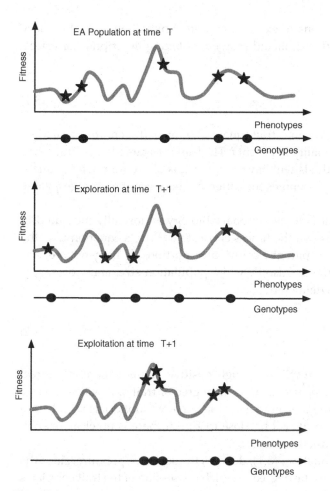

Figure 8.4
Exploration versus exploitation in evolutionary search. Circles denote the population of five genotypes; stars represent their fitness. This assumes a perfect correlation between genotype and phenotype space such that genotypes produce phenotypes directly above them in the diagram.

the fitness landscape. An exploratory strategy encourages even further spreading, with no extra emphasis on points near the middle (where a high-fitness genotype was discovered at time T). In contrast, the exploitative strategy shifts more resources (i.e., genotypes) to the areas that have already yielded high fitness values. In terms of EA genetic operators, the explorative strategy may have employed more crossover to produce new genotypes that are distant from the parents, while exploitation might have slightly mutated clones of the two best individuals from time T to produce neighboring genotypes at $T + 1$.

To formalize the contributions of exploration and exploitation to evolutionary progress, De Jong (2006, 160–162) includes an enlightening description of Price's classic equation:

$$\Delta Q = \frac{\text{Cov}(z,q)}{\bar{z}} + \frac{\sum z_i \triangle q_i}{N\bar{z}}. \tag{8.1}$$

Here, ΔQ is the change in some trait/quality, Q, over the entire population, while q_i is the value of that quality in parent i. The trait is assumed to be graded, to some degree, so that individuals will have more or less of the trait, with q_i quantifying that amount. Also, $\triangle q_i$ denotes the difference between the average q value in parent i's children and q_i.

As mentioned earlier, fitness in evolutionary biology is normally measured in terms of reproductive success, so the fitness correlate in Price's equation is z_i, the number of children to which parent i contributes portions of its genotype; $\bar{z}$ is the population average of the z_i. Also, N is the population size, and $\text{Cov}(z, q)$ is the covariance of z and q, as defined by

$$\text{Cov}(z,q) = \sum_{i=1}^{N} \frac{(z_i - \bar{z})(q_i - \bar{q})}{N}. \tag{8.2}$$

In a nutshell, the covariance will be a high positive value if there is a direct relation between reproductive success and the degree of trait q. It will be a large negative value if the two are inversely related, i.e., many individuals with high q values produce few offspring. It will be close to zero if there is no clear relation between q values and reproductive success.

The first term in Price's equation, (8.1), denotes the selective pressure, since the covariance of q and z indicates the degree to which possession of the trait correlates with reproductive success. When selection pressure is low, *any* value of q could yield just as many offspring as any other value, which would be reflected in a low $\text{Cov}(z, q)$ value. Conversely, when selection pressure is high, certain traits (in Price's model, high values of q) give a definite reproductive advantage over others, reflected in many (high-q, high-z) pairs in the population, and thus a high $\text{Cov}(z, q)$ value.

The second term in equation (8.1) captures the degree to which children *differ* from their parents with respect to trait q. It therefore represents population variability from generation to generation, the opposite of inheritance.

Clearly then, Price's equation shows that the rate of evolutionary change is directly proportional to the combination of selective pressure and variability. If both are high, evolutionary change should be swift, but if either is low, the rate is reduced.

Consider each of these options. If variation is high but selection low, then many new genotypes (i.e., many diverse q values) are produced, but all receive approximately equal priority in terms of reproduction. Hence the population will not deviate much from some average value of q. But if selection is high and variation is low, then an initial population (of presumably similar q values) will not diversify. Thus, selection, no matter how strong, will have very little to *favor*, since all q values will be similar.

For evolution to attain optimal but hard-to-reach areas of a search space, it needs to progress through many rounds of ruthless filtering of high-variance gene pools. Each filtering round moves the population's center of gravity to a slightly higher point, from which many exploratory moves are taken by the genetic operators, with only the best such moves resulting in useful new points, to which evolution shifts its focus and produces new feelers in search space.

The assumption of homogeneous initial q values normally holds in nature but not in EAs, which are often initialized with random genotypes. However, even with a diverse gene pool as a starting point, evolution will stagnate under conditions of high selection and low variance. That is , high selection will initially favor one or a few of the slightly better genotypes, causing many of the new offspring to have q values within a tight range. From there, exploration will be quite slow because of low variation. In effect, selection forces the population to converge to homogeneity so early in evolution that the favored individuals probably only represent local maxima in the search space; evolution then gets stuck there because of low variation.

8.4   Representations for Evolutionary Algorithms

In most cases, the problem domain and target computational machinery for an EA will influence the choice of phenotype, which should mean something in that domain. For instance, if the goal is to design a finite-state automaton for controlling a chemical process, then the phenotype will probably include states and the transitions between them. The choice of genotypic syntax will depend upon the phenotypes and on the EA designer's preferences. Genotypes may be close (or identical) to phenotypes and thus very problem-dependent, or they may be very generic. For example, bit vectors are very generic genotypes, while real numbers arranged into rows and columns are more specific and probably map directly to real-array phenotypes.

EA researchers often differentiate between two general classes of genotype-phenotype mappings, direct and indirect, although details of the distinction often vary. These are essentially synonymous with direct/indirect *encodings* or *representations*; the genotype represents the phenotype, since the former re-creates the latter when processed by the mapping function.

A direct EA representation is one in which the genome can be divided into several components (i.e., genes), where each such segment independently encodes some aspect of the phenotype (e.g., a trait) and there is a one-to-one relation between each gene and trait. For example, in the bin-packing problem discussed earlier, each gene is a bit string that encodes the container for a particular ball. These are independent, since the placement of any ball (a trait) does not affect the ball placement encoded by another gene. In terms of genetic operations, any mutation to one gene will affect only its own translation, not that of other genes. So independence has both developmental and evolutionary significance.

If the genome for a direct mapping always encodes a value for each phenotypic trait, then the representation is *complete*; these representations typically involve fixed-length genomes. On the other hand, a *partial* direct encoding has genomes that only denote some traits of the phenotype (e.g., the weights of only some connections in a neural network); partial direct genomes in the same population will often have different lengths. In either case, the genes still determine traits independently.

Complete encodings can take advantage of genome position to simplify their syntax, whereas partial encodings cannot. For example, if the genome encodes the open/closed settings for N gates in a flow-routing problem, then under a complete encoding, the kth bit on the chromosome can always denote the state of the kth gate; and the whole genome can have a constant length of k bits. Alternatively, with a partial representation (where, for example, all unmentioned gates are assumed closed), each gene would require both a gate index and a binary state.

Indirect mappings can be either *expanded* (i.e., full length) or *generative*. Like a complete direct mapping, the expanded indirect genome has one entry for each trait, but independence is no longer guaranteed. Hence, one gene-to-trait translation can depend upon another gene/trait. For the bin-packing problem, one possible expanded indirect genome has the same number of genes and bits as does the direct representation, but if the ith gene encodes the integer j, then the semantics is "place the ith object in the first bin that can accommodate it, beginning with the jth bin, ascending thereafter, and possibly looping back to the beginning of the bin list." Clearly, the placement of low-index objects can affect the decoded bins of higher-index objects, and single mutations can have widespread consequences for the phenotype: the final placement of all objects.

Finally, a generative indirect representation guarantees neither a one-to-one mapping between genes and traits nor independence of gene translation. Typically, these involve relatively small genotypes that produce much larger phenotypes. For the bin-packing problem, a generative representation may encode a few parameters that control key decisions in a packing algorithm. For instance, assume an algorithm that takes an object (of size S) and the remaining space in each bin as input. It then computes the ratio of S to available space for each bin and then sorts these ratios in descending order. It then removes all ratios greater than 1 to produce a priority list of length L (where $L \leq K$, the number of bins). The evolved parameters of this EA are K probabilities, which are normalized to sum to 1. The kth such value indicates the probability of choosing the bin that yields the kth largest ratio (after all ratios above 1 have been removed).

Since each object can produce a priority list of a different length (because of the filtering of containers with insufficient space), the algorithm must also remove from K the $K - L$ probabilities of the insufficient containers and renormalize the remaining L values. Then, by spinning a simulated roulette wheel with L sectors whose size distribution reflects these L probabilities, the algorithm can stochastically choose among the L possible bins. Since K is normally quite a bit less than M (the number of objects) for interesting bin-packing problems, the length K genotype produces much larger (size M) phenotypes.

These categories (direct, expanded, and generative) permit the practical comparison of genotypes with respect to the amount of effort required to convert them into phenotypes. However, a universally useful classification of EA representations seems far-fetched, since many thousands of EA applications (with specialized representations) exist, and different perspectives can be important in analyzing them.

While this formulation keys on the relation between genotype and phenotype, other classifications focus on one or the other. For example, De Jong (2006) uses the syntax of the genotype as the defining property to produce four main classes:

- *Fixed-length linear objects* Simple vectors of values, with each value denoting a gene and each vector having the same length and retaining that length throughout the EA run.

- *Fixed-size nonlinear objects* Potentially complex data structures, some of which are easy to linearize, such as n-dimensional arrays and trees, and others that require more work, such as connected graphs with loops. These do have a fixed size, so crossover is generally straightforward to implement. However, the sections that get swapped during crossover are not necessarily effective, modular, or meaningful building blocks.

- *Variable-length linear objects* Vectors of values whose lengths are not all equal and may vary during the course of the run. Here, mutation is easy but crossover can pose a challenge.

- *Variable-length nonlinear objects* These are the most difficult, since the complex objects can also vary in size, making crossover potentially difficult both syntactically and semantically.

Another alternative classification is based on the semantics of the information that the genotype contributes to the phenotype. It consists of two main genotypic categories:

- *Data-oriented* These encode several data values whose use in the phenotype may vary but does not include actual data manipulation or program control.
- *Program-oriented* These encode explicit data-processing and control information—supplementary data may also be encoded—to form the kernel of an executable program at the phenotypic level.

This separation closely mirrors two distinct representational semantics found in the EA community:

- *Collections of parameters* These are often variables in optimization problems, used by genetic algorithm (GA), evolutionary strategy (ES), and evolutionary programming (EP) researchers.
- *Computer programs* Used in the field of genetic programming (GP).

This perspective, along with the direct/indirect classification of mappings, is central to this book's discussion of EA representations.

8.4.1   Data-Oriented Genotypes
The classic example of a data-oriented genotype is a list of parameters, encoded either as a bit string or as an array of integers or reals. These parameters may represent anything from dial settings for a factory controller to variable values in a function optimization problem to weights for a neural network to room numbers for an exam scheduler. The examples are endless, and in many cases an EA is the perfect tool for the job.

At a lower level there are a host of relatively standard syntactic representations for genotypes and their corresponding mappings functions. Six of these are shown in figure 8.5.

At the center of figure 8.5 the classic, direct-encoding bit vector (common to genetic algorithms) is a simple list of bits, subsequences of which are translated into phenotypic traits such as integers or real numbers. Mutation is a simple bit flip.

One potential weakness with the classic bit vector representation is the mediocre correlation between genospace and phenospace. There is no guarantee that a small (large) change in a genotype will result in a correspondingly small (large) change in the phenotype. Consider the genotype 011 for phenotype 3. By flipping *every* bit, i.e., by making the maximal change to the genotype, the neighboring phenotype, 4, is formed. So a large change in genotype space produces a small change in

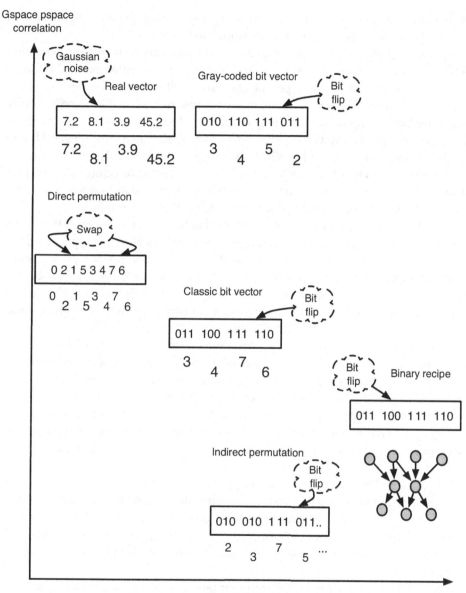

Figure 8.5
Six of the many data-oriented representational approaches in evolutionary computation. These vary with respect to the distance between genotypes and phenotypes, i.e., the developmental effort (x-axis) and the correlation between genospace and phenospace (y-axis). Genotypes (rectangles), phenotypes (under rectangles) and mutation operators (clouds) are shown. (*Top left*) Representations that traditionally require the least computational overhead. (*Bottom right*) Representations requiring more resources.

phenospace. On the other hand, by mutating only one bit, the most significant, in 111, the phenotype jumps dramatically from 7 to 011 = 3.

In short, the classic bit vector representation has only a mediocre correlation between genospace and phenospace, and involves a modest amount of effort to convert genotypes into phenotypes, i.e., to convert bit strings to integers or reals.

To combat the correlation problem, many EA researchers use Gray coding. Gray codes are binary encodings for integers that are designed to have a high correlation. Hence, given any Gray-coded bit vector, V, that maps to integer M, any single-bit mutation to V will produce an integer close to M. The conversion of Gray-coded bit vectors to integers is nearly as easy as the decoding of normal bit strings, so classic and Gray-coded bit vectors appear at similar locations along the x-axis of figure 8.5, but the Gray codes are much higher on the y-axis.

Many EA practitioners use representations that have both high correlation and a simple conversion process. Real vector genomes are typical of this type. Here, the genome is simply a list of real numbers, so the genotype and phenotype are essentially the same. Mutation is performed directly on the reals, not on their binary representations. The standard form of mutation is to perturb the original value by a small amount, which is chosen from a normal distribution with a mean of 0 and a problem-dependent standard deviation.

Many applications, such as the traveling salesman problem (TSP), involve solutions that are permutations of a set of integers. For these, a direct permutation representation is often appropriate, wherein the kth gene encodes the index of the kth city in the tour. These also employ a list of numbers (in this case, integers), where genotype and phenotype are identical. Mutation involves the simple swapping of two integers in the list. Crossover of permutations requires a bit more effort in order to avoid duplications and omissions of city indices in the children.

Expanded indirect representations of permutations are also possible, such that the genotype is a bit string that translates (with some effort) into a valid permutation. For example, if the kth gene has value j, this entails that the kth tour member is the jth element in the list of unassigned cities. This representation must be used with caution, since the genospace-phenospace correlation is very weak, as shown in figure 8.5.

Finally, many generative indirect representations involve complex developmental routines that severely compromise the genospace-phenospace correlation. For example, bit strings may be converted into circuit layouts or neural network topologies (as depicted at the lower right of figure 8.5). These are much more difficult representations to handle, but they enable evolution to solve intricate design problems.

A Robot Example To illustrate the wide variety of representations for any given problem, consider a situation where the link between genotype and phenotype is less obvious. In this example, a robot must be evolved to enter a potentially dangerous area and outline the periphery of all suspicious objects with warning markers. The first prototype might be a simple office robot that must differentiate between red (dangerous) and blue (harmless) objects and learn to pick up and place blue objects around red ones to mark the hazard, as depicted in figure 8.6.

Assume that the robot has eight color sensors, four for blue and four for red, with a sensor of each type for each of the four directions: front, back, left, and right. The red sensors are denoted RF, RB, RL, RR, while the blue are BF, BB, BL, BR. Sensors give simple binary readings, so, for example, BB = T (i.e., True) means that a blue object is detected directly in back of the robot. In addition, two sensors detect whether the robot is carrying a red or a blue object. These are denoted RC and BC and are also binary.

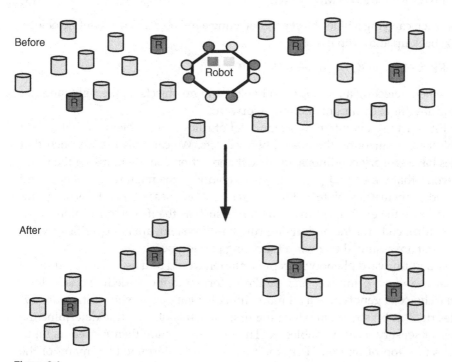

Figure 8.6
A robotic task of placing blue (light gray) warning markers around dangerous red (R) objects, where the robot has eight color sensors, four each for red and blue (dark and light circles on the robot's exterior) and two carry sensors (shaded squares in the robot's interior)

The robot has six possible actions:

- MF   Move forward.
- MB   Move back.
- TL   Turn left 90°.
- TR   Turn right 90°.
- PICK   Pick up the object in front of the robot.
- DROP   Put the carried object down in front of the robot.

If the robot were designed by hand, then some of the rules might be

$$BC \wedge RR \Longrightarrow TR, \tag{8.3}$$

that is, when carrying a blue object and detecting a red object on the right, turn to the right.

$$BC \wedge \neg BB \wedge \neg RB \wedge RF \Longrightarrow MB \wedge DROP, \tag{8.4}$$

that is, when carrying a blue object and detecting a red object in front but no objects in back, back up and drop the object.

$$RL \wedge \neg RR \wedge \neg BR \wedge \neg BC \wedge \neg RC \Longrightarrow TR \wedge MF, \tag{8.5}$$

that is, when detecting a red object on the left and no objects on the right, and not carrying anything, turn right and move forward.

The EA must use evolution to design a set of rules that enable the robot to find red objects and surround them with blue objects. We can bias the EA such that all rules have sensory conditions on the left and action combinations on the right, but beyond that we should probably give evolution free reign to discover useful sense-and-act heuristics. Note that this representation does not qualify as program-oriented, since the explicit control commands such as the if-then ones, along with kernel control code for the underlying rule-based system, are not specified by the genome but are assumed to govern the acting phenotype.

The straightforward phenotypic representation would be a list of rules, using the same sense-and-act primitives and similar in format to the preceding ones. However, the choice of genotype is a bit more difficult. One option, shown in figure 8.7, encodes rules as bit strings in which the first three bits of each rule determine the number of sensory input variables, K. The next $5K$ bits are then parsed as input variables (see top of figure). Then, the final six bits determine how many of the six possible actions will be performed, with opposing actions such as TL and TR simply canceling one another out if both are true.

This representation is reasonably compact and flexible, since few bits are wasted, and rules can have varying lengths. However, it is highly susceptible to lethal

Sensory input gene (5 bits)

T or F? (1)	Red or blue? (1)	Direction or carry? (1)	Direction (2)

Action gene (6 bits)

Turn left? (1)	Turn right? (1)	Go forward? (1)	Go backward? (1)	Pickup? (1)	Drop? (1)

Rule segment (5K+9 bits))

Number input terms = K (3 bits)	K input genes (K x 5 bits)	Actions (6 bits)

Translating a rule segment

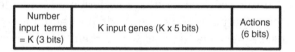

Complete genotype: set of rules

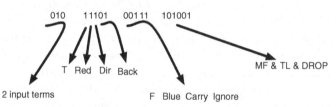

Figure 8.7
Flexible genotype representation for the robot controller. Rules consist of K input terms, where $0 \le K \le 7$ is encoded by the first three bits (the count bits) of the rule and a six-bit action vector. The dynamic coding of K implies that rules differ in bit length, and a small change to the count bits of one rule can alter the interpretation of all succeeding rules. The translated sample rule is RB $\wedge$ ¬BC $\implies$ MF $\wedge$ TL $\wedge$ DROP.

mutations, particularly those of the length-3 count bits. If one bit of one count changes, this alters the translation of the rest of the genome. Hence, a good rule set could, with one bit flip, become a terrible one. In other words, variation would be high, but at the cost of low inheritance. Of course, mutations to other bits would provide more local changes and strike a friendlier balance between variation and inheritance.

One way to provide some insurance against lethal variations is to enable rule codes to spread themselves around the genome, without necessarily bordering one another. Figure 8.8 (top) illustrates the basic idea. Here, as in real biological genomes, various tags indicate the beginning of a coding segment. For simplicity, assume the tag 11111 denotes the start of a new rule segment. The genotype-to-phenotype parser simply searches the genome from left to right until it hits a 11111 tag. It then parses the bits that immediately follow the tag as a rule (in the same manner as shown in figure 8.7). From the end of this rule, it continues to the right in search of another tag. Between the end of a rule segment and the next tag, many unused bits may lie. These perform a similar function to introns in genetics, since they help isolate gene segments and thus reduce their interaction with other segments. For instance, if the count of one rule increases, the unused filler bits can be used to extend the rule segment without interfering with the downstream rules. They also increase the probability that randomly chosen crossover points will occur between rule segments rather than within them, thus preserving linkage inside the segment, i.e., the segments tend to be inherited as units.

Even with tags and filler bits, a potentially unwanted source of variation still exists within the rule segments. If the count bits mutate to a higher number, then the new inputs will be taken from the old action segment, and some of the new actions will come from the trailing filler bits. Similar problems occur if the count decreases. This annihilates potentially useful action combinations (which evolution may work so hard to produce). It makes more sense to preserve the actions and existing sensory input conditions of a rule and to use the filler bits as the source of any additional input conditions. A simple reordering of the rule bits facilitates this

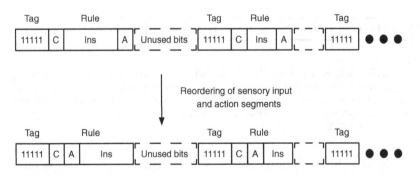

Figure 8.8
(*Top*) Genotype representation for a robot controller that helps prevent lethal mutations by allowing rule segments to be separated by free space (unused bits). The tag 11111 indicates the start of a rule segment; C, Ins, and A denote the count, sensory input, and action bits, respectively. (*Bottom*) A slightly modified encoding in which action bits occur directly after the count bits and thus become protected from the side effects of count changes.

straightforward reduction in variation: now the action bits come directly after the count bits and before the input bits, as shown in figure 8.8 (bottom).

Yet another bit vector genotype for the robot problem is possible; this one removes all potential interrule interactions stemming from the side effects of genetic operations. Consider that each of the ten possible sensory inputs can take on one of three values in a rule: true, false, unimportant. This gives a total of $3^{10} = 59,049$ possible preconditions for a rule, hardly a difficult size for today's computers. To represent any number between 0 and 59,048 (or 1 and 59,049) requires $\lceil \log_2 59,049 \rceil = 16$ bits. Thus, as shown in figure 8.9 (top), a rule could be represented by a 16-bit precondition selector, which chooses among the 59,049

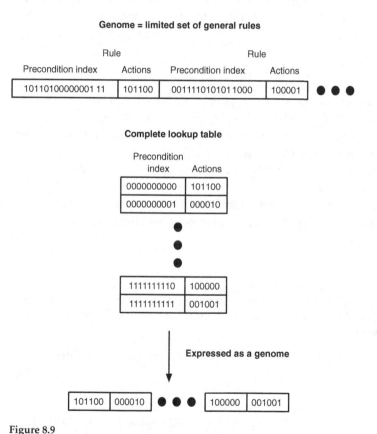

Figure 8.9
Additional genotypes for the robot problem that use indices into the collection of all possible preconditions. (*Top*) The genome handles only some of the sensory input situations, so the index of each scenario is required. (*Bottom*) All 1,024 scenarios are accounted for, so scenario indices are implicit in the genome; and index length can be reduced from 16 to 10.

unique preconditions, followed by a 6-bit action vector identical to those used in the previous three representations.

Finally, if a complete lookup table for all possible sensory input scenarios is desired, then one can ignore the *unimportant* option for each variable. This yields a much smaller total of $2^{10} = 1,024$ scenarios. Associating six action bits with each scenario gives a complete action strategy and uses only $6 \times 1,024 = 6,144$ bits. Such a strategy is show in figure 8.9 (bottom). Though theoretically possible, these enumeration-based encodings become impractical for domains involving either many inputs or many distinct values for each input. The current robot example, however, lies well within their scope.

For the lower encoding strategy of figure 8.9, precondition bits are not needed in the genome, since they are implicitly represented by the indices of each six-bit action vector. For example, the precondition for the fourteenth action vector (using zero-based indexing) is 0000001110 (i.e., 14 in binary), which denotes the conjunction:

$$\neg RF \wedge \neg RB \wedge \neg RL \wedge \neg RR \wedge \neg BF \wedge \neg BB \wedge BL \wedge BR \wedge RC \wedge \neg BC, \tag{8.6}$$

under the assumption that for genotype-to-phenotype translation, the sensor variables are ordered as follows: RF, RB, RL, RR, BF, BB, BL, BR, RC, BC.

Clearly, this representation explicitly encodes an action strategy for each of the 1,024 specific sensory scenarios. Hence, it can be described as an *extensional* strategy, since the extension of a concept is typically defined as the set of all individual members. The earlier versions are *intensional* in the sense that single rules are lists of only the important attributes—unless they explicitly mention all ten input sensors on their left-hand sides—from which extensions must be computed.

For example, the rule

$$RB \wedge \neg BC \implies MF \wedge TL \wedge DROP \tag{8.7}$$

is intensional, with an extension consisting of $2^8 = 256$ detailed rules to account for all possible values of the eight sensory variables that are *not* explicitly mentioned on the left-hand side of equation (8.7). One such rule is (core intensional components in boldface)

$$\mathbf{RB} \wedge \mathbf{\neg BC} \wedge RF \wedge RL \wedge RR \wedge BF \wedge \neg BB \wedge \neg BL \wedge \neg BR \wedge \neg RC \implies \mathbf{MF} \wedge \mathbf{TL} \wedge \mathbf{DROP}, \tag{8.8}$$

and another is

$$\mathbf{RB} \wedge \mathbf{\neg BC} \wedge \neg RF \wedge \neg RL \wedge \neg RR \wedge \neg BF \wedge BB \wedge BL \wedge BR \wedge RC \implies \mathbf{MF} \wedge \mathbf{TL} \wedge \mathbf{DROP}. \tag{8.9}$$

Although, in theory, these two representations, intensional and extensional, provide the same *expressibility* (they can represent the same behavior rules), it is clear that a general rule such as (8.7) would have a very slim chance of arising in the lookup table, since its complete extension (all 256 cases) would need to appear. In other words, the table would need to have 256 identical action vector entries (of MF $\wedge$ TL $\wedge$ DROP) to account for all scenarios in which RB is true, BC is false, and the other eight sensory variables take on any possible truth value combination.

In this and many other situations, the choice between an intensional and an extensional representation must be considered. Extensional representations can often by molded into fixed-size genomes with none of the intergene dependencies (termed *epistasis* in biology) that plague many of the more flexible intensional representations. With low epistasis, the danger of one mutation completely reorganizing the phenotype decreases significantly. Hence, heritability, variation, and selection can cooperate to evolve useful problem solutions. However, the extensional approaches hinder the emergence of useful general rules. They can also produce very large genomes, which require more evolutionary time to improve.

Regardless of these differences, a key feature of all the preceding representations is that they allow random bit flip mutations and recombinations to produce genotypes that are guaranteed to be translatable into legal phenotypes. The balance between variation and heritability of these genotypes with respect to the parent genotypes will vary, but they will develop into understandable phenotypes regardless of the extent of mutation and crossover.

This feature insures that the basic bit vector genetic operators of mutation and crossover can be programmed once and used for all the preceding genotypic representations and many more. For each new bit vector representation, only a new translation/development module must be written to convert the genotypes to legal phenotypes. Thus, at least at the genotype level, an EA appears quite representation-independent: the same generators of variation can work on genotypes that encode rule sets, neural networks, arrays of control variables, and many other phenotypes.

In effect, the genetic operators embody the hypothesis-generating intelligence of an EA. If these are designed to work on any bit string, regardless of the phenotype it encodes, then their extreme generality and domain-independence trade off against a complete lack of understanding concerning strategic changes most likely to improve solutions, i.e., a complete lack of semantic knowledge.

Although all genotypes are represented by bits at some layer of the computer, the true level of the genotype for evolutionary computation is defined by the genetic operators and the knowledge that they use to mutate and recombine genotypes. The genotype becomes equivalent to the phenotype in those cases where the genetic operators incorporate the high-level semantics of the phenotype to manipulate genotypes.

In the robot example we could elevate the genetic operators closer to the phenotypic level by devising mutation rules such as the following:

- Switch any blue sensor reading, such as BF, to the corresponding red one, RF, or vice versa, in the precondition of a rule.
- Switch any movement to the opposing movement, i.e., MF to MB, and TR to TL.
- If the rule includes a positive red sensory reading and no mention of a blue object being carried, then supplement the actions with a move away from the red object and remove any move toward it.

Crossover rules might include conditions like these:

- When combining partial precondition lists of two rules, resolve any contradictions such as $RL \wedge \neg RL$ by randomly choosing one of the two terms.
- When combining partial action lists of two rules, resolve pairs of opposing actions by randomly keeping one and deleting the other.

In this case, it would be safe to say that the genetic operators are working directly on the phenotype, although this book always differentiates the genotype and phenotype regardless of the simplicity of the developmental process. Although the use of high-level genetic operators can lend advantageous direction to search, it may also bias evolution too strongly, precluding the discovery of optimal phenotypes that might require the creative forces of synthetic evolution over the rigid guidance of fundamental engineering principles.

Unfortunately, there is no standard answer for the proper level of abstraction for genetic operators. Problem-independent operators have obvious advantages, including both code reuse and providing evolution with maximum search flexibility. However, extremely complicated search problems sometimes require the bias of higher-level operators. As with much of evolutionary computation, the crafting of representations and genetic operators is more art than science. And as such, it is one of the more interesting, creative, and challenging aspects of evolutionary problem solving.

A Neural Network Robot Controller   Although evolutionary algorithms constitute fundamental approaches to Bio-AI, the preceding robot controllers employ a classic GOFAI representation: a set of if-then rules. This is one way to (relatively seamlessly) blend the two AI paradigms, and because one can evolve just about any aspect of an AI system, this type of hybrid is commonplace. In contrast, evolving artificial neural networks (EANNs) have deeper roots in Bio-AI.

Consider the ANN of figure 8.10, again used to control the cylinder-moving robot. This receives a binary vector of length 10 on its input layer, one bit for the

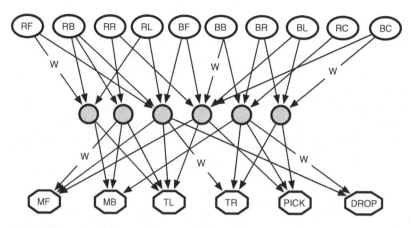

Figure 8.10
Sample ANN for controlling the robot of figure 8.6. Only some connections are shown, and all connections have weights (*W*).

Boolean value of each sensory precondition. These values are weighted and fed forward to the hidden layer and then the output layer, with the activation levels of the output neurons determining the actions taken by the robot. For example, all output nodes above a given threshold may trigger their corresponding action, with conflicts (i.e., turn left and turn right) resolved in favor of nodes with higher activation levels.

The network may contain other connections, such as feedback links from the output to the hidden layer, but such details are irrelevant for the current discussion. The main point is that several parameters (i.e., the connection weights along with a few factors associated with the activation functions) will determine its decision-making *logic*. In theory, given a sufficient number of hidden nodes, this ANN can realize any of the preceding robot rule sets. However, the precise combination of these parameters needed to mirror a given rule set can be very difficult to design by hand; this search for a proper parameter vector often calls for an evolutionary algorithm.

Given a fixed ANN topology, the size of this parameter vector is also fixed. Hence, the EA chromosome is simply a long array of values (using one of the many direct encodings mentioned earlier), with no need for *padding* to prevent the widespread disruption of semantic content by a single mutation. However, the EANN research community often has problems with crossover: either is omitted it completely, or it is restricted it to certain chromosomal locations (Nolfi and Floreano, 2000).

Under the common view of a neuron as a detector of particular patterns (on its afferents), the weights on all incoming arcs combined with the properties of a node's activation function constitute a module that essentially defines each detector. These are the building blocks of the EANN chromosome, and once effective subset vectors arise, a good EA will maintain and combine them, although splitting and merging have more deleterious effects. Hence, many EANN systems include constraints such that crossover only occurs between (not within) these modules.

These ANNs serve as intensional representations of a rule set. To derive the extension, simply input all 1,024 sensory cases and record the resulting outputs. The beauty of ANNs is their ability (when combined with the right search technique) to formulate parameter vectors that perform complex mappings from inputs to outputs. The rule-based extensions of these mappings often consist of many specific rules whose combination embodies both general and case-specific behavior, akin to human knowledge. However, it is typically not the same rule set that a human engineer would formulate, and thus the knowledge content is difficult to interpret.

In figure 8.11 the given weights, firing thresholds, and activation functions are those that a person might use to mimic the following four robot-controlling if-then rules:

- $(RR \lor RL) \land BC \Longrightarrow DROP$.
- $RF \Longrightarrow MB$.
- $\neg RF \Longrightarrow MF$.
- $BF \Longrightarrow PICK$.

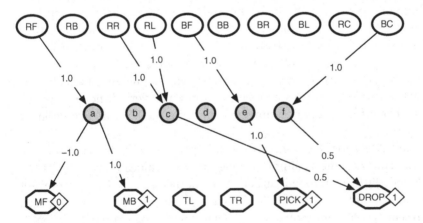

Figure 8.11
Weights (labels on edges) and firing thresholds (numbers in diamonds) that provide the same functionality as the set of four if-then robot rules given earlier. All firing thresholds in the middle layer are 1.0, and all undrawn connections are assumed to have weights of zero. The activation function for each neuron is a step function that produces a 1 if the sum of weighted inputs is greater than or equal to the firing threshold, and a 0 otherwise.

For example, the first rule is realized by the nodes RR, RL, BC, c, f, and DROP (along with the links between them). Remembering that all input layer nodes produce only 1's and 0's, the weighted input to node c will be 0, 1, or 2. Since c has a firing threshold of 1, it will fire when one or both of RR and RL are active: when their disjunction is true. Node f is simpler and fires if and only if BC is true. Finally, the output node DROP, with incoming weights of 0.5 and a threshold of 1, will only fire when both c and f have fired; it detects their conjunction.

Note that the weights shown are just a small fraction of the total network; all others are presumably zero. This constitutes the purest and simplest continuous-valued parallel to the set of four discrete rules. However, there are an infinite number of continuous alternatives to the discrete solution, each containing many more nonzero weights and nonunitary firing thresholds. A search algorithm operating in this continuous space could home in on any of them and return it as an optimal result.

Figure 8.12 displays one such network in which the dotted connections provide noisy additional consequents to each of the four rules. However, each has a small weight that a minor firing threshold change can counterbalance. For example, RB and BB provide noise to node c; neither alone nor in combination can they fire c, but they can make it more difficult for RR or RL to activate c. Lowering c's threshold to 0.9 effectively negates this noisy inhibition. A corresponding threshold reduction is also needed at DROP to maintain the integrity of the original rule.

The plethora of functionally equivalent solutions allows evolutionary search to find something useful, though probably not the needle-in-a-haystack solution of figure 8.11. This requires very little special treatment in the chromosome, other than the preceding (not completely necessary) restriction of crossover points to

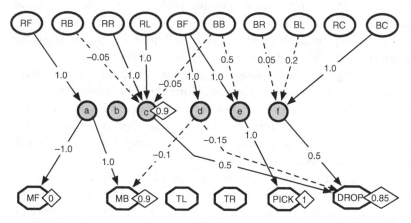

Figure 8.12
Functionally equivalent network to that of figure 8.11, but one whose behavior is much less transparent

module boundaries. To bias search toward the purest continuous solutions with, for example, a minimum of nonzero weights and nonunity firing thresholds would require a fitness function with various penalties for impure values or complex restrictions upon mutation and crossover. In short, EANNs can find intensional solutions using relatively simple genotypes and genetic operators because of the nature of this solution-rich search space. ANNs may not be easy to understand, but useful ones need not be difficult for a search algorithm to find.

8.4.2   Program-Oriented Genotypes

In the robot example, the genotype encodes either a set of if-then behavioral rules or a collection of ANN parameters. Thus, the genotype forms the basis for a rule-based or ANN-based controller. However, the genotype does not contain the if-then structure nor the neural integrate-and-fire architecture; those decisions are made by the EA designer, who interprets bit segments as if-then rules to be run by a rule-based system kernel or as weights and thresholds to govern an ANN.

However, in program-oriented genotypes a significant amount of the computational control knowledge resides in the genotype itself and is thus open to manipulation by genetic operators. Hence, an if-then may mutate into an if-then-else or a while loop or a case statement. As discussed earlier, John Koza coined the term *genetic programming* (GP) for this type of EA.

In most GP applications the genotype and phenotype are nearly identical. Hence, the genotype resembles a piece of computer code, although it must often be slightly repackaged to actually run. This means that the genetic operators must be designed to manipulate code, not just bit strings, such that new genotypes are syntactically valid code segments, not random gibberish. For many programming languages, designing such genetic operators would be an arduous task indeed. Imagine randomly swapping the lines of two C++ or Python programs. The odds of the child programs actually running are diminishingly small.

Fortunately, the LISP programming language provided the perfect substrate for Koza's groundbreaking forays into GP. The prefix format of LISP commands, combined with the recursive structure of LISP programs, facilitates a simple tree-based representation of LISP code. Trees are then easily mutated and recombined such that with the enforcement of a few minor constraints, genotypes can be randomly combined to produce viable new code trees.

As a simple example, consider a curve-fitting problem of finding a mathematical function (of a single variable X) that best matches a set of data points: $(x_1, y_1), (x_2, y_2), \ldots, (x_n, y_n)$. This function may be complex, but it can probably be decomposed into the recursive application of many simple operators, such as addition, subtraction, multiplication, and division. Hence, these will be the four primitive operators: $+, -, *$, and $/$. In GP terminology, these are the *function set*.

Computer programs (and mathematical functions) also need variables and constants; these are called *terminals* in GP. For a single-variable curve-fitting problem, the terminal set must contain X along with a few numerical constants, such as $(-3, -2, -1, 0, 1, 2, 3)$.

Figure 8.13 shows a typical GP program in both tree form and as a linear list. The list is straightforward to execute when the *arity* (i.e., number of arguments) of each function is known, and the code is written using prefix notation (i.e., operators come before their operands).

When a randomly generated GP tree (call it GPT) combines these functions and terminals, and is then wrapped within a lambda expression to form (lambda (X) GPT), it will compile and run, provided that the following constraints have been respected:

- When building trees, all functions must have all their argument slots (i.e., child nodes) filled by either terminals or subtrees with functions as their roots.

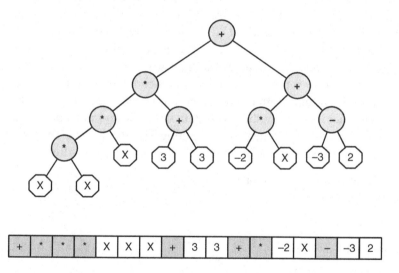

(lambda (X) (+ (* (* (* X X) X) (+ 3 3)) (+ (* -2 X) (- -3 2)))))

Figure 8.13
(*Top*) GP tree representing the function $6X^3 - 2X - 5$. Functions are denoted by shaded circles, terminals by octagons. (*Middle*) A linearization of the same function using prefix notation. (*Bottom*) LISP code for the same function. To produce an executable piece of code, the lambda (X) wrapper creates an unnamed function with one argument (X) and the GP tree as the function's body.

- *The closure property*   All functions must be defined so that all their argument slots accept any of the terminals and any of the possible outputs from any of the functions.
- Functions such as division that would normally crash on certain values, such as a 0 denominator, must be rewritten to handle those cases and output a value that is an acceptable input to all functions.

Usually, to satisfy the closure property, all GP functions for a given problem domain are designed to work with the same type of data, such as real numbers or Booleans. If arithmetic and logical functions are combined in a program, then it is common to rewrite the Booleans so that they interpret positive input arguments as true and nonpositive values as false, and output a 1 for true and a -1 for false. Functions such as division are rewritten to output a 0 if division by zero is attempted. Similarly, a log function may output a 0 if given a negative input. Together, these constraints insure that random combinations of functions and terminals will run without error. Of course, this says nothing about the actual quality (a.k.a. fitness) of the programs.

The standard genetic operators for GP are mutation and crossover. Mutation involves the replacement of a random subtree with a newly generated subtree (not necessarily of the same size). Standard GP crossover, depicted in figure 8.14, merely swaps random subtrees of two GP trees. The closure property guarantees that the results of mutation and crossover are valid programs.

When working with a linear representation of a prefix-coded tree, mutation and crossover work similarly in that subtrees are replaced and swapped, respectively. However, subtrees are not inherent in the linear data structure and must be found using a simple trick (Koza, 1994) that begins by selecting a random spot in the code vector and then moving forward while applying a basic counting procedure:

1. $i =$ random index into the code vector V.
2. count $= 0$; subtree $S = \emptyset$.
3. While count $\neq -1$, do:
   a. count $=$ count $-1+$ arity$(V(i))$.
   b. Append $V(i)$ onto end of S.
   c. $i = i + 1$.
4. Return S.

Here, the arity of a function is the number of arguments that it takes, while the arity of a terminal is 0. Figure 8.15 shows a simple example of subtree hunting in the program from figure 8.13. Once subtrees are found, they are swapped by

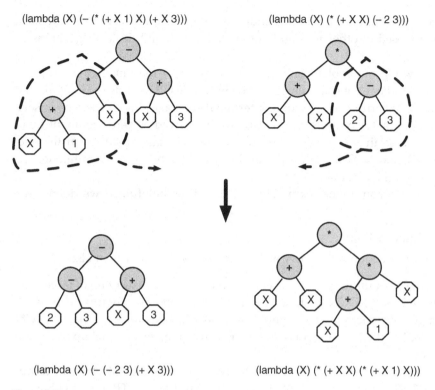

(lambda (X) (– (* (+ X 1) X) (+ X 3)))

(lambda (X) (* (+ X X) (– 2 3)))

(lambda (X) (– (– 2 3) (+ X 3)))

(lambda (X) (* (+ X X) (* (+ X 1) X)))

Figure 8.14
Crossover via subtree swapping in genetic programming. (*Top*) Subtrees (outlined by dotted lines) of two parents are exchanged. (*Bottom*) Child genotypes resulting from the subtree swap.

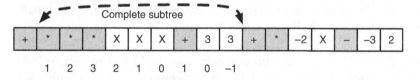

Figure 8.15
Using the simple counting procedure to find a random subtree. Numbers below each code element denote values of the *count* variable after that particular item has been processed. The leftmost spot is randomly chosen, and the algorithm than proceeds to the right. When the count reaches −1, the traversed segment constitutes a well-formed subtree.

removal and insertion into their code arrays. Since the two swapped subtrees may vary in size, code vectors for linear GP must have flexible dimensions.

The preceding example illustrates the representational flexibility of GP: given a few primitive functions and terminals, a wide variety of complex functions can be crafted by evolution. These open-ended design possibilities are the trademark of GP as opposed to other EAs and automated search and design algorithms in general.

To see how GP can evolve computer programs other than mathematical functions, consider the robot example. A relatively simple GP permits the evolution of quite complex control code. Assume a terminal set consisting of the ten sensory inputs and six motor outputs. These can also be viewed as zero-argument functions. In this case, the most obvious primitive functions are logical, not arithmetic, so AND, OR, and NOT are appropriate choices, with the former two taking two arguments and the latter taking one.

In addition, a conditional expression such as IF is helpful, so we define two versions:

- IF2(condition, action)
- IF3(condition, action, alternative action), corresponding to if-then-else

Finally, a block-building construct enables the sequential execution of large code segments. In LISP, PROGN (code sequence, code sequence) serves this purpose. To create longer sequences of code, simply nest the PROGNs; for example, (PROGN TL (PROGN MF PICK)) performs a left turn, forward move, and pickup operation in sequence.

So the complete function set is AND, OR, NOT, IF2, IF3, PROGN, and the terminal set is RF, RB, RL, RR, BF, BB, BL, BR, RC, BC, MF, MB, TL, TR, PICK, DROP.

To satisfy the closure property, each logical operator returns T (true) or F (false), while all six motor commands output T. IF2, IF3, and PROGN are defined (as in standard LISP) to return the output value of the final argument that gets evaluated. For example, IF3 returns the output of its alternative action in cases where the condition is false, and PROGN returns the return value of its second code sequence.

Evolution will invariably produce programs that are inefficient or that, at least on the surface, make little sense; but they do run. For example, (IF3 ML RF BB) uses a movement command in the condition spot. Since ML always returns true, the net effect is to move left and then read the front red sensor but do nothing contingent on its value. Not surprisingly, GP programs are often very hard to interpret.

Figure 8.16 shows a GP program of similar functionality to the four sense-act rules shown earlier for the ANN robot controller. This code would run on each timestep of the fitness assessment simulation.

Although a more complete comparison of GP and the other EAs appears in textbooks (Banzhaf et al., 1998; De Jong, 2006), the essential difference is representational: the GP evolves complete algorithms with the great majority of data-processing and control decisions determined by evolution, not the user. In theory, nothing else separates GP from the rest of EA. All general discussions of fitness landscapes, fitness assessment, selection strategies, and so on, need not specify the type of EA, although each community (GA, ES, EP, GP) has different consensus preferences.

8.5   Bio-AI's Head Designer

Just as the natural world depends upon evolution, Bio-AI relies heavily upon evolutionary algorithms. Though other Bio-AI tools, such as ANNs, swarms (Bonabeau et al., 1999), and Lindenmayer systems (Lindenmayer and Prusinkiewicz, 1989) have, strictly speaking, no necessary tie to EAs, they all invoke emergent processes whose outcomes defy accurate prediction. Hence, the task of hand designing local behaviors to generate target global patterns often becomes rather daunting, as does the formulation of intelligent, solution-building tools. In many cases, the trial-and-error combination of primitive components is the only feasible approach, with EAs being a popular search algorithm for the job.

As proven mathematically by Holland (1995), EAs skillfully manage the combination of exploration and exploitation, though the true optimality of this balance is highly representation-dependent, pertaining mainly to direct encodings. Unfortunately, many evolutionary design problems require indirect encodings, moving us well outside the realm of guaranteed feasibility; crafting genotypes, phenotypes, and their mappings becomes a very creative endeavor. As shown, different syntactic choices can reduce phenotypic sensitivity to mutation such that inheritance

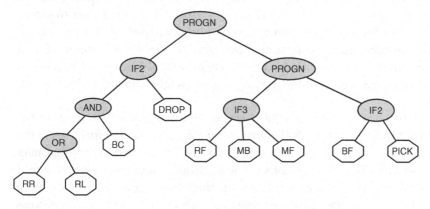

Figure 8.16
Sample GP program for the robot problem that is equivalent to the four rules and ANN shown earlier

remains relatively high, and the EA can methodically exploit a promising region of the search space. However, many of these options add a lot of extra baggage to the genotype in order to preserve the generality and expressibility of an intensional phenotype. Switching to an extensional phenotype alleviates many of these problems, but at the expense of expressibility, since certain general traits become highly unlikely, though not impossible, to evolve. In restricted domains, all preconditions, of both general and specific nature, can be enumerated, thus relieving many potential problems at the genotype level while maintaining intensionality. However, these scale very poorly.

Artificial neural networks manifest quantitative reasoning engines at a level well below that of rule-based systems. They can serve as generators of extensional rule sets, but the genotypes used to produce ANNs can employ direct or generative encodings. By evolving ANN parameters (e.g., weights, thresholds) the EA can retain a simple chromosome with direct encoding and standard genetic operators, while producing a structure that exhibits both general and highly context-dependent behavior without the need to enumerate all possible rules when defining the genotype (as in figure 8.9). As described in chapter 12 on evolving neural networks, the direct encoding of all ANN parameters also becomes impractical for complex domains, thus creating the need for a second level of generators: genomes that encode the key parameters of structures that generate ANN parameters.

Genetic programming conveniently supports the evolution of both intensional rule sets, by allowing myriad combinations of primitive elements in preconditions and actions, and the control logic itself. It epitomizes the spirit of Bio-AI by combining basic building blocks into an infinity of emergent designs. GP raises genotypes to the phenotypic level, a decision that would usually require a host of special-purpose genetic operators and impose a strong search bias just to insure syntactic validity. However, the homogeneous syntax of functional programs (in concert with features such as closure) allows generic, tree-based, mutation, and crossover mechanisms to produce a wide variety of legal phenotypes. GP can even evolve programs that generate the topologies and weight sets for ANNs (Gruau, 1994).

As in GOFAI, representational choice governs the success or failure of design search in Bio-AI. What confounds the issue is the extra representational level of the genotype. Whereas GOFAI usually employs hand-crafted phenotypes (that may adapt via learning), Bio-AI drops a level and searches in a purely syntactic world to find genotypes that translate into successful phenotypes (that may also undergo learning). This is nature's way, not necessarily AI's; but a long string of EA success stories, reviewed by Banzhaf et al. (1998) and Floreano and Mattiussi (2008), testify to the wisdom of this general approach to computational creativity.

9 Evolution and Development of the Brain

9.1 Neurons in a Haystack

The human and mouse genomes both consist of a little over 3 billion base pairs, which produce approximately 21,000 and 23,000 genes, respectively. Interestingly, a rice plant has one order of magnitude fewer base pairs but 28,000 genes. These numbers, huge in their own right, represent the tips of enormous icebergs (search spaces). Assuming that all combinations of base pairs are possible, though many are surely lethal, a sequence of length 3 billion is a point in a space of $4^{3,000,000,000}$ possibilities, which is approximately equivalent to 166 followed by 1.8 billion zeros. By comparison, the estimated age of the universe, in seconds, is 43 followed by 16 zeros.

How in the world (or universe) could evolution find the needles in the haystack that are (or have been) the DNA codes for viable phenotypes? Neo-Darwinism clearly explains how evolution selects good over bad genotypes, thereby filtering inferior designs; and modern genetics reveals how the good genes get passed from parents to children. But what about variability? How does nature generate this collection of thriving phenotypes when the odds of generating any one of them would appear to be so infinitesimal as to discourage even a repeat lottery winner from investing in nature's grand prize drawing?

The third-order emergence of the brain (as best understood by evolutionary neuroscience) provides an intriguing account of gradual complexification, producing nervous systems capable of progressively more sophisticated behavior. This shows how evolutionary adaptation has achieved impressive cognition but no indication that it *should*: it gives no hints as to forces that might stack the deck in the brain's favor. In short, Neo-Darwinism sheds little light on whether brains are evolutionarily inevitable or impressive conquests of intimidatingly long odds. If it is merely the latter, then AI has little to gain from a deep exploration of the evolutionary paper trail, since brains would be the contingent outcomes of the lengthy history of life on earth, a history that not even the most die-hard bio-inspired hacker would hope to re-create in simulation. However, if general principles (not just historical contingencies) underlie the brain's evolutionary emergence, then these might prove invaluable for AI and the automated search for sophisticated machine intelligence.

Many of the more promising general principles involve tight interactions between evolution and development, as reviewed in books such as *Evolving Brains* (Allman, 1999), *Principles of Brain Evolution* (Striedter, 2005), *Developmental Plasticity and Evolution* (West-Eberhard, 2003), and *The Plausibility of Life* (Kirschner and Gerhart, 2005). This chapter investigates several of these phenomena in light of their relevance to AI.

9.2   The Neuromeric Model and Hox Genes

Though different species follow different developmental pathways, they exhibit remarkable similarities at an intermediate point known as the *phylotypic stage* (Wolpert et al., 2002). So the search for general principles of neural evolution and development might profitably begin there. Bergquist and Kallen (1953) noticed that all vertebrate embryos have a similar elongated, segmented hindbrain during the phylotypic stage, as shown in figure 9.1. The ringed segments, termed *neuromeres*, are modular zones of high cell division (to form neurons) and radial migration (of neurons to their proper layer). Later, Puelles and Rubenstein (1993) found that this pattern encompasses the midbrain and forebrain as well.

The hindbrain neuromeres develop into brain regions such as the cerebellum and pons, which are tightly tied to sensory and motor systems, while the midbrain and forebrain segments become areas such as the basal ganglia, hippocampus, and prefrontal cortex, all of which are involved in high-level cognitive processes.

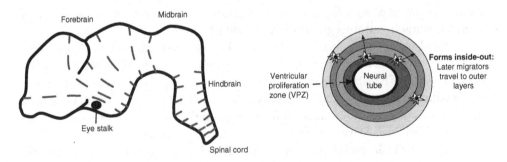

Figure 9.1
(*Left*) Phylotypic vertebrate embryo composed of many neuromeres. (*Right*) Multiple layers of a neuromere through which neurons migrate radially to their proper locations.

Hence, the neuromeric model provides the perfect developmental scaffolding for the brain's emergent control heterarchy (figure 9.2).

Genetic evidence shows that homeobox, (or Hox), genes control brain segmentation, just as they control the subdivisions of the body's central axis (Puelles and Rubenstein, 1993; Allman, 1999; Striedter, 2005). In fact, the vertebrate brain archetype (figure 9.3), looks remarkably similar to a body plan. The evolutionary complexification of the brain begins with the simple addition of more modules,

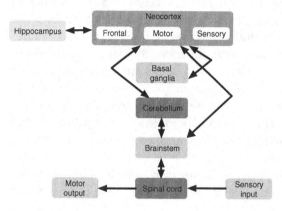

Figure 9.2
Control levels in the brain, which form more of a heterarchy than a strict hierarchy: interactions are too abundant to claim that any one region fully directs another, though higher-level reasoning is known to involve the upper areas, while purely reactive behavior only requires the lower components

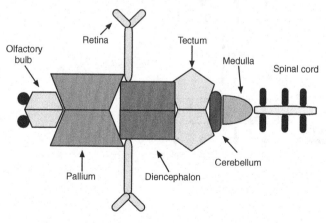

Figure 9.3
Overhead view of the vertebrate brain archetype. Note the similarity to an animal's body. Adapted with permission from Striedter (2005, 66).

like extra scoops on top of an ice cream cone. Folding these modules to fit them inside the skull complicates brain anatomy, particularly that of primates and other mammals, whose large cortices obscure the view of other regions. However, at the coarse level, the basic pattern of complexifying brain anatomy is one of simple linear addition along with differential growth of each segment. The relevance of this spatial patterning for AI is questionable: computational success often depends upon modularity, but proximity relations among modules are usually more of a hardware than a software issue.

The more useful principle involves the homeobox genes and the manner in which they support complexification. Classic work by Ohno (1970) shows the prevalence of gene duplication in evolution, and work by several Nobel laureate geneticists (summarized by Allman, 1999) illustrates the importance of *duplication and differentiation* of genetic material to the gradual emergence of complex phenotypes. A shown in figure 9.4, a copied gene (or gene group) gives evolution the flexibility to *experiment* with new functionalities (by mutating the copy) while retaining the original functionality. This is a low-risk route to complexification that begins with the production of an evolutionarily neutral copy that can eventually morph into a selectively advantageous trait.

The homeobox, a sequence of 180 DNA base pairs, provides a textbook example: it has duplicated and differentiated several times during evolution, with all

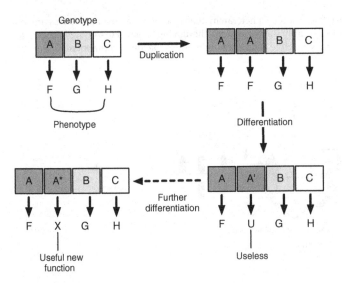

Figure 9.4
Duplication and differentiation of genetic material. Gene A, which produces functionality F, duplicates, yielding a redundant source of F. Subsequent mutations to the copy produce A' (useless) and A* (useful), while maintaining F via the original A gene. Thus, the phenotype complexifies without risking the loss of F.

mutated copies aligned in sequence along the chromosome. During development only some of these Hox regions become activated in particular body or brain segments, but as shown in figure 9.5, the activation sequence mirrors the segment topology: neighboring segments involve the activation of neighboring Hox regions. Each new Hox derivative provided by evolution yields new and varied segments along the phenotype (Allman, 1999).

Once again, the spatial aspects of Hox evolution and developmental expression may serve a limited utility in AI—in the automated design of physical structures (Bentley, 1999)—but the general principle of duplication and differentiation supports a wide range of simulated evolutionary discoveries. For example, Koza et al. (1999) employ this principle in genetic programming (GP): modules are copied then allowed to mutate in order to diversify phenotypic behavior. Similarly, Stanley and Miikkulainen (2003) recognize its importance in artificial developmental systems, and Federici and Downing (2006) fully leverage duplication and differentiation in the evolution and development of artificial neural networks. More generally, the theory of neutral complexification (Kimura, 1983) and its potential for improving search (particularly in rugged landscapes) has drawn considerable attention in the evolutionary computation community (Yu and Miller, 2006; Wu and Lindsay, 1995).

9.3 Neural Darwinism and Displacement Theory

For Bio-AI, the neuromeric model combined with duplication and differentiation constitute a promising scheme for evolving gradually complexifying controllers composed of neural regions. Displacement theory (DT) (Deacon, 1998)

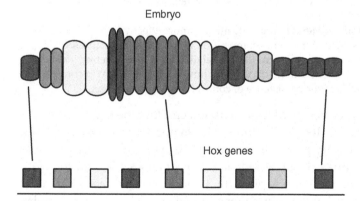

Figure 9.5
Hox groups are aligned on chromosomes in the same order as they are activated in different body and brain segments

complements this scheme by explaining the sizing and connectivity of these regions, and via a process that renders the coevolution of bodies, brains, and brain regions as a much less improbable accomplishment.

The basis of DT lies in *neural Darwinism* (Edelman, 1987; Edelman and Tononi, 2000), also known as *the theory of neuronal group selection* (TNGS). It is now known (Sanes et al., 2011) that considerably more neurons are produced during development than survive into adulthood. Neural Darwinism essentially frames this as a *survival-of-the-best-networkers* competition. Neurons must first grow axons to physically search for target structures (e.g., neurons or other types of cells), some of which can only accept a small number of afferents. Then, the source and target must exhibit correlations in activity patterns to firm up the link. A neuron pair may initially fire randomly but eventually becomes entrained. Failure of either search process (regarding physical contact or behavioral coordination) can spell death for the source or target. Thus, the formation of active connections is paramount to survival.

DT expounds on TNGS by proposing that the networking competition during early development enables brains to *scale to fit* the body's sensory and motor apparatus. In short, primary sensory and motor areas of the brain are sized according to their immediate inputs or outputs, respectively. Secondary region sizes derive from those of the primary regions, and deep cortical structures grow or shrink to fit their input and output sources. Figure 9.6 conveys the essence of TNGS and DT. Note the expansion of the three neuron groups along the path from the largest sensory input, S1, to the largest motor output, M2, and the decline of groups B and D, which lose the competition for C's dendrites and C's axons, respectively. As Deacon (1998) explains,

> So although genetic tinkering may not go on in any significant degree at the connection-by-connection level, genetic biasing at the level of whole populations of cells can result in reliable shifts in connection patterns. . . . Relative increases in certain neuron populations will tend to translate into the more effective recruitment of afferent and efferent connections in the competition for axons and synapses. So, a genetic variation that increases or decreases the relative sizes of competing source populations of growing axons will tend to *displace* [my emphasis] or divert connections from the smaller to favor persistence of connections from the larger. (207)

Developmental neuroscience clearly supports and employs the key tenets of DT. For example, Fuster (2003) documents the earlier maturation of posterior brain regions (used in early sensory processing) and the late maturation of frontal areas such as the prefrontal cortex (PFC). Striedter (2005) combines DT with the work of Finlay and Darlington (1995) to explain the trend of greater neocortical (and especially PFC) control of lower brain regions in higher organisms. First, Finlay

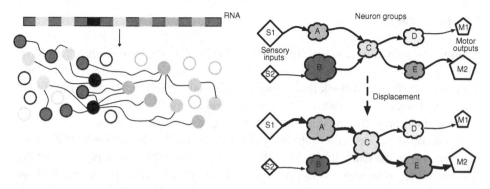

Figure 9.6
(*Left*) Theory of neuronal group selection, wherein genes govern the production of neurons whose ultimate survival depends upon their ability to establish active efferent and afferent connections. Uncon-
nected neurons (open circles) will perish. (*Right*) Displacement theory, in which neuron groups with
good networking possibilities expand while others shrink. Sizes of sensory, motor, and neuron group
icons represent relative sizes of the corresponding neuron pools.

and Darlington show that *late equals large* in neurogenesis: larger brain regions
are those that mature later in development. Second, a key corollary of Deacon's
DT is that *large equals well-connected*: big brain regions send many axons to other
regions and thereby exercise significant control over them. Together, these show
how small changes in developmental timing (in higher mammals) have enabled
the frontal regions to mature later, hence grow larger, and hence exhibit greater
control over a wide variety of cortical areas. And greater frontal control correlates
well with behavioral sophistication, as illustrated by comparisons of manual dex-
terity vis-à-vis frontal control of motor areas in mammals such as cats, monkeys
and humans (Nakajima et al., 2000).

Together, these theories paint neurogenesis as a process in which genetics determines the neuromeric structure, the basic properties of neurons in different layers of the neuromeres, and the maturation speed of neural regions. But the final
sizes of these regions and their interconnectivity arise from self-organizing activity, wherein neural regions competitively search for targets: they compete for the
opportunity to cooperate (i.e., fire in a coordinated fashion).

DT agrees with theories of *concerted evolution* of brain regions, which argue for
the interdependence of developing neural modules, in contrast to *mosaic* theories,
which champion independence (Striedter, 2005). In general, the notion of scaling
to fit reduces the long odds of brain-body coevolution; in DT the body evolves
and the brain grows and wires to accommodate it. This casts a large part of brain
formation as an extensive physical search process.

9.4   Facilitated Variation

Around the turn of the twenty-first century, two biologists, John Gerhart and Marc Kirschner, published two key papers (Kirschner and Gerhart, 1998; Gerhart and Kirschner, 2007) and a book (Kirschner and Gerhart, 2005) motivating and detailing their *Theory of Facilitated Variation*, the means by which variation has been enabled/facilitated by an evolutionarily emergent set of constraints.

The account begins with *evolvability* (Kirschner and Gerhart, 1998), which they characterize as the capacity to generate heritable, selectable phenotypic variation. It is one thing to produce variation, but quite another to yield variations that are viable, can be inherited, and offer natural selection something to grab onto and use to differentiate successes from failures (and thereby maintain evolutionary transition). Evolvability has two key prerequisites: robustness and adaptability. A robust system tolerates perturbation by, in effect, damping their internal consequences. Genetically, this means that mutations rarely have lethal consequences: genomes can buffer the effects of random changes to a few base pairs. Conversely, a system is adaptive if it can easily achieve internal modifications to combat external change. So the genetic version of adaptivity lies in the ability to achieve significant phenotypic change with minimal genotypic change: if the environment changes drastically (such that the available phenotypic plasticity cannot cope), then within a few generations, a genotypic modification will arise to produce a more viable phenotype.

Facilitated variation postulates several key mechanisms for achieving robustness and adaptability. In a nutshell, the theory states that structures and mechanisms that are modular, structurally connected via *exploratory growth* and functionally coupled via *weak linkage*, will produce species that are robust to both genetic and environmental change but can also adapt by easily transitioning to new phenotypes.

Nascent biological mechanisms that exhibit these key properties are termed *core processes* and are believed to have reached a stable evolutionarily status, such that the vast majority of phenotypic change (whether minor or extensive) now involves modifications to the interaction topologies of these processes but not to the mechanisms themselves. Thus, these core elements have an evolutionarily analogous status to the major body plans that arose during the Cambrian explosion (Gould, 1989). Some of the oldest core processes, whose origins date back 3 billion years, include energy metabolism, membrane formation, and DNA replication. Those aged closer to 2 billion years include meiosis, contractile activity, and microfilament and microtubule formation, while 1 billion years have passed since the emergence of intercellular signaling pathways, cell adhesion, and apical-basal cell polarization. Finally, anterior-posterior and dorsal-ventral axis formation, along with complex regulatory processes, are a mere 550 million years old, having arisen 10 or 20 million years prior to the Cambrian explosion of body forms.

They are the building blocks of life, but unlike the building blocks of typical complex-system agents championed by Holland (1995), these do not map uniquely to specific phenotypic traits or exhibit relative simplicity compared to the emergent complexity of the entire organism. Holland's other building blocks (Holland, 1992), at the genotype level, are compact regions of the chromosome and thus not likely to be broken apart by crossover. In contrast, the genetic basis of facilitated variation's core processes may be spread throughout a genome that has gradually evolved to support and preserve these fundamental mechanisms.

Systems built upon core processes derive their complexity from *both* the intricacy of the primitive elements and their interaction networks. Facilitated variation involves complex regulatory networks among sophisticated components whose *pairwise interactions are* simple, a crucial prerequisite to the whole enterprise. Termed *weak linkage*, the principle that signals between core processes are simple compared to intraprocess activity; this is arguably the theory's strongest supporting pillar of robustness and adaptability.

Facilitated variation embodies several aspects of complex systems. First, robustness and adaptivity cover both ends of the classic edge-of-chaos, power law distribution, as shown in figure 9.7, where the x-axis plots the degree to which a phenotype changes in response to a single random genetic mutation, and the y-axis charts the frequency of each amount of phenotypic change. Evolvability entails a power law distribution over this space, wherein most mutations lead to no significant phenotypic change (the high left side of the power law curve), but occasionally, a single mutation can produce a major (useful) jump in design space (the curve's nonzero tail on the right). Thus, the graph displays robustness on the left and adaptability on the right.

Second, exploratory growth, so vital to the emergence of coordinated interactions among a multitude of components, is a textbook example of Mitchell's fourth principle of complex systems: parallel exploration and exploitation among the individual agents (M. Mitchell, 2006). Thus, search is fundamental to facilitated variation, not merely as an abstract perspective from which to interpret the emergence of a properly linked collection of components but as visible processes occurring within individual cells and regions, processes in which many trial structures arise but only a few persist.

9.4.1 Modularity

Basically, modular structures are those in which the intracomponent activity is more extensive than its intercomponent counterpart. Structurally, in a neural network, for example, the vast majority of synapses onto a neuron of module M would

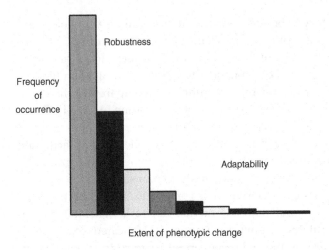

Figure 9.7
Combination of robustness and adaptability embodied in facilitated variation. Each phenotypic change
is assumed to arise from similar small genotypic perturbations.

come from other neurons of M, with only a small minority arising outside of M.
This is modularity at the phenotypic level. As summarized in figure 9.8, many
levels of modularity exist. On the chromosome a modular gene is one whose component base pairs are co-located and thus difficult to separate by the perturbing
effect of crossover. A genotype-phenotype mapping can exhibit varying degrees
of modularity, inversely proportional to the pleiotropic interactions among genes.
For example, in a modular situation, gene A is only involved in the production
of trait X, while in a less modular (more pleiotropic) scenario, it would also play
a role in the production of traits Y and Z. In the former case, a mutation of A
would only affect X, while the latter case would entail more widespread change.
Finally, modular phenotypic structures (such as cortical columns, brain regions, or
tightly interacting subsets of a neural network) can typically restrict a perturbation's extent to the module itself.

   Modularity is a widely recognized factor in the emergence of sophisticated systems (Holland, 1992; Holland, 1995; Simon, 1996), whether via evolutionary or
other adaptive means. Once structures or mechanisms are consolidated and isolated (to some degree) from external influences, their probability of disruption
declines and their potential for self-modification without global repercussions
increases: they enhance both robustness and adaptability.

9.4.2   Weak Linkage
The lack of strong interregional activity is not only a hallmark of modularity but also a prerequisite to facilitated variation. As shown in figure 9.9,

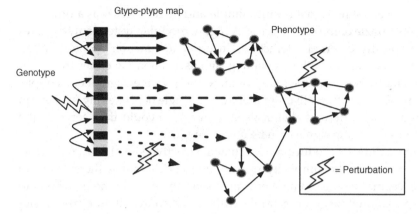

Figure 9.8
Three examples of modularity in an evolutionary context: (*Left*) Locations of genes on the chromosome. (*Middle*) Genotype-phenotype mapping. (*Right*) Structures constituting the phenotype. In each case, a perturbations amidst a modular subsystem should only cause local damage.

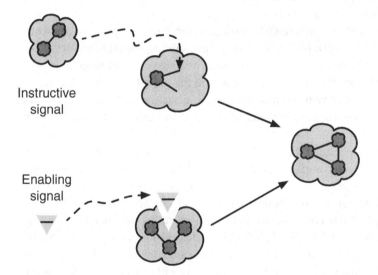

Figure 9.9
Difference between a complex instructive signal and a simple enabling signal

a key biological distinction exists between *instructive* and *enabling* signals. The former is characterized by core processes that require considerable external assistance for their normal operation: the prerequisite external signals are complicated, possibly housing large chunks of the complete process. Thus, the process itself exhibits less modularity because of the intricate external communication.

Conversely, an enabling signal is quite simple and merely serves as a tiny piece in the complete puzzle of the core process. A highly modular system only requires these low-complexity external interactions in order to function properly. Weak linkage entails a good deal of these enabling signals, originating either from simple individual messages or simple pieces of more complex messages. For example, with chemical messages, the active signal may be a single benzene ring, which can be found on a wide variety of chemicals, any of which could thus serve as the messenger. Thus, enabling signals can be quite general.

Whereas the ability to produce a particular instructive signal may only arise a few times in nature, the simplicity of enabling signals makes them easier to generate. Hence, many other components may send them, an immediate boon to robustness and adaptability. For example, if the *usual* source of an enabling signal weakens or vanishes, an alternative source might already exist or easily arise, aiding robustness. Similarly, a network of interacting processes can more easily reconfigure (into a new phenotype) in the advent of a simplified signaling protocol, whereas complex communication would overconstrain the situation, giving few (if any) alternative arrangements.

In fact, Gerhart and Kirschner (2007) consider most core components to house the entire interaction network *plus an inhibitor*; the enabling signal then blocks the inhibitor, thereby disinhibiting the core process. The signal is thus selectable only for its ability to block the inhibitor, not for its role in the actual core process. Hence, the core process can evolve without a corresponding change to the signal, and the long odds of coevolution are avoided such that viable variations can more readily evolve.

Weak signaling can produce situations in which the control of a core process can shift between internal and external sources, which would allow an environmentally induced mechanism to become innate or vice versa. As shown in figure 9.10, an enabling signal might originally stem from an external source, but because of its simplicity it could easily become internally produced via a simple genetic mutation. West-Eberhard (2003) provides numerous examples of this phenomenon, including sex determination by temperature (instead of genes) in turtles and snakes, cricket wing length controlled by either crowding conditions or genes, and foraging strategies in fruit flies determined either innately or by starvation/satiation levels.

Thus, a phenotypic change of a *plastic* nature could smoothly become innate, providing a selective advantage in environments where the particular change had become ubiquitously desirable. Furthermore, this internalization of a signal source helps explain how natural selection could favor weak linkage (and other aspects of facilitated variation). Assume that various species benefit from lifetime flexibility owing to evolved weak linkage between the environment and core components. If simple mutations could easily internalize these linkages to the genome, then

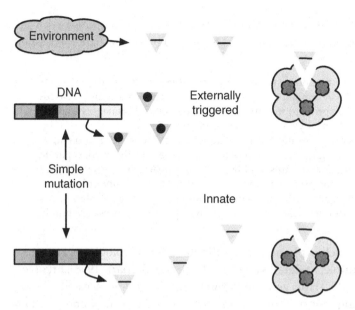

Figure 9.10
Shifting the control of a core process from environmental to internal via a simple mutation that allows the organism itself to produce the enabling signal

there would be a steady influx of weakly linked, internal-signaling networks to the gene pool. In short, there would be no shortage of genomes that exhibited facilitated variation, and some would presumably have a selective advantage for individual organisms, particularly those living in unpredictable environments (where renewed external triggering of certain core components could become desirable). Thus, many genotypic adaptations may indirectly arise from selective pressure favoring phenotypic flexibility.

Neurons are classic examples of weak-signaling core components, since they all employ the same communication currency: action potentials. In addition, signal transmission across synapses involves neurotransmitters, of which there are only a few dozen well-conserved varieties such as glutamate, GABA, and acetylcholine. Furthermore, postsynaptic terminals often house receptors for several of the neurotransmitters, thus further increasing their communication possibilities. Essentially, if a new neuron type were to arise from the mutation of a contemporary variety, it would almost certainly sprout some of the same receptors and thus be immediately prepared to talk to the rest of the brain.

9.4.3 Exploratory Growth
Along with the ability to easily communicate with one another, the core components of facilitated variation can *find* each other using trial-and-error exploratory

search processes whose high energetic costs trade off (very favorably) with an amazing level of developmental flexibility. As explained by Gerhart and Kirschner (2007),

> Examples include the formation of microtubule structures, the connecting of axons and target organs in development, synapse elimination, muscle patterning, vasculogenesis, vertebrate adaptive immunity, and even behavioral strategies like ant foraging. All are based on physiological variation and selection. In the variation step, the core process generates not just two output states, but an enormous number, often at random and at great energetic expense. In the selective step, separate agents stabilize one or a few outputs, and the rest disappear. Although that agent seems to signal the distant process to direct outputs to it, it actually only selects locally via weak linkage among the many outputs independently generated by the process. Components of the variation and selection steps of the process are highly conserved. (8285)

In some cases, many such explorations coordinate to form biological structures. A most impressive example, as detailed by Kirschner and Gerhart (1998), involves the development of limbs. Hox genes determine the locations of cartilaginous points in the developing embryo, and exploration does the rest. Namely, bones form to join these points, muscles and tendons grow to link up the bones, motor neurons sprout axons to innervate the muscles, and blood vessels grow to feed them. Furthermore, within each neuron, selection occurs among the many axons, only a few (if any) of which manage to find a target dendritic tree; the rest wither away. As discussed earlier, neural Darwinism and displacement theory explain the search processes wherein neural subpopulations attain sizes and connection patterns in a *grow-to-fit* manner.

In their seminal developmental biology text Wolpert et al. (2002) describe the use of filopodia—thin cytoplasmic extensions—to pull mesenchyme cells along during migration:

> When filopodia make contact with, and adhere to, the blastocoel wall, they contract, drawing the cell body toward the point of contact. Because each cell extends several filopodia, some or all of which may contract on contact with the wall, there seems to be a *competition* [my emphasis] between the filopodia, the cell being drawn toward that region of the wall where the filopodia make the most stable contact. The movement of the primary mesenchyme cells therefore resembles a *random search* [my emphasis] for the stable attachment. (281–282)

The authors cite a similar behavior for neural crest cells, precursors to the entire peripheral nervous system (along with other parts of the body) and travelers of many long and diverse migratory routes. Kirschner and Gerhart (1998) view neural crest cells as the epitome of an exploratory process, with variation stemming from the ability to follow many paths and differentiate into numerous cell types, and selection performed by the chemical signals in various compartments—often defined by local signal-chemical portfolios, not physical barriers—during the early stages of development.

Crest cells originate from different rhombomeres of the developing brain, with each having a characteristic set of active Hox genes (Sanes et al., 2011), as discussed earlier. These give each migrating crest cell a general identity, which then determines the target region for its wandering search. Striedter (2005) proposes a similar mechanism for the formation of laminae and topological maps in the brain.

Laminae are simply parallel layers of cells, classically illustrated by the six-layered mammalian cortex. The key point is that neurons from one layer of a region will often target neurons of a particular lamina of another region. If each layer has a characteristic portfolio of chemical signals, then growing axons *read* that chemical signature and internalize it to the extent that it affects the target regions they seek.

As discussed in chapter 7, a topological map is a connection pattern between two regions wherein nearby neurons in layer A have targets in layer B that are also neighbors. The general concept extends beyond the nervous system to sensory inputs and motor outputs. For example, nearby fragments of the visual field are normally handled by nearby neurons in the retina, which, in turn, feed signals to neighboring neurons in the thalamus, which maintain these spatial correlations in sending signals to the visual cortex. Similar sequences of maps exist in the auditory system, where the *environment* consists of sound frequencies (see chapter 10). Again, Striedter (2005) cites evidence of chemical signatures biasing axonal target selection to form the initial maps, with experience-based synaptic change fine-tuning them afterwards.

Laminae and topological maps are critical components of neural systems. They have several functional advantages, including modularity, wiring efficiency (since axons between these regions tend to run in parallel, exhibiting much less criss-crossing than with random connections), and generalizability. The last stems from interregion correlations, which help insure that similar sensory situations promote similar neural firing patterns that invoke similar motor responses. This allows organisms to handle novel situations with behaviors that were successful in similar contexts. Thus, exploratory growth produces efficient neuroanatomical structures that support adaptive behavior.

Figure 9.11 illustrates another contribution of exploration to robustness and adaptability in a neural context. The former occurs when the C neurons are abnormally displaced, but axons from the A neurons still find them: the perturbation is countered via exploration, enabling standard A–C connections to form. Similarly, adaptability occurs when a single mutation changes the affinity of A or C neurons such that B neurons become the dominant source of afferents to the C population, thus yielding a new neuronal topology and potentially a new phenotype.

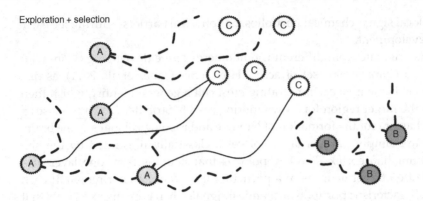

Exploration + selection

Figure 9.11
Exploratory axonal growth processes leading to the formation of neural topologies. Dashed lines represent exploratory axons; solid lines, explorers that found targets.

In all these examples DNA encodes only the exploratory processes, not the resulting patterns of connectivity. Again, the genome embodies a recipe, not a blueprint.

Analogously, the evolutionary algorithm for finding Steiner trees (chapter 3) uses a chromosome that, like the Hox genes, only indirectly encodes locations of special (Steiner) points. Kruskal's algorithm determines the best way to connect them. To add more biological realism to that example, evolution would need to discover Kruskal's, Prim's, or some similar connection-generating algorithm as a core process.

The practical implication of exploration and its facilitation of variation is an important take-home lesson: the production of novel phenotypes does not require concerted change to many parts of the genome, a very low-probability combination of events, but rather a single change to a factor affecting an early phase of development. The rest just grow to fit the altered context.

In short, nature has not endowed core processes with foresight, but with *persistence*. A multitude of parallel explorations combined with weak linkage allow many core elements to eventually find connection points. From this, networks emerge, and with them, the possibilities for more patterned, predictable, and intelligent behavior.

9.4.4   Emerging Topological Maps
Exploratory growth, when combined with Hebbian learning in the form of basic spike-timing-dependent plasticity (STDP; see in chapter 10), provides an emergent account of topological map formation, as displayed in figures 9.12 and 9.13. First, similar concentration gradients in the two regions will bias the searching axons

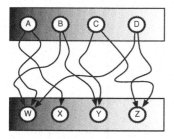

Figure 9.12
Developmental emergence of a topological mapping between the upper and lower layers, assuming that concentration gradients (shading) mirror one another. Note the biased but imperfect topology, best illustrated by the errant link from D to W and the missing link from B to X. Based on (Striedter, 2005, 189).

to target corresponding neurons in the lower layer, though *errors* may abound. Despite the errant nontopological connections, experiences in the real world can easily fine-tune the map, as shown in figure 9.13. In each case, an abstract object activates the upper (presumably sensory) neurons, with can, in turn, team up to fire the lower-level neurons. For example, in scenario 1 the object activates A and B, which stimulate W.

The local actions in this emergent phenomenon are the following STDP mechanisms:

- A synapse strengthens when pre- and postsynaptic neurons fire simultaneously or with presynaptic activity slightly preceding postsynaptic.

- A synapse weakens if the presynaptic but not the postsynaptic neuron fires, or if postsynaptic firing precedes presynaptic firing by a small amount.

Thus, in scenario 1, the A–W and B–W links strengthen, while the A–X and B–Y links weaken (because of lack of postsynaptic firing). Similarly, D–W weakens in scenario 2, as does C–Y, since neither W nor Y spikes. Only in scenario 3 does D–W strengthen, and this would seem to be a more atypical sensory situation (i.e., one that stimulates A and B but then hops over C and stimulates D) given the general continuity of the sensory world. Also note that D–W weakens in scenario 4, since A and B suffice to activate W, while D activates slightly later (since the object moves left to right). Thus, the presynaptic node (D) activates after the postsynaptic node (W), leading to further synaptic depression.

Note that a right-to-left movement of the same object would not have a completely opposite effect. D would activate before A and B, but D alone would not suffice to activate W, which could not fire until at least B did (and possibly not until both A and B fired, since some of D's input to W will have leaked away by the time

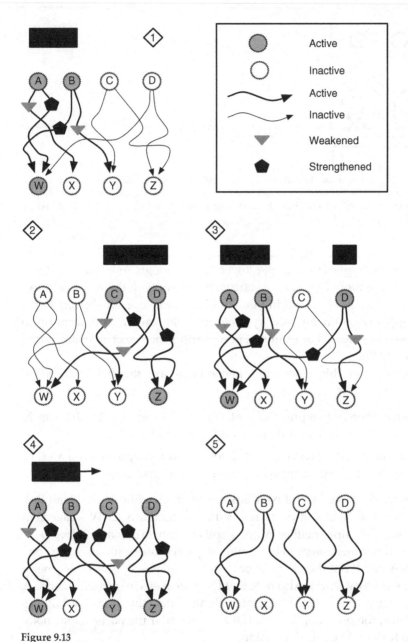

Figure 9.13
Once formed by development, the topological map is fine-tuned by experience. Black bars denote stimulation in the sensory (i.e., visual) field corresponding to neurons A–D. Each scenario 1–4 results in some neurons activating; it is assumed that lower-layer neurons require at least two upper-layer inputs (within a short time window) to activate. Synaptic tuning follows basic Hebbian learning and STDP principles. Scenario 5 depicts the network after many rounds of learning, whereby those synapses that many experiences would tend to weaken have disappeared.

the object stimulates B). Thus, although the D–W link would strengthen (since D fires just before W), the B–W link would not weaken but only get stronger. In fact, the details of STDP entail that B–W (and quite possibly A–W) would strengthen even more than D–W, since its spikes would be closer in time to (yet still precede) W's spike. Thus, in the overall competition to activate W, D would again lose ground to B (and possibly to A as well).

Furthermore, any strengthening of A–W and B–W relative to D–W can quickly set a positive feedback in motion: any co-occurrence of A and B could fire W almost immediately, so even if D fires a few milliseconds after A and B, it will be *too late* (i.e., after W fires), thus compounding the depression of D–W.

In short, the regularity of the real world will, in all likelihood, provide an overwhelming number of stimuli that strengthen the connections shown in scenario 5 while weakening nontopological links such as D–W, which only improve on very large or discontinuous stimuli. Hence, the combination of coarse, imperfect chemical gradients, Hebbian STDP principles, and the continuity of the real world suffice to promote the emergence of the brain's topological maps.

9.4.5 Deconstraining Evolution

The net effect of modular core processes, weak linkage, and exploratory growth is a deconstraint of evolution. Now, many genotypic and phenotypic changes become viable because perturbations are constrained to remain within a modular area, but those with nonlocal consequences are nonlethal because of the flexibility inherent in weakly linked interaction topologies and exploratory growth mechanisms. Just as it is easier to innovate with a group of open-minded individuals, all self-confident in their own abilities but willing and eager to interact with many others, in many constellations, so, too, can nature more readily produce novel phenotypes when its core components are stable, complex, and competent but easy to interface with others.

As a simple example of how constraints affect a search space, consider the Boolean logical constraint in equation (9.1), where the goal is to find an assignment of truth values to the four Boolean variables such that the entire expression is true. There are $2^4 = 16$ possible states in the search space, and as shown in table 9.1, nine of them satisfy this constraint.

$$(a \vee \neg b) \wedge (c \vee d). \tag{9.1}$$

Notice what happens when the second disjunction is supplemented with a b (equation (9.2)). This makes the disjunction easier to satisfy, which in turn

Table 9.1
The nine viable solutions (boldface) to the constrained logical expression of equation (9.1)

Possible States			
0000	**0001**	**0010**	**0011**
0100	0101	0110	0111
1000	**1001**	**1010**	**1011**
1100	**1101**	**1110**	**1111**

Note: Binary vectors denote truth values for a–d; 1 = true, 0 = false.

Table 9.2
The fourteen viable solutions (boldface) to the deconstrained logical expression of equation (9.2)

Possible States			
0000	**0001**	**0010**	**0011**
0100	**0101**	**0110**	**0111**
1000	**1001**	**1010**	**1011**
1100	1101	**1110**	**1111**

Note: Binary vectors denote the truth values for a–d; 1 = true, 0 = false.

deconstrains the entire expression, allowing more points in the search space to satisfy it, as shown in table 9.2.

$$(a \vee \neg b) \wedge (b \vee c \vee d) \equiv (a \vee c \vee d). \tag{9.2}$$

This simple example has direct parallels with biological deconstraint. Anytime a core process is modified to accept an additional type of enabling signal, it effectively supplements its disjunction of salient signals, thereby allowing a wider variety of components to serve as afferents; and this increases the set of viable interaction topologies. When a design space has more feasible solutions, the probability that an emergent process can stumble onto one of them can only increase.

But if evolution has become deconstrained, then how has nature managed to navigate the daunting design space of living things? Though weak linkage and exploratory growth reduce constraint, the core processes, now well locked in for the past billion or so years, greatly increase it. In essence, the evolutionary emergence of the core has shifted nature's evolutionary search to a small corner of the space of possibilities. Given the chemistry and physics of earth, that corner may be a strong attractor, or it could be one of several viable regions. Either way, without very powerful intervention, life as we know it is basically stuck there. Weak linkage and exploratory growth then deconstrain *that region of search space*. Thus, in terms of search space terrain, these features help nature make the most out of the corner into which the core processes have corralled life.

9.5   Representation and Search in Neural Evolution and Development

By now, the fact that Darwinian evolution is an impressive trial-and-error search process should need little elaboration: it finds amazing designs both in nature and in Bio-AI by essentially tossing options out there and letting selection filter them out. The role of search in development has traditionally received less attention. However, researchers such as Edelman, Deacon, Kirschner, and Gerhart, though they did not discover these mechanisms, have highlighted the importance of exploratory actions combined with selection during the growth of neural structures.

In order for AI to incorporate the essence of exploratory growth, a study of the representations underlying it could prove helpful. In nature, of course, that representation is DNA, which contributes to life's robustness and adaptability in many ways. For our purposes, the essential facts about DNA are the following:

- Triplets of DNA bases, called *codons*, constitute amino acids, which are the building blocks of proteins but also have other uses in the body.

- A *gene* is a long continuous segment of DNA bases that codes for compounds of functional significance in the organism, typically polypeptides or RNA chains.

- *Functional genes* directly express functionally significant compounds, while *regulatory genes* produce signaling compounds that promote or inhibit the expression of other genes, whether functional or regulatory. These signals serve an enabling (or more often, inhibiting) role but not an instructive role. As indicated earlier, many such signals can arise from the genome or the organism's environment.

- *Junk DNA* comprises segments not known to express anything of functional or regulatory significance.

First, the coding scheme for amino acids houses many redundancies: only 2 of the 20 amino acids stem from unique codons, while all others have between 2 and 6 different codon sources. This enhances robustness, since many mutations may alter base pairs without changing the resulting amino acids (and proteins).

Another form of insurance against genetic change is junk DNA, which seems to provide buffer zones between expressed genes such that crossover during meiosis has less chance of splitting a coding gene in two: the crossover points often occur within the junk DNA.

In general, the exact positions of genes on the chromosome have little significance; interactions between genes (for example, a regulatory gene R and a functional gene F) happen through an intermediary chemical signal, S, that R produces, not via direct contact between the R and F segments. So gene R may be close to or far from F on the chromosome. This supports gene duplication, since copied genes need not bump existing genes from important, predefined chromosomal locations. The modularity of genes combined with the lack of location specificity facilitates changes to the gene composition of a chromosome: additions and subtractions should not interfere with the other genetic interactions. Only when a gene differentiates can it alter functionality.

All told, these features of DNA and the general (and evolutionarily successful) arrangements of it (i.e., genotypes) lend ample support to both phylogenetic and ontogenetic search processes. For evolution, the robustness of nascent genotypes to genetic change allows nature to experiment with novel genotypes, many of which are neutral, since they produce the same phenotypes (and or equally fit phenotypes) as their predecessor chromosomes. Evolution can thus explore design space with much lower risk than if every mutation yielded new phenotypes.

The combination of control and functional genes into a *genetic regulatory network* (GRN) forms a mill of novelty. Without regulation, every functional gene would presumably be ubiquitously expressed (or vestigial), and new phenotypes would require the addition of new functional genes. The presence of regulatory genes provides an exponential number of possible phenotypes from the *same* set of functional genes, some of which may be on (expressed) and others repressed at any particular time in evolution or any spatiotemporal point in development. In fact, cell types are largely differentiated by their subset of active functional genes.

This means that at any point in evolution or development, anything with a chromosome has many options, with regulatory signals (from both the environment and the cell itself) determining the outcome: the state of the GRN. From the evolutionary perspective, there are many GRNs in many different states that constitute an organism; in development, the focus is often on a particular cell and the status of its GRN. For example, Kirschner and Gerhart (2005) discuss neural crest cells and their ability to migrate within the embryo and then eventually transform into their final cell type based on signals in the extracellular environment. In essence, these cells search for a fitting location and for the appropriate biochemical behavior from a long list of possibilities. The options exist all the time; they just require the proper signals to activate. Thus, the exploratory search underlying so much of development depends upon the flexibility of the GRN to *read* ambient signals and *react* accordingly.

In fact, Kirschner and Gerhart have considerable evidence that functional genes (like core processes) have been conserved over millions of years, possibly all the way back to the Precambrian era and beyond. Variation has become the sole province of the regulatory genes, which, like the automatic steering of a car, can affect change with minimal effort (because of weak linkage).

From a representational perspective, GRNs take the direct-versus-indirect distinction to a higher level. Their indirect representations require a complex translation process, but one that can be characterized by two stages: (1) the formation of all products of the individual genes (such as neurons), followed by (2) myriad interactions between those products (many mediated by weak linkage) to produce the phenotype. GRNs add considerable variability to stage 1 by expressing only a subset of the genes, but a subset that is sensitive to, and thus adapted to, its environment. This changes the game completely; and while nature has mastered this innovation, Bio-AI researchers have only just begun to investigate.

9.6   The Bio-Inspiration

What biological inspiration should AI draw from the evolution and development of the brain? Clearly, many of the core processes (such as energy metabolism, exploratory axonal growth, and anterior-posterior axis formation) demand a (computer) module or two in simulations designed to accurately mimic biology. But how many of them are vital to the emergent design of an artificial intelligence? Does a computational system need a metabolism, a procedure for gradually growing a connection between two hardware or software modules, or a means of differentiating top from bottom? These are clearly more relevant for an embodied AI system, but lower-level core processes such as cell adhesion and contractile activity may have few implications for AI.

9.6.1   Genetic Regulatory Networks

Good AI systems are often flexible at the phenotypic level. They use sophisticated machine learning algorithms to adapt. However, if, as argued earlier, AI wishes to take advantage of indirect encodings and the genotype-phenotype distinction in the search for intelligent systems, then genotypes may need enhancements to *facilitate* useful variation. Nature eventually found these enhancements in GRNs, but whether AI should employ bio-inspired abstractions of these representations (or pursue entirely different, engineering-based routes) remains unclear.

GRNs have been used in many ALife systems, often producing impressive low-level emergence, such as a single cell that repeatedly divides to produce a multi-cellular clump that then performs a physical task, such as pushing a box (Bongard, 2002), or a similar seed cell that divides and differentiates into a spatial form that matches a target pattern, such as a French or Norwegian flag (J. Miller and Banzhaf, 2003; Federici and Downing, 2006).

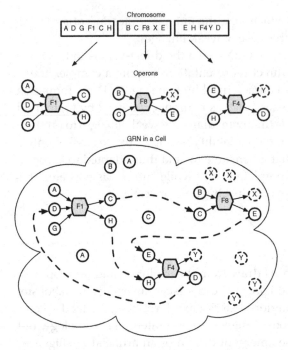

Figure 9.14
Basic genetic regulatory network (GRN) as implemented by several Bio-AI researchers (Bongard, 2002; Eggenberger, 1997a; Reisinger and Miikkulainen, 2007) and described in more detail in chapter 12. The evolutionary computation chromosome encodes a group of operons, each of which takes several input substances and can produce one (or more, in some implementations) outputs. The relations between inputs and outputs then form a network, with the cell serving as the repository for all substances, which are analogous to proteins in biology.

In these and many other Bio-AI systems, the GRNs resemble that of figure 9.14, where an evolutionary algorithm chromosome encodes several modules known as *operons*, each of which works similarly to an if-then rule: when its preconditions are satisfied, the consequent is performed. For example, if the concentrations of all substances mentioned in the precondition exceed particular thresholds, then a product substance is produced (whose amount and possible location of production may also be specified by the operon). Unlike the if-then rules of classic GOFAI expert systems, the operons often run in parallel, not series, to continually update the concentrations of substances in the local environment, i.e., the cell.

Systems employing GRNs normally consist of many cells, which may exchange substances either directly or via an external milieu. The full portfolio of substance concentrations often determines a cell's type, which then maps to a functional component of the AI system. For instance, the cell may correspond to an artificial neuron or a body segment or even an innervated body part. Special substances may also promote cell division or death, axonal migration; or it may translate into receptors (on the cell membrane) for other substances. The general operon-based model supports a wide range of applications, all of which empower the genome with considerably more flexibility than in traditional evolutionary computation.

Instead of the artificial chemicals of GRNs, some Bio-AI systems employ grammars (i.e., sets of rewrite rules) as the basis for simulated development (Stanley and Miikkulainen, 2003; Floreano and Mattiussi, 2008). This exhibits less biological realism but often creates more lifelike structural facsimiles of both plants and animals, as best exemplified by L-systems (Lindenmayer and Prusinkiewicz, 1989) and their many applications (Floreano and Mattiussi, 2008). Both the GRN-driven (chemical-based) and grammar-based developmental models have these advantages:

- *Scaling* Small genotypes suffice to generate large phenotypes, and the compressed genotype defines a more manageable search space.

- *Symmetry* Large phenotypes produced from the repeated execution of small genotypes typically have recurring structures, offering a selective advantage for many tasks, such as those involving locomotion.

- *Robustness* Reexecution of the developmental procedure (later in the life of a system) can often repair a damaged phenotype or, in general, help to maintain a stable, functioning state.

Regardless of these advantages, artificial evolutionary developmental systems have yet to achieve convincing success on anything beyond quite simple tasks, though simplicity in the eye of the observer often belies fascinating emergence at the evolutionary and developmental levels. Finding effective developmental recipes for growing complex physical structures or their control systems has proven to be an extremely daunting task, despite the help of artificial evolution in exploring the genotype search space. Researchers always return to the same question: What aspects of natural development can actually contribute to improving artificial design, rather than being merely scientifically interesting to implement?

9.6.2 Conserved Core Processes

The most appropriate bio-to-tech transfer might be at a very high level of abstraction. For example, Bio-AI can benefit from the general concept of conserved core processes, which evolutionary search employs as building blocks that because of weak linkage, can be combined in many ways. Specialists in the area of genetic programming (GP) understand this well. Their evolutionary simulations begin with a set of primitive modules/functions that can be combined into an exponential number of problem-solving systems. Search occurs within the space of these combinations, but the primitives are hand-designed and fixed.

9.6.3   Weak Linkage and Tags

GP also incorporates weak linkage to varying degrees. Original versions of the concept (Koza, 1992) employed the *closure property*, discussed earlier, to insure that the outputs of any module could be accepted as the inputs of any other module. This allowed the full complement of module combinations to at least run (i.e., not crash the computer) though not necessarily to produce useful results. Later versions (Montana, 1995; Koza, 1994) introduced strongly typed modules that only accept certain types of arguments. This restricted GP search but in no way detracted from the creativity of GP design. The field exploited these and other constraints to make quantum leaps from toy problems to real problems to the automatic design of devices that rival or surpass those of modern engineers (Koza et al., 1999; Koza, 2003).

Perhaps no other AI researchers have embraced facilitated variation more than Lones and Tyrrell, who employed weak linkage and classic GP modularity in their *enzyme genetic programming* (EGP) system. In this work the genome encodes modules but says nothing about their interaction topology, whereas standard GP employs genomes that fix the topology. Each module includes an interface whose syntax reflects the module's computational semantics, in much the same way that an enzyme's 3-D shape (its interface) implicitly mirrors its functionality; the interface is not an arbitrary signal nor receptor. The complete topology then emerges from a search process in which modules attempt to find others with compatible interfaces.

EGP's interfaces are sophisticated versions of the more general concept of *tags*, which have received more attention in the EA community. In fact, Holland (1995) includes tags in a short list of key elements of a complex adaptive system, and they are the cornerstone of his general model of signal boundary systems (Holland, 2012). A tag is simply a syntactic structure (typically a string of bits or other symbols) attached to a component that is visible to other components and typically determines the degree to which different components interact. Unlike EGP's interfaces, tags generally do not reflect the behavioral semantics of the component, although they may eventually evolve to correlate in some way with the internal activity. Or, to achieve deception in situations where agents are best suited by avoiding interactions (e.g., with predators), tags may evolve to signal something totally different than the component's actual behavior. Either way, tags influence interactions and allow interaction topologies to emerge during development rather than being predetermined by the genome. This allows more *natural* topologies to form in which components have affinities for one another, and many such local compatibilities increase the odds of global coherence.

Basically, tags allow good designs to fall into place once proper combinations of core elements and tags arise. In this case, the inspiration from facilitated variation entails hand-designing the core modules but allowing the search process to determine the quantities of components along with their tags. The exploratory process is then embodied in the algorithm that components use to search for neighbors in the emerging topology.

Depending upon the matching criteria, tagging can support weak linkage. For example, if one agent's output tag must perfectly match another's input tag, then this obviously restricts interaction. But if local compatibility requires only, say, a 50 percent match, then more relations can form.

Matching criteria also have a strong effect upon the robustness of a tag-based representation. Figure 9.15 plots the match degree of two binary tags (of length 200) as a function of the bitwise mutation rate. The two tags are originally identical, but, naturally, as the mutation rate rises, matching declines. However, the algorithm for computing matches varies for the three plots, and each yields a different curve. A summary of these metrics follows.

The Hamming distance, H_{ab}, between bit strings a and b is simply the number of nonmatching bits in the two strings. For example, if $a = 110111$ and $b = 010101$, then $H_{ab} = 2$, since they differ only at two positions: 1 and 5. The match degree

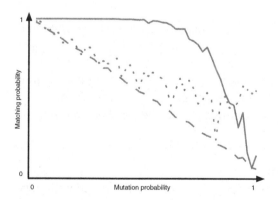

Figure 9.15
Varying degrees of robustness (to bit mutation) for different tag-matching metrics. The x-axis denotes the probability of bit mutation; the y-axis records the matching degree of two tags that were originally identical. Matches are computed using three different algorithms based on Hamming distance, M_H (dashed line), the integer value of the bit string, M_I (dotted line), and the special chemically inspired *streak* model, M_S (solid line).

based on Hamming distance, M_H, is defined as $M_H = 1 - \frac{H_{ab}}{n}$, where n is the string length.

The integer metric for matching, M_I, compares the integer values of each bit string, with smaller differences yielding higher matching. Formally, $M_I = 1 - \frac{|I_a - I_b|}{2^n}$, where I_a and I_b are the integer values of strings a and b, respectively.

Finally, the streak metric, M_S, stems from

- the length (s) of the longest contiguous segment of perfect bitwise equality,

- the length (d) of the longest segment of perfect bitwise inequality, and

- the comparative odds of each such streak, where odds are the inverse probability.

For example, if $n = 10$, the probability, when comparing two n-length strings, that four contiguous bits match is $\frac{7}{2^4}$: the probability of a four-bit match is $\frac{1}{2^4}$, and there are 7 spots where such a streak could begin. Thus, the general probability for a k-bit match is

$$p_k = \frac{n - k + 1}{2^k},$$

and this is the same as the probability of finding k consecutive bits that do not match. The streak match degree is then

$$\frac{\frac{1}{p_s}}{\frac{1}{p_s} + \frac{1}{p_d}} = \frac{p_d}{p_s + p_d}. \tag{9.3}$$

M_S is high for the combination of large s and small d, since these translate into small p_s and large p_d. In other words, a high match is one where there is a long (low probability) matching streak but no lengthy nonmatching streak. As shown in figure 9.15, M_S yields the most desirable behavior: it tolerates a good deal of mutation, though certain well-placed mutations (e.g., in the middle of the maximum equal or unequal streak) can perturb the match values between the strings, which may alter the emerging interaction topology.

9.6.4   Interacting Adaptive Mechanisms

Another relevant, though often overlooked, aspect of contemporary evolutionary developmental theory involves the bidirectional interactions between evolution, development, and learning. Bio-AI researchers often implement each adaptive phase in lockstep: first generate a new chromosome using genetic operators, then convert it into a phenotype using a developmental routine, and then run the phenotype in an environment and permit small behavioral tweaks based on experience. However, these mechanisms interact in many ways. For example, development often continues well into adolescence and even adulthood, thus overlapping with (and often blurring the distinction between itself and) learning. More important for our discussion, the effects of phenotypic and developmental change can affect evolution.

The prevalence of equivalences of genetic and environmental factors in producing phenotypic traits allows evolution to find GRNs that confer an immediate selective advantage (in an unstable environment) and then to co-opt this phenotypic flexibility for evolutionary purposes by allowing gene products to (permanently) enable or disable various operons to stabilize phenotypes in more predictable environments. Thus, the evolutionary search for GRNs that equip phenotypes with effective lifelong search capabilities also accelerate evolutionary search. This relation is easily confused with Lamarckianism (Lamarck, 1914), a (largely) disproven evolutionary theory that proposed the (relatively immediate) genetic transmission of phenotypic change. However, it is best exemplified by the Baldwin effect (Baldwin, 1896; Weber and Depew, 2003), a very plausible, indirect mechanism wherein the emergence of lifetime plasticity can have a selective advantage in a new (but predictable) environment, followed by genetic hardwiring (via mutation) to handle these conditions, which offers a better solution in situations where lifetime plasticity has a significant cost to the organism.

This bidirectional relation between evolution on the one hand and development or learning on the other has received considerable interest in Bio-AI (Turney et al., 1997; Bala et al., 1996; Mayley, 1996; Downing, 2001), as has the more general combination of a broad-scale search process (such as evolution) to find general strategies and a focused process of adaptivity for tuning those strategies during the course of their deployment (Eiben and Smith, 2007; Nolfi and Floreano, 2000; Michalewicz and Fogel, 2004).

9.6.5 Solving Network Problems with Fruit Flies

An interesting (and very bio-inspiring) example of search and emergence during development involves the sensory organ precursor (SOP) cells (Sanes et al., 2011) of *Drosophila's* (fruit fly's) nervous system. As depicted in figure 9.16, SOP cells, which become sensory bristles, originally differentiate from a homogeneous population of proneural cells. Interestingly, this differentiation creates a population in which every non-SOP cell is adjacent to at least one SOP, and no SOP cells are adjacent. This pattern emerges from chemical signaling in which SOPs inhibit neighbor cells (from becoming SOPs) by emitting the protein Delta, which interacts with another protein, Notch, to prevent SOP formation. Although the details of these chemical processes are beyond the scope of this book, their essence has

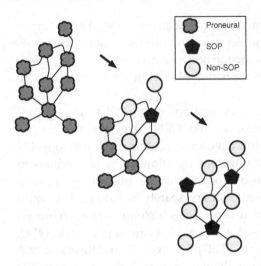

Figure 9.16
Emergent segregation of *Drosophila* proneural cells into SOP and non-SOP varieties. All non-SOPs con-nect to at least one SOP, and no two SOPs are linked.

been abstracted into a very elegant, fully distributed solution to a vexing problem in computer networks (Afek et al., 2011).

The minimal independent set (MIS) problem involves finding a set of *local leader* nodes in a network such that all nonleaders are adjacent to a leader, and no leaders are adjacent to one another. Computer scientists and mathematicians struggled with distributed algorithms for solving this problem for decades before Afek et al. (2011) looked to *Drosophila* for a solution. The resulting algorithm high-lights the role of localized search (within each node) in leading to the global solu-tion: the MIS.

The procedure/simulation begins with a network of N nodes, many connections between them, and two empty pools, one for leaders (L) and one for nonleaders (NL). D is the maximum number of neighbors for any node in the network, and M is an integer constant.[1] Each step of the simulation involves two signaling rounds for each node (n) that has not yet been assigned to either of the pools. These nodes presumably execute this procedure synchronously, with the same step being con-currently executed by all nodes.

1. $p = \frac{1}{D}$.
2. If $p \geq 1$, exit.
3. Repeat $M\log_2 N$ times:

a. Round 1

(1) With probability p do:

(a) Broadcast message B to all neighbors(n).

(b) state(n) ←1.

(2) If n receives B in round 1, then state(n) ← 0.

b. Round 2

(1) If state(n) = 1 (i.e., n broadcast B but did not receive B in round 1):

(a) Add n to L.

(b) Broadcast B to all neighbors(n).

(2) If n receives B in round 2, add n to NL.

4. $p \leftarrow 2p$.

5. Go to step 2.

Notice that each unassigned node makes a simple stochastic decision on each pass through round 1: whether or not to broadcast B, which constitutes declaring its intention to become a leader node. These declarations, when met by similar messages from a neighbor, simply cancel one another: neither becomes a leader (on this pass). Initially, these declarations occur infrequently, but each round through the outer loop doubles p, leading to more declarations among the remaining uncommitted nodes. Thus, each node exhibits a probabilistic *try-and-try-again* behavior until either its eagerness is rewarded (thus gaining entrance to L) or its passivity is punished by an eager neighbor, thus relegating it to NL. The individual search behaviors are quite simple, but the end result is the emergent solution to a complex problem.[2]

9.7 Appreciating Add-Hox Design

One glance at an anatomical brain map, with its lobes, gyri, sulci, and vesicles, is usually enough to convince an AI researcher to keep her day job at the keyboard. More detailed figures—of layered cortical columns and their mishmash of connections—only bolster the impression that although a good deal of brain activity seems computational, the organ has evolved a form that is more historically contingent on the Precambrian era than the Industrial Revolution. Brains emerged for survival on a merciless planet, not for designing steam engines and solving scheduling problems. So indeed it is easy to write off the engineering potential of many neuro-inspirations.

However, when viewed as multiple modular extensions of a segmented body by the duplication and differentiation of the same Hox complex that governs so much of development, brain anatomy appears less daunting and more generalizable to a wide range of problems requiring hierarchical control. When the *grow-to-fit* search processes forming networks of muscles, vessels, axons, and dendrites are fully exposed on the drawing board, the coevolution of controller and body seems more inevitable than enigmatic. This awareness—that evolution eventually crafted an omnipotent strategy that can, with only small tweaks, generate an exceptionally diverse menagerie of morphologies and regulators—is one of the most valuable sources of bio-inspiration that mammalian evolution has to offer. One search process, evolution, configured the primitives for another search process, development, which produces phenotypes capable of even more (mental and physical) search.

So aside from the Darwinian trilogy of inheritance, variation, and selection (and their roles in design search), and the well-documented machine learning contributions of large networks of simple neuronlike processors, the key motivation for AI researchers to study brains and their evolution is the general system properties that help produce useful variation: modularity, duplication and differentiation, weak linkage, and exploratory growth. The details of their realization may vary across application domains, but the general concepts deserve careful consideration when equipping a search algorithm with tools for discovering innovative forms of intelligence.

10 Learning via Synaptic Tuning

10.1 A Good Hebb Start

Neural networks are often viewed as functions or mappings from activation states of input neurons to those of output neurons, with the intelligence behind this transformation distributed over the myriad connection weights. Comprehensive logical descriptions of these functions are often extremely difficult to formulate. However, the learning process through which these mappings emerge can follow very simple intuitive local rules for synaptic modification, many inspired by that classic principle of Hebb (1949): *fire together, wire together*.

These local rules appear to underlie a good deal of biological learning and are sufficient to produce sophisticated input-output relations in both natural and artificial neural networks. From their simplicity arises complex knowledge representations. This chapter investigates some of these local learning schemes and how they support various search processes that lead to the emergence of complex global network patterns and, in general, intelligent behavior.

First, however, we consider the most classic and widespread form of ANN adaptivity: supervised learning via the backpropagation algorithm. Though it embodies little emergence and even less biological realism, it employs a local search method that enables ANNs to learn a wide range of complex mappings, thus making thousands of practical contributions across numerous disciplines.

10.2 Supervised Learning in Artificial Neural Networks

Supervised learning scenarios arise reasonably seldom in real life. They are situations in which an agent (e.g., human) performs many similar tasks, or many similar steps of a longer task, and receives *frequent and detailed feedback*: after each step, the agent learns whether her action was correct and which action should have been performed. In real life this type of feedback is not only rare but annoying. Few people wish to have their every move corrected.

In machine learning (ML), however, this learning paradigm has generated considerable research interest for decades. The classic tasks involve classification, wherein the system is given a large data set, with each case consisting of many features plus its characterization or class. The system uses these examples of feature-to-class associations to construct a general-purpose mapping from feature vectors to hypothesized classes.

For example, the features might be meteorological factors such as temperature, humidity, and wind velocity, while the class could be the predicted precipitation level (high, medium, low, or zero) for the next day. The supervised learning classifier system must then learn to predict future precipitation when given a vector of meteorological features. The system receives the input features for a case and produces a precipitation prediction, whose accuracy can be assessed by a comparison to the known precipitation attached to that case. Any discrepancy between the two can then serve as an error term for modifying the system (so as to make better predictions in the future). Classification tasks are a staple of machine learning, and feedforward ANNs trained by backpropagation are one of several standard tools for tackling them. Since its conception in the mid 1970s (Werbos, 1974) and formalization and popularization a decade later (Rumelhart et al., 1986), backpropagation has accounted for a large percentage of all successful ANN applications.

The basic idea behind backpropagation learning is to gradually adjust the ANN's weights so as to reduce the error between the actual and desired outputs on a series of training cases. The presence of these desired outputs demarcates the task as *supervised*, while the strategy of searching in the direction of decreasing error falls under the category of *gradient descent* search methods.

Each case is typically a pair, (d_i, r_i), indicating an element of the mapping between a domain and a range for some (yet unknown) function. By training the ANN to reduce this error, one effectively discovers the function.

Figure 10.1 summarizes the basic process, wherein training cases are presented to the ANN one at a time. The domain value, d_i, of each case is encoded into activation values for the neurons of the input layer. These values are propagated through the network to the output layer, whose values are then decoded into a value of the range, r^*, which is compared to the desired range value, r_i. The difference between these two values constitutes the error term.

The challenge in gradient descent is to figure out which combination of weight changes (to the many hundreds or thousands of synapses) will actually reduce the output error. Intuitively, we know that if a positive change in a weight will increase (decrease) the error, then we want to decrease (increase) that weight. Mathematically, this means that we look at the derivative of the error with respect to each

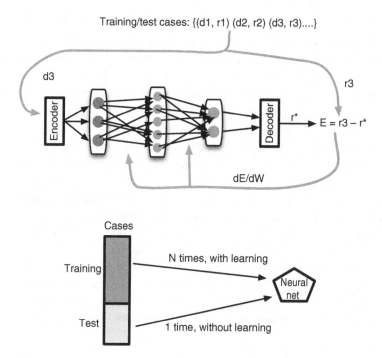

Figure 10.1
Backpropagation algorithm. (*Top*) During training, cases are encoded and sent through the network. The resulting outputs are compared to desired outputs to compute an error term, E, which is then used to compute changes to the weights of the network so as to reduce E. (*Bottom*) Data is initially partitioned into training and test sets; the former passes through the network many times, inducing weight change each time; the test data passes through just once, without learning, to assess the networks general ability to handle data on which it has not been trained.

weight: $\frac{\partial E}{\partial w_{ij}}$, which represents the change in the error given a unit change in the connection weight from neuron j to neuron i.

Once we find this derivative, we update w_{ij} via the following equation:

$$\Delta w_{ij} = -\eta \frac{\partial E_i}{\partial w_{ij}}. \tag{10.1}$$

This essentially represents the distance times the direction of change. The distance, η, is a standard parameter in neural networks, often called the *learning rate*. In more advanced algorithms this rate may gradually decrease as training progresses. This equation is easy and captures a basic intuition: decrease (increase) a weight that positively (negatively) contributes to the error. Although the calculations for $\frac{\partial E}{\partial w_{ij}}$ are beyond the scope of this book, they are available in most neural network textbooks (Haykin, 1999; Arbib, 2003).

If one updates all the weights using the same formula, this amounts to moving in the direction of steepest descent along the error surface, as shown in figure 10.2. This constitutes a form of local search, since the entire weight vector is employed to compute output values, the basis for evaluation. However, unlike many of the local search methods discussed earlier, backpropagation involves a very intelligent modification of each weight, based on the detailed calculation of $\frac{\partial E_i}{\partial w_{ij}}$, a process involving information from all over the network and thus not local. Furthermore, since the desired outputs and the errors that they spawn are easily interpreted as global directives as to the behavior of the system, the training of ANNs via backpropagation hardly qualifies as emergent.

Thus, one of the biggest success stories in Bio-AI, the backpropagation algorithm, is only loosely inspired by neuroscience and the emergent nature of intelligence. Still, it performs a sophisticated search in a space of complex mappings, often yielding solutions when other techniques fail. This earns respect not only from engineers and other technical problem solvers but also from brain researchers, who may see no scientific value in the learning algorithm itself but do appreciate the final result: a proof-of-principle neural network that can perform a particular mapping from sensory inputs to motor outputs. The existence of

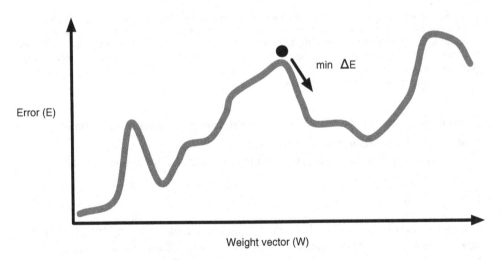

Figure 10.2
Gradient descent as search in an error landscape. The choice is among alternative changes to the ANN's complete weight vector. The change that gives the maximum decrease (arrow) from the current location (dot) is preferred. Note that in this case the maximum decrease, a *greedy* choice to move to the right, does not lead in the direction of the global error minimum, which resides at the far left.

such a network can help justify a neuroscientist's claim that, indeed, a collection of neurons (with particular connection strengths between them) suffices to perform a particular mental function.

Backpropagation is not without its shortcomings, which include a tendency to get stuck in local error minima and a common need for extremely many repeat presentations of an entire training set, both of which can be argued to have some precedence in the brain but rarely with the same negative consequences as in ANNs. Hence, an ongoing challenge in ANN research is to find other sources of neuro-inspiration that can translate more directly into problem-solving success. Many such attempts begin at the level of the synapse, and its modification by predominantly local Hebbian means.

10.3 Hebbian Learning Models

The key insight of Hebb (1949), that stronger synaptic bonds form between neurons that fire together, underlies numerous learning rules for artificial neural networks, a few of which are discussed here. These rules are typically *local* in that synaptic change hinges solely on the behaviors of the presynaptic and postsynaptic neurons, not on any global error signals (as in classic supervised ANNs). This locality is appealing from a biological angle, since the brain appears to employ Hebbian learning, and from an ALife perspective, wherein complex system functionality *emerges* from the myriad local interactions of simple components in the complete absence of a global controller.

Unfortunately, the naive application of local rules across repeated ANN learning rounds can quickly lead to explosive weight increases and decreases, thus driving networks to states of extreme instability or deeply entrenched stagnation. Only by invoking additional mechanisms, which typically lack biological plausibility or strict locality, can the modeler rein in these networks, forcing them into behavioral regimes exhibiting an ample mix of stability and adaptivity.

In the following discussion u_i and v denote activity of a presynaptic and postsynaptic neuron, respectively. They typically represent either the last activation level or the difference between the current firing frequency of the neuron and its time-averaged firing rate. The term *neural output* is used as a general reference to u_i or v, without any assumption about the exact nature of the physical variable.

Also, the term *long-term potentiation* (LTP) refers to a strengthening of a weight, while *long-term depression* (LTD) denotes a weakening. The long-term aspect of each change stems from the biological uses of LTP and LTD, where this type of synaptic change persists for hours, days, or longer. In ANNs there is no such guarantee of the duration of the change.

The classic Hebbian learning rule is simply

$$\triangle w_i = \lambda u_i v, \tag{10.2}$$

where λ is a positive real number (often less than 1) representing the learning rate. Equation (10.2) captures the basic proportionality between weight change and the correlation between the two neural outputs, as illustrated in figure 10.3.

When u_i and v represent absolute or fractional firing rates, as they often do, they are never negative. Thus, the right-hand side of equation (10.2) is always positive, and weights increase without bound. In fact, a positive feedback occurs, since positive firing rates produce weight increases, which insure that the influence of u_i upon v gets stronger, which tends to raise v's firing rate, which then elevates the weight, and so on.

One can simply put an upper bound on weight values, but then all weights eventually reach this limit and the ANN has no diversity and hence no interesting information content.

Another attempt to relieve the problem is to view u_i and v either as membrane potentials or as the differences between current and average firing rates. Both of these interpretations allow u_i and v to be positive or negative. Thus, $u_i v$ will often be negative, and weights should decrease as well as increase.

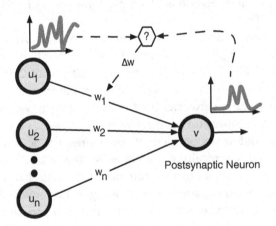

Presynaptic Neurons

Figure 10.3
Learning in neural networks. The comparison of presynaptic and postsynaptic firing histories determines changes to the connection weight between the two neurons. Here, weights have a single subscript, denoting the presynaptic neuron. Graphs show the neuron's membrane potential as a function of time.

The latter interpretation has spawned several useful learning rules in which presynaptic or postsynaptic (or both) firing rates are relativized to a normal level. For example, the weight-updating expression of equation (10.3), known as a *homosynaptic* rule, normalizes only the postsynaptic output to a threshold, θ_v:

$$\triangle w_i = \lambda(v - \theta_v)u_i. \tag{10.3}$$

Notice that this rule can exhibit long-term potentiation LTP or LTD depending upon whether v is above or below θ_v, which typically represents an average firing rate (computed and updated over the course of a simulation). The term *homosynaptic* refers to the fact that if v is above (below) threshold, then *all* presynaptic neurons that have a nonzero firing rate will experience LTP (LTD) along their connection to v. That is, all presynaptic neurons that fire will see the same type of change.

In contrast, equation (10.4) constitutes a *heterosynaptic* rule, since when $v > 0$, only those presynaptic neurons that fire above threshold θ_i will experience LTP, while the others—even those that do not fire at all—will experience LTD:

$$\triangle w_i = \lambda v(u_i - \theta_i). \tag{10.4}$$

A popular learning scheme that only normalizes the postsynaptic neuron but requires both pre- and postsynaptic firing is the BCM (Bienenstock, Cooper, and Munro) rule (Bienenstock et al., 1982):

$$\triangle w_i = \lambda u_i v(v - \theta_v). \tag{10.5}$$

Rules that normalize both pre- and postsynaptic neurons include

$$\triangle w_i = \lambda(v - \theta_v)(u_i - \theta_i) \tag{10.6}$$

and

$$\triangle w_i = \lambda v(v - \theta_v)u_i(u_i - \theta_i). \tag{10.7}$$

10.3.1 Weight Normalization

According to the pure Hebbian learning rule of equation (10.2), when u_i and v are non-negative numbers, the weights will increase without bound. One seemingly obvious fix is to use activation functions whose outputs can be negative, for example, the hyperbolic tangent function described earlier. Unfortunately, this is not enough to avoid positive feedback and the ensuing unstable increase in weights (See appendix B for details.)

To avoid problems of weight vector instability that occur with so many of the popular learning rules, more direct methods of weight restriction are available.

One of the most direct mechanisms is to simply normalize the weights associated with a particular neuron. Here, the options include

- *which* weights to normalize, for instance, weights on all incoming (or outgoing) links to (from) each neuron;
- *how* to normalize, whether by
    - dividing each weight by the sum of the weights or the sum of the absolute values of the weights (in cases where weights can be positive or negative) or
    - subtracting the same quantity from each weight.

Another common approach is to randomly initialize the weights and then, for each node, to sum either the incoming or outgoing connection weights. Assume that incoming weights are chosen, whose initial sum for node i is σ_i. Then, after each round of learning-based weight change, equation (10.8) would be used to renormalize each incoming weight such that the input sums for each node equal the originals.

$$w_{ij} \leftarrow \frac{\sigma_i w_{ij}}{\sum_{j=1}^{n} | w_{ij} |}. \tag{10.8}$$

Although computationally expensive, these methods do have the desired effect of preventing runaway weights.

A simpler and computationally cheaper method is the Oja rule (Oja, 1982), which includes a *forgetting* or *leakage* term that involves the weight itself:

$$\triangle w_i = \lambda v(u_i - v \mid w_i \mid) = \lambda u_i v - \lambda v^2 \mid w_i \mid. \tag{10.9}$$

Note that the first term in this rule is purely Hebbian, while the second term involves leakage that is proportional to both the magnitude of the current weight and the postsynaptic firing level. This second term combats the standard positive feedback problem with the earlier rules: when neurons fire at high rates, synapses tend to strengthen, which helps neurons to fire even harder in the future. This tendency for postsynaptic firing to increase weight leakage is relevant to spike-timing-dependent plasticity (STDP), discussed later in this chapter.

As detailed by Oja (1982), the Oja rule stems from a simplification of the Taylor series expansion of the following (standard) weight normalization scheme, wherein each new weight value is divided by the Euclidean sum of all new weights:

$$w_i \leftarrow \frac{w_i + \lambda v u_i}{\left(\sum_{i=1}^{n} [w_i + \lambda v u_i]^2 \right)^{0.5}}. \tag{10.10}$$

Thus, the beauty of the Oja rule is its ability to achieve the same result as weight normalization, but without actually performing that computationally expensive process after each round of learning. In short, weight normalization emerges from repeated applications of the local update rule.

The Oja rule has nice theoretical properties, including the ability to perform principal component analysis (PCA), though it lacks complete biological plausibility. However, its stability is easily proven (Dayan and Abbott, 2001), and thus it is a popular learning mechanism for ANNs designed to solve complex engineering problems by predominantly local learning mechanisms.

10.4 Unsupervised ANNs for AI

All the preceding mathematical formulations of Hebb's rule provide the basis for unsupervised learning: the tuning of synapses despite the lack of external feedback from an instructor or from the environment in general. The network itself must extract meaningful relations (and build useful higher-level structures) from the sensory input data. These may be invariant patterns in the data or key variants that allow the system to differentiate important classes of inputs. Either way, the system learns these concepts by itself.

In many cases, the result of unsupervised learning is a set of classes or clusters into which each previous input scenario falls. These facilitate generalization, since new input cases can be ushered into the proper cluster and then handled with the action associated with that group. This ability to generalize behavior is critical for the survival of living organisms and essential for the success of AI systems in complex environments (where all possible input scenarios cannot possibly be enumerated and planned for ahead of time).

Many ANNs used in AI research abstract away a good deal of the biological detail while retaining the essence of Hebbian learning. These models have legitimate utility for many of AI's unsupervised learning problems while providing further instances of search and emergence in a neural context. In many of these cases, the emergent global patterns stem from local competitive or cooperative interactions among artificial neurons. In general, the basic Hebbian notion of firing together and wiring together has a very cooperative connotation: neurons working in concert will tend to promote one another's activity. Or, neurons can be viewed as competitors in trying to excite a common postsynaptic neuron or to become the main postsynaptic detector of a presynaptic pattern. These mechanisms take center stage in the following ANNs.

10.4.1 Hopfield Networks

Hopfield networks (Hopfield, 1982) are a popular species of ANN that despite their simplicity capture two of the brain's key functions: storage and retrieval of distributed patterns. They do this in a completely unsupervised manner by recognizing correlations among neural firing histories and modifying weights to record those relations.

In the brain many forms of information appear to be distributed across large populations of neurons. As discussed in chapter 7, this distributed (or *population*) coding has several advantages:

- *Storage efficiency*   In theory, k neurons with m differentiable states can store m^k patterns.

- *Robustness*   If a few neurons die, each pattern may be slightly corrupted, but none is lost completely.

- *Pattern completion*   Given part of a pattern, the network can often *fill in* the rest.

- *Content-addressable memory*   Patterns are retrieved using portions of the pattern (not memory addresses) as keys.

Hopfield networks take advantage of population coding to store many patterns across a shared collection of nodes, with each node representing the same aspect of all stored patterns and each connection weight denoting the average correlation between two aspects across all stored patterns.

Figure 10.4 displays an auto-associative Hopfield network, where *auto* implies that the correlation is between aspects/components of the *same* pattern. In this case, the components are simply small regions of the image, with components a and b representing small regions centered at distinct locations of the image plane. An auto-associative network encodes the average correlations (across all stored patterns) between all pairs of components.

In figure 10.4 the correlation between components a and b is computed for each of the two images. It is positive in the upper left image, since a and b both contain some black color, but it is negative in the upper right image, since a contains black but b does not. Thus, the left image contributes +1 to the average correlation between a and b, while the right image contributes −1. If there are P patterns to store, then P such correlations will be averaged for every pair of components. This correlation analysis of all P patterns constitutes the *training phase* of the Hopfield algorithm. It is where the learning occurs.

Note that this training reflects Hebbian learning at a coarse level: when two components are highly correlated in a pattern, then the link between their nodes in the ANN will be strengthened. Alternatively, a negative correlation, i.e., one component is on (black) while the other is off (white), leads to a weight reduction.

The Hopfield network (a small portion of which appears in the middle of figure 10.4) has a *clique* topology, meaning that every node is connected to every other node. Each node represents a component, and each arc weight denotes the

Training

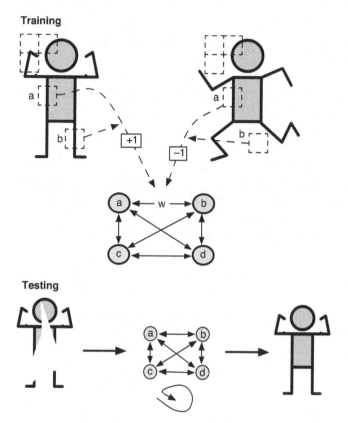

Testing

Figure 10.4
Procedures for an auto-associative Hopfield network. (*Top*) Training (learning) phase. (*Bottom*) Testing (pattern retrieval) phase.

average correlation between the components of the arc. In Hopfield networks the arcs are bidirectional, since a correlation is a symmetric property.

Once trained, the auto-associative Hopfield network can be employed to retrieve one of the P original patterns when given only a portion (or corrupted version) of it, as shown at the bottom of figure 10.4.

The retrieval process begins by *loading* the partial pattern into the network: components that are on in the partial pattern will have their corresponding ANN nodes set to a high activation level, while those that are off will engender low activation levels. The network is then run, with nodes summing their inputs and using their activation functions to compute new output levels. This continues until either the network reaches a quiescent state (i.e., no nodes are changing activation levels) or a fixed number of update steps have been performed. The final activation levels of the nodes are then mapped back to the image plane to create the output pattern. For example, if node f has a high activation level, then the f component of the image will be painted black.

Similar to an auto-associative net, a hetero-associative net records average correlations between *pairs* of patterns. These are useful for associating one pattern with its typical successor pattern in a sequence. Hence, when given the predecessor pattern, the network can *predict* the successor.

10.4.2   Basic Computations for Hopfield Networks

In Hopfield and many other associative networks, the learning phase is a one-shot batch process in which all patterns are analyzed and all correlations averaged. The weights of the network are then set to those averages and never modified.

A typical learning (i.e., weight assignment) scheme for Hopfield networks is

$$w_{jk} \leftarrow \frac{1}{P} \sum_{p=1}^{P} c_{pk} c_{pj}, \tag{10.11}$$

where P is the number of patterns, c_{pk} is the value of component k in pattern p, and $c_{pk} c_{pj}$ is the local correlation in pattern p between components k and j.

For a hetero-associative network, the corresponding scheme is

$$w_{jk} \leftarrow \frac{1}{P} \sum_{p=1}^{P} i_{pk} o_{pj}, \tag{10.12}$$

where P is now the number of pattern *pairs*, i_{pk} is the kth component of the predecessor (input) pattern of pair p, and o_{pj} is the jth component of the successor (output) pattern of pair p. The product of the two components is the local (k, j) correlation for pair p.

Once all weights have been computed, the net can be run by loading input patterns and updating activation levels. For discrete Hopfield networks, where firing levels are either +1 or −1, the following activation function is common:

$$c_k(t+1) \leftarrow sign(\sum_{j=1}^{C} w_{kj} c_j(t) + I_k), \tag{10.13}$$

where C is the number of components (for example, 64 in an 8-by-8 image plane), $c_k(t+1)$ is the activation level of the kth component's neuron at time $t+1$, w_{kj} is the weight on the arc from node j to node k, and I_k is the original input value for component k. The I_k term insures that the original bias imposed by the input pattern has an effect throughout the entire run of the network.[1]

Figure 10.5 illustrates the basic training and test procedure for Hopfield networks.

10.4.3 Search in Hopfield Networks

The process by which a running associative network gradually transitions to quiescence has a searchlike quality. In this case, a state is a vector consisting of the activation levels of each neuron in the ANN, and the biases that push it toward

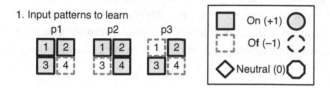

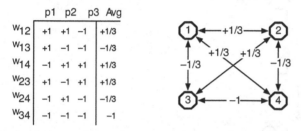

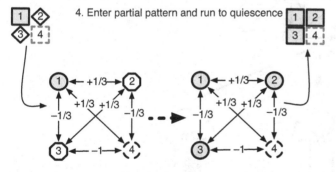

Figure 10.5

Procedure for encoding patterns in a Hopfield network and retrieving them via partial patterns. Steps 1–3 involve computing the average correlations between all pairs of components (i.e., pixels) in the input patterns and using them as weights in the network. Step 4 shows the encoding of a simple partial pattern. All nodes then update their activation levels *simultaneously* (synchronously), to attain the next state, which turns out to be stable: further updates will not change any activation levels. The final state corresponds to pattern p1.

equilibrium are the connection weights, whose values reflect the network's history
of pattern learning.

Unfortunately, stable states do not necessarily correspond to any of the original
P input patterns. They may be *spurious*, i.e., they map to patterns that were not part
of the training set.

Figure 10.6 illustrates this problem. The original input pattern (*top*) is loaded
into the auto-associative network, but nothing guarantees that the quiescent state
will map to the correct pattern instead of a spurious one.

Hopfield (1982) quantified the search for quiescence as a search for minima on an
energy landscape by defining a function for mapping ANN states to energy levels.
The general Hopfield learning procedure (equation (10.11)) and activation function
(equation (10.13)) then insure that low-energy states are those corresponding to

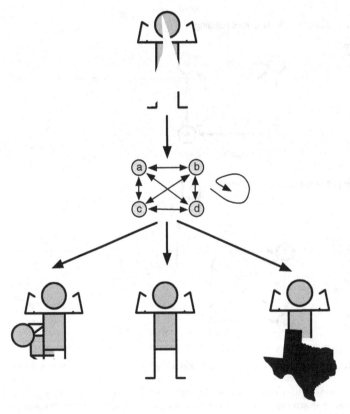

Figure 10.6
Correct (*bottom middle*) and spurious (*bottom left/right*) outputs of a Hopfield network when given a
corrupted input pattern (*top*)

patterns on which the net was trained. However, nothing prohibits some spurious patterns from achieving locally minimal energy as well. Here, a local minimum is defined as a state, M, such that any state M* that is created by updating all activation levels of M *once* has higher energy than M.

Hopfield's energy function is

$$E = -a \sum_{k=1}^{C} \sum_{j=1}^{C} w_{jk} c_j c_k - b \sum_{k=1}^{C} I_k c_k, \tag{10.14}$$

where c_k is the activation level of component k's neuron, I_k is the original input value loaded into the kth component's neuron, and a and b are positive constants.

Although complex in appearance, equation (10.14) is actually quite intuitive. Consider the term $w_{jk} c_j c_k$, and assume that $w_{jk} > 0$, meaning that the jth and kth components had, on average, a positive correlation in the training patterns. Next, consider two cases:

- sign(c_j) = sign(c_k) means that the jth and kth components are both on or both off. Hence, they are positively correlated in the current state of the ANN. This agrees with w_{jk}, which also indicates a positive correlation between components j and k, since $w_{jk} > 0$. Hence, there is no conflict, only agreement, between c_j, c_k, and w_{jk}. This is reflected in the fact that $w_{jk} c_j c_k > 0$. Since the summations in equation (10.14) are preceded by negative factors, each pair of activation values that agrees with the corresponding weight will contribute negatively to the total energy, where *lower* total energy means *more* local agreement.

- sign(c_j) $\neq$ sign(c_k) means that components j and k are negatively correlated in the current state. This disagrees with the positive weight between them and is reflected in the fact that $w_{jk} c_j c_k < 0$. Hence, this pair of components will contribute positively to the total energy.

There is a similar set of cases when $w_{jk} < 0$; in those, agreement is signaled when sign(c_j) $\neq$ sign(c_k).

Given Hopfield's energy function, one can now sketch an *energy landscape* for an associative network, where network activation states map to energy levels. Figure 10.7 illustrates a possible landscape for the hypothetical situation of figure 10.6. Note that the two spurious patterns occupy local minima and are thus *deceptive* quiescent states for Hopfield search.

Just as Simon's ant runs along trajectories determined by the beach's terrain, and Simon's expert draws conclusions based on the history of her experience, Hopfield's ANN runs downhill along a landscape sculpted by the network's history of learned patterns.

The emergent nature of Hopfield networks is certainly debatable. Though only neuron-neuron interactions govern the emergence of global patterns, the clique

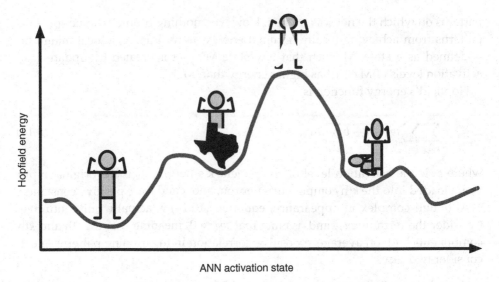

Figure 10.7
Hopfield energy landscape in which the initial corrupted pattern from figure 10.6 occupies a high-energy point, and the correct pattern resides at the global minimum. Spurious activation states correspond to local minima in the landscape, minima to which the Hopfield network could easily quiesce.

topology of the network seems to stretch the definition of *local*. Still, no single group of neurons runs the show; the combination of all low-level activation levels and weights determines the final outcome. Furthermore, it is important to note that the repeated application of the local update rule (equation (10.13)) leads to the gradual emergence of the low-energy global state defined by equation (10.14).

10.4.4   Hopfield Search in the Brain
The behavior of real brains is believed to exhibit many of the same properties as Hopfield networks:

- The strength of synaptic connections between two neurons (or populations of neurons) often reflects the degree to which the receptive fields of those neurons are correlated, where the *receptive field* of a neuron is that part of the sensory space (e.g., a small region in the upper left quadrant of the visual field) for which the neuron appears to be a detector.

- Certain activity patterns appear to be *stable attractors*, i.e., quiescent states, to which real neural networks eventually transition. These attractors may represent salient concepts in the brain.

Consider the duck rabbit flip-flop picture of figure 10.8. Most people cannot view this as both a rabbit and a duck simultaneously; rather, the interpretations seem to

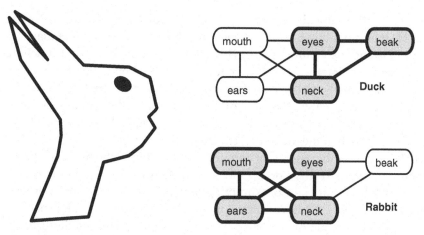

Figure 10.8
(*Left*) Flip-flop figure that resembles both a duck and a rabbit, depending upon perspective. (*Right*) Two copies of the same neural network, with nodes coding for properties of the duck-rabbit figure. Thick lines and shading indicate activity of neurons and synapses. This is inspired by a similar example, of the Necker cube, in a popular book by Pinker (1997).

alternate, particularly if one stares at the picture for a long time. This is evidence that both interpretations represent stable attractors in memory but that there is some overlap between the neural states corresponding to each, as shown by the abstract neural networks on the right of the figure.

The components of each attractor stimulate one another, as shown by the thick links in figure 10.8, in much the same way that highly correlated nodes in a Hopfield network excite one another because of their high connection weights. However, with overlap, when one attractor is active, some of its components, such as the "eyes" and "neck" nodes, also stimulate portions of competing attractors.

In addition, neurons are known to *habituate* to stimuli, meaning that they tend to reduce their AP production in the presence of a continuous stimulus. To see this, try staring at something for a few minutes and feel how hard it is to keep your thoughts focused on that particular item; the mind tends to wander.

Hence, a stable attractor such as the duck pattern will habituate while simultaneously lending some stimulation to the "mouth" and "ears" nodes. Eventually, the balance of firing power shifts and "rabbit" becomes the main interpretation, until it, too, habituates and the duck returns.

10.4.5   Competitive Networks
In certain types of ANNs the nodes compete for activity such that the most active nodes can both inhibit other nodes from firing and localize learning to only their

own connections, typically the incoming (afferent) links. In the brain, topologies in which neurons inhibit many of their neighbors (whether immediate or
slightly more distant) are commonplace. This often serves the important function
of filtering noise, such that the final stable pattern consists of only the neurons that
detect meaningful signals. Other competitions lead to useful structural isomor-
phisms between aspects of the physical world and regions of the brain specialized
to handle those properties. These *topological maps* are beautiful examples of how
the brain encodes much of the inherent structure of the physical world.

Figure 10.9 illustrates the essence of competitive learning in the brain. In many
brain regions, particularly the cortex, the main (often excitatory) neurons are
known as *principal cells*. In close proximity are one or several types of interneurons, which are typically inhibitory. Active principal cells tend to stimulate nearby
interneurons, which then inhibit all their postsynaptic targets. This manifests feedback inhibition, wherein the activation of one (or a few) neurons in a layer quickly
leads to the inhibition of the rest of the layer. Thus, the principal cells can be viewed
as competing, and the neuron that is most active simultaneously with a subset of
input neurons can become a detector for that input pattern.

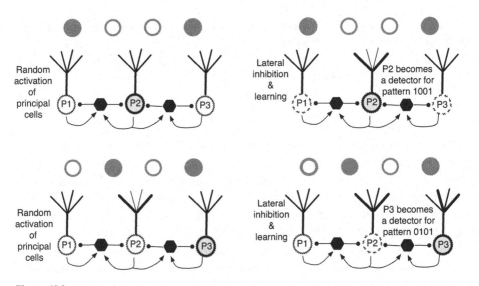

Figure 10.9
Competitive learning in a layer of neurons. (*Top left*) Principal cells fire randomly, and neuron P2 fires
approximately coincident to the first and fourth input neurons. (*Top right*) P2 inhibits its neighbor neurons, thus insuring that very few neurons in its layer fire together. The synapses between the active
inputs (shaded circles) and P2 strengthen by Hebbian learning (thick line segments). P2 thus becomes
a detector for input pattern 1001. (*Bottom left/right*) A similar process occurs with P3, which becomes a
detector for 0101.

In practical applications of competitive ANNs, the focus is on the weights of the input arcs to each output node, n_i. The vector of input weights to n_i, $\langle w_i \rangle = \langle w_{i1}, w_{i1}, \ldots, w_{im} \rangle$, typically represents a prototype of the patterns that n_i is (or has learned to become) specialized to detect. In this sense, each output node represents a class or category, and input patterns can be clustered according to the output node that they maximally stimulate.

Figure 10.10 illustrates a standard topology for artificial competitive networks, with one input and one output layer. The output nodes represent categories/classes that essentially compete to capture the different input vectors, with each such vector falling into the category whose prototype it most closely matches.

The generic competitive ANN algorithm is quite simple. Patterns are repeatedly presented to the input layer and activations recorded on the output layer. The output node with the highest activation on pattern P *wins* and has its input weights adjusted so that its prototype more closely resembles P. The update formula for the weights into winning node n_i on input pattern P is then

$$w_{ij} \leftarrow w_{ij} + \eta(P_j - w_{ij}), \tag{10.15}$$

where P_j is the *j*th value of pattern *P*, i.e., the value loaded onto input neuron *j*. After many epochs (i.e., presentations of the entire training set) the cases often become clearly segregated into classes, with the prototype vector of each class

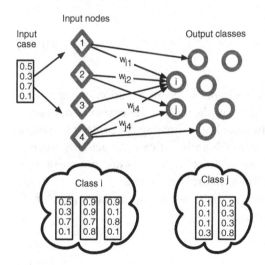

Figure 10.10
Topology of a competitive network, with an input layer and an output layer. Each output node represents a class, whose prototype is given by the input weights to that node. Through learning, input cases become associated with different output nodes and thereby become clustered into instance sets (clouds) of the classes represented by those nodes.

located close to the middle of the subspace delineated by its cases. In this way, competitive networks function as clustering mechanisms for complex data sets.

In practice, many competitive networks are not really neural networks at all. They are simply lists of prototype vectors that are matched to case vectors. The vector with the closest match, using a standard Euclidian distance metric for n-dimensional space, is updated in the direction of the case.

However, one can capture these same dynamics in an actual neural network that involves a few key components:

- Many inhibitory links between all output units
- Excitatory links between output units and themselves
- Normalized case and prototype vectors

The first two features constitute a *Maxnet*, which insures that when all output nodes are activated, they will compete to shut each other off while stimulating themselves. After a transitory period the network settles into a state where the only neuron with a nonzero activity level is the one that originally had the highest activation level.

Maxnets are easy to implement, but care must be taken to properly set the (fixed) weights on the inhibitory and self-stimulating arcs, ϵ and θ, respectively. For example, if the network consists of m output neurons, then the following settings work well:

$$\theta = 1, \ \epsilon \leq \frac{1}{m}. \tag{10.16}$$

Weight Normalization in Competitive Networks   Normalization of input and prototype weight vectors (though slightly complicated) is the key to understanding how equation (10.15) achieves its purpose: how adjusting the prototype weight vector of output node n to more closely match an input case, C, actually insures that n fires harder on the next presentation of C. This serves as one more example of the importance of weight normalization in neural networks.

First consider the basic Euclidean distance between an input pattern vector, P, and the weight vector for output neuron i, $\langle w_i \rangle$:

$$\sqrt{\sum_{j=1}^{n}(P_j - w_{ij})^2} = \sqrt{\sum_{j=1}^{n} P_j^2 - 2P_j w_{ij} + w_{ij}^2}. \tag{10.17}$$

Now, if input vectors and prototype weight vectors are in normalized form, then their unit length is 1. Hence,

$$\sum_{j=1}^{n} P_j^2 = 1 = \sum_{j=1}^{n} w_{ij}^2. \tag{10.18}$$

Combining equations (10.17) and (10.18),

$$\sqrt{\sum_{j=1}^{n}(P_j - w_{ij})^2} = \sqrt{\sum_{j=1}^{n} P_j^2 - 2P_j w_{ij} + w_{ij}^2} = \sqrt{2 - 2\sum_{j=1}^{n} P_j w_{ij}}. \tag{10.19}$$

Thus, to minimize the distance between P and $\langle w_i \rangle$, one should maximize $\sum_{j=1}^{n} P_j w_{ij}$, which is just the sum of weighted inputs to neuron i. Hence, the prototype vector with the best match to the input case (i.e., that which is closest to it in Euclidean space) will have the highest sum of weighted inputs and thus will have the highest activation level.[2]

Another way of looking at this is that each normalized vector represents a unit vector in n-dimensional space. The term $\sum_{j=1}^{n} P_j w_{ij}$ is the dot product of these vectors, and with unit vectors the dot product is equal to the cosine of the angle, ϕ, between the vectors. Then, $\cos \phi$ is large, i.e., approaches 1, if and only if ϕ approaches 0, i.e., the vectors are similar. Hence, a good match between the normalized input case and prototype is indicated by a large dot product: $\sum_{j=1}^{n} P_j w_{ij}$.

In summary, if the application allows a normalization of input patterns and weight vectors, and if computational resources permit output neurons to participate in Maxnet competitions to determine the largest activation level, then competitive networks can be implemented as true ANNs that learn via equation (10.15) and allow a population of dedicated pattern detectors to arise from local interactions.

10.4.6   Self-Organizing Maps

The neural networks that are perhaps most synonymous with emergence are self-organizing maps (SOMs), or *Kohonen nets* (Kohonen, 2001) . These employ a dynamic mixture of competition and cooperation to produce topological mappings between two spaces, e.g., a perceptual field and a neural layer, two neural layers, or a neural layer and an actuator field.

In many competitive maps the spatial relations between the output neurons have no significance, and thus it makes no sense to discuss the *neighbors* of a neuron. However, these relations have meaning inside the brain, since the firing of a neuron in a region of the brain will often have consequences for nearby neurons.

When neurons are viewed as detectors of various phenomena, whether visual, olfactory, auditory, or tactile, it is very often the case that nearby neurons in the

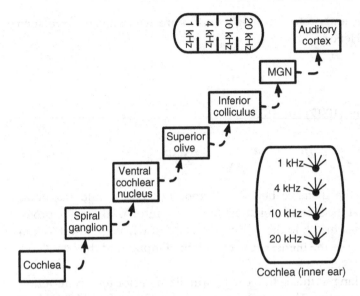

Figure 10.11
Tonotopic maps. Topological mapping between the space of sound frequencies and seven successive layers of the auditory processing system in the mammalian brain. The correlation between frequencies and neuron locations is preserved through all seven layers. Based on Bear et al. (2001, 368–373) and Kandel et al. (2000, 1127).

brain serve as detectors for similar stimuli. These local similarities of preferred stimuli are abundant in the brain. Figure 10.11 shows one impressive example from the auditory cortex, where a long chain of processing layers maintains a correlation between neural neighborhoods and sound frequencies.

The essence of a topological map is the isomorphism between two spaces, at least one of which is a population of neurons. The two spaces are isomorphic when components that are close (distant) in one space map to components that are close (distant) in the other space. This basic idea is shown in figure 10.12, where the top mapping is not isomorphic, but the bottom one is.

SOMs operate similarly to standard competitive networks in that a case vector is presented, and the best-matching prototype is modified in the direction of the case. However, Kohonen added an interesting twist: neurons have a meaningful spatial relation to other neurons, and when an output neuron *wins* a case, it *shares the prize* with its neighbors in that both winner and neighbors adjust their prototypes toward the case. This neighborhood of sharing changes during the training phase, typically beginning with a large radius that decreases as training progresses and prototypes specialize toward particular subsets of the case patterns. Over time, this sharing results in an isomorphic mapping in which the prototypes of nearby

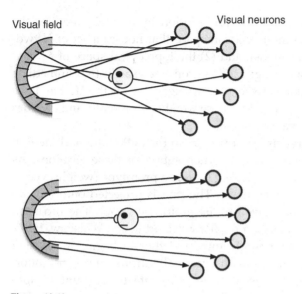

Visual field   Visual neurons

Figure 10.12
(*Top*) Typical initial situation for the application of a self-organizing map (SOM), with poor correlation between the two spaces: the visual field and the collection of neurons. (*Bottom*) A well-correlated
(i.e., isomorphic) mapping between the two spaces.

(in neuron space) neurons are more similar to one another than to those of distant
neurons. In addition, SOMs typically have a dynamic learning rate (η in equation
(10.15)) that begins large and decreases during the run.

The net result is a convergence of the Euclidean and topological neighborhoods
of the neurons in the SOM (see figure 10.12). The significance of this convergence
is multifaceted, including physical factors such as reduced total *wiring* between
brain layers, measured in terms of the lengths of axons and dendrites.

In terms of brain function, a key advantage of topological maps is that if situation A_1 maps to neuron (or more likely neuron population) N_1, presumably via a
learning process, then situations similar to A_1, such as A_2 and A_3, would map to
neighbors of N_1, which would elicit similar behavior. Thus, the topological map
allows the brain to reuse and generalize its behaviors to situations similar to those
that it has explicitly learned. The neighbors of N_1 may not share all of its functionality, which would seem appropriate, since A_2 and A_3 may require similar but not
exactly the same actions as N_1 supports.

SOMs have been applied to a wide range of clustering and unsupervised learning problems. One of the more creative of these (Durbin and Willshaw, 1987)
involves the traveling salesman problem (TSP), in which the goal is to find the
shortest circuit that includes every point (e.g., city) in a particular collection

(Garey and Johnson, 1979). Framing TSP for solution by an SOM requires the formalization of two spaces: (1) the Euclidean space that houses a set of N two-dimensional points (e.g., city locations), and (2) the topological space of M neurons, which is a one-dimensional ring: each neuron has exactly two immediate neighbor neurons. The connection between the two spaces is via the afferent links of the neurons; each neuron has two links whose weights represent x and y values of the neuron's prototype city location.

The N city locations then serve as input vectors to the SOM, and each neuron competes to be the winning detector of a certain subset of those locations. As shown in figure 10.13, each neuron of the ring houses a prototype (weight) vector that maps to a point on the Cartesian plane. The ring is projected onto the city space, with each neuron's location given by its prototype vector. The projected neuron ring then serves as scaffolding for a TSP tour (Durbin and Willshaw, 1987). When these prototype vectors are randomly initialized, the ring begins as a twisted mess, but after many rounds of training, with a gradually shrinking ring neighborhood (of weight-update sharing) and declining learning rate, the ring unfolds into a circuit (bottom right of the figure), providing the backbone of an optimal TSP tour. From competition and cooperation emerges a global solution.

10.5   Spikes and Plasticity

The earlier mathematical models of Hebbian plasticity assume that the presynaptic and postsynaptic outputs represent firing *rates*, which are sufficient for many ANNs (such as Hopfield and Kohonen nets) and their application scenarios, wherein neural outputs typically represent general activity levels, not actual action potentials, i.e., *spikes*. However, in the brain, the basic Hebbian model also holds at the level of spike trains. At the firing rate level, the model indicates that neurons with simultaneously high activity tend to develop strong synaptic interconnections, i.e., the model explains correlation-based learning. But dropping one level to that of the spike times of individual neurons indicates how the brain may learn causal relations among stimuli and thus become adept at predicting future from present stimuli.

10.5.1   Prediction

Several prominent neuroscientists have championed the ability to predict as a fundamental function of all brains, from insects to humans (Llinás, 2001; Hawkins, 2004). The basic argument for prediction's unique importance begins with sensorimotor behavior, where the processing speed of sensory inputs simply cannot keep pace with the rate at which the coupled agent-environment system changes in many (often intense or life-threatening) situations. In calculating the next action,

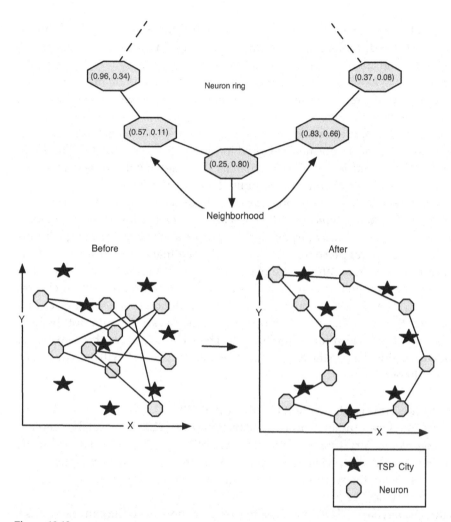

Figure 10.13
Use of a self-organizing map to solve the traveling salesman problem (TSP) using the method of Durbin and Willshaw (1987). (*Top*) The neuron topology is a ring, with each neuron's weight vector denoting a location in two-dimensional space. (*Bottom*) City locations (stars) and neurons (octagons) shown on the Cartesian plane, with coordinates corresponding to the locations of the cities on a scaled map and to the weight vectors of each neuron.

the nervous system does not have time to interpret all sensory data. Instead, it uses predictions of future states to guide action, and only when predictions conflict with reality must the system realign its expectations with the sensory present. Neuroscientists (D.M. Wolpert et al., 1998; Llinás, 2001) generally agree that the brain needs

predictive abilities similar to those found in control theory (such as Kalman filters) to handle the different time scales involved in sensing, moving, subconscious
action-choice, and conscious decision making.

As described in Downing (2007a), based on a similar distinction made by Squire
and Zola (1996), a prediction involving an explicit expression of knowledge is
termed *declarative*, whereas one in which no such expression occurs but the agent's
behavior clearly indicates anticipation of a particular situation is *procedural*. The
former tends to involve information stored in the cortex (Hawkins, 2004) and hippocampus (Gluck and Myers, 2000), whereas the latter involves the cerebellum
(D.M. Wolpert et al., 1998) and basal ganglia (Houk et al., 1995).

Assume the dictionary definition of *predict*: "to declare or indicate in advance"
(Merriam, 2003). Then, a typical predictive-learning scenario involves an organism
that witnesses event A *followed by* event B. It learns this correlation, along with the
all-important temporal precedence of A over B, such that in the future, when the
organism observes A, it can declare (e.g., via language) or indicate (via various
anticipatory actions) that B will soon occur. Thus, if a neuron (or neural cluster)
represents A, its activation should lead to the stimulation of a neuron or cluster
representing B; but the linkage need not (and in many cases should not) be bidirectional: B's cluster need not stimulate A's. The trick is to learn this temporally
dependent relation despite the fact that the neurons for A and B will both be highly
active within overlapping time windows.

10.5.2 Spike-Timing-Dependent Plasticity (STDP)

Several neurological studies, summarized by Song et al., (2000), reveal the presence
of two complementary tuning curves for synaptic strength, both based directly
on Δt, the time of the presynaptic spike minus that of the postsynaptic spike.
Figure 10.14 shows the relation of Δt to synaptic potentiation (when $\Delta t \leq 0$) and
to depression (when $\Delta t > 0$).

On the surface, basic STDP would seem to explain how the brain can learn causal
relations: if neuron A fires just before neuron B, then the connection from A to B
is potentiated, and future firing of A should stimulate B, thus predicting the event
that B represents. Also, since B fires after A, the $B \to A$ link would actually weaken,
thus reducing the chance that the future occurrence of B's event would predict A's
event. However, neurons rarely fire single action potentials but many per second,
often so many that the 40 ms windows overlap to such a degree that there is no
obvious pairing of spikes with which to assess before and after relations.

For example, in figure 10.15, each neuron fires eight times within a 50 ms time
frame, with neuron A's three pulses (labeled U) leading off the frame. Since these
spikes all come prior to neuron C's first triple (labeled X), neuron A would seem
to be the predecessor (at least within this one time frame). However, by the end of

sequence Y, neuron C has fired more times (six) than neuron A (five). To further confuse the issue, the pairing of pre- and postsynaptic pulses quickly becomes nontrivial. For example, is the second spike of X a successor to the second (or third) spike of U, or is it a predecessor to the first spike of sequence V? Maybe it should be treated as both? Maybe all relations between all pre- and postsynaptic spikes (within a 40 ms window of one another) should factor into the calculation of the synaptic change, $\triangle s$? In fact, many methods for comparing spike trains (see van Rossum (2001) for several examples) consider all such pairings.

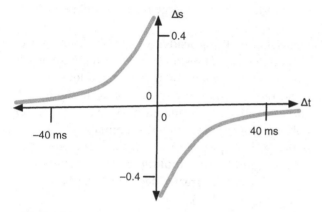

Figure 10.14
Spike-timing-dependent plasticity (STDP) involves the dependence of the change in synaptic strength ($\triangle s$) upon the timing difference between pre- and postsynaptic spikes ($\triangle t$). The time window for these dynamics is roughly 40 ms, and the maximum magnitude of change is roughly 0.4 percent of the maximum possible synaptic strength/conductance (Song et al., 2000).

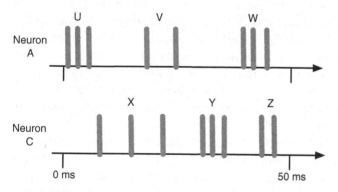

Figure 10.15
Two hypothetical neurons firing at 160 Hz, or eight spikes per 50 ms time window

But how might STDP arise from a pair of neighboring neurons with complex spike trains? Surely the brain does not explicitly use each pairing of pre- and postsynaptic spikes to govern synaptic modification, as do many computational neuroscience algorithms.

One intriguing possible explanation arises from the STDP model of Song et al. (2000). As depicted in figure 10.16, the pre- and postsynaptic spikes in this model each lead to the modification of various factors (P_a, P_b, and M), and these affect synaptic conductances (w_a, and w_b), but at no point are spike times compared (to produce a $\triangle t$). Still, the algorithm yields STDP as in figure 10.14; furthermore, it achieves weight stabilization without explicit normalization of conductance values. In short, STDP and weight stability emerge from a small set of local interactions.

In this procedure, note that presynaptic firing leads to increases in the corresponding memory terms (P_a or P_b). If neuron A fires several times in quick succession, as shown in figure 10.15, then P_a increases without suffering a lot of decay. If postsynaptic neuron C fires shortly thereafter, the update of w_a is significant, whereas a longer delay (i.e., a more negative $\triangle t$, (as in figure 10.14) entails greater decay of P_a and thus a less sizable increment to w_a when C eventually does fire.

On the other hand, if C fires many times just prior to a spike from neuron A, then C's activity raises M. If A fires right after this buildup of M, w_a experiences a large decrease, whereas a longer delay entails greater decay of M and a smaller decrease in w_a (similar to the far right figure 10.14).

In both these scenarios the memory variables and their effects upon the weights produce the fundamental STDP relations of figure 10.14 but without the explicit comparison of numerous pre- and postsynaptic spike times. This exhibits an interesting form of emergence in that LTP and LTD constitute *temporally* global states (i.e., those of long duration) that arise from interactions that are spatially and *temporally* local. In figure 10.16 note that none of the update equations involve time stamps: variables change based on the current values of other variables, not on the details of their histories. Although P_a, P_b, and M give lumped indications of previous spiking activity, they are not detailed memories; their interactions with other variables do not require comparisons of a temporally nonlocal nature (as when comparing spikes across large time windows). All weight modifications stem from the current values of a few leaky integrative memories—mechanisms that reappear in several guises throughout this book.

In addition to the classic STDP produced on individual synapses, this model also leads to an emergent competition among synapses for conductance strength. Those presynaptic neurons whose action potentials consistently precede postsynaptic spikes are rewarded with potentiation, while those whose spikes tend to come after the downstream action potential become depressed. Furthermore, randomly

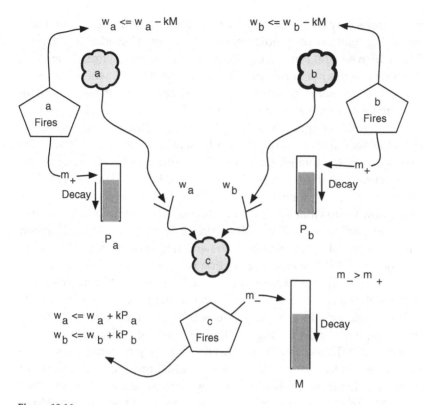

Figure 10.16
Interactions in the algorithm of Song et al. (2000) depicted for a circuit of two presynaptic neurons (a and b) and one postsynaptic neuron (c). The key presynaptic *memory* parameters, P_a and P_b, are incremented (by m_+) when neurons a and b fire, respectively, while the memory for postsynaptic firing (M), increases (by m_-) when c fires; typically, $m_- > m_+$. All three memories experience continuous decay. The parameter k represents the maximum attainable synaptic weight/conductance. Presynaptic firing initiates decreases in the corresponding weights (w_a and w_b), while postsynaptic firing raises the weights. The differentials in replenishment and decay between the presynaptic and postsynaptic memories result in the ultimate increase or decrease of each synaptic weight in a manner that mirrors the relations in figure 10.14.

firing afferents also yield synaptic depression because of a critical element of the model: $m_- > m_+$. Since a randomly firing neuron's spikes will precede and succeed a postsynaptic spike with equal probability, its net synaptic change would be negligible if m_- were to equal m_+. Thus, the model insures that only temporal precedence of presynaptic spikes can produce potentiation, while all else leads to depression; the weight vector of the postsynaptic neuron becomes tuned to the predictive presynaptic neurons and ignores all others. Thus, the tuning of synapses to capture salient afferents is another emergent aspect of neural learning.

Presynaptic neurons may cooperate as well as compete in their implicit attempts to control the firing pattern of a postsynaptic neighbor. Consider a situation (figure 10.17) in which several presynaptic neurons in a group (G) fire at the same time, t, while a lone presynaptic competitor (C) fires at time $t + 2d$, where d is just a few milliseconds. Furthermore, assume that these combined G+C inputs stimulate the postsynaptic neuron (P) to fire at time $t + 3d$. STDP entails that the C-P synapse will strengthen to a greater degree than the G-P synapses. But all weights should increase, as will the net effect of G and C upon P. Hence, in the future, P may fire earlier than $t + 3d$. In fact, although the G-P weights may not be as strong as the C-P weight, the cumulative effect of the G neurons could be sufficient to excite P to threshold before time $t + 2d$, at, for example, time $t + d$. This modification to the relative firing order now means that the G-P weights should continue to increase, but the C-P weight will decline, since C now fires after P. In short, by firing together, the G neurons seize control of P, while the solo neuron C, though initially spiking in a perfect position for LTP, loses the competition with G for influence upon P.

As emphasized by Song et al. (2000), their STDP model manifests emergent weight normalization via two distinct firing modes of the postsynaptic neuron, P. In one mode P receives considerable input—its activation function is saturated—and thus fires in response to the average excitation from the entire population of presynaptic neurons, G*. This normally entails a regular firing pattern that is insensitive to small changes in the firing patterns of individual presynaptic neurons. In this situation assume that many of these neurons were to correlate with P and thus have rising weights; the risk of explosive weight gain would seem high. However, at some point, a small subset, G' of G∗ would be sufficient to evoke a spike from P. All members of G* − G' would then lose the pre-before-post relation to P and would thus experience weight decreases on their synapses to P. So the total synaptic strength has an emergent upper bound.

In the second firing mode P's total input is much less (and closer to the firing threshold) such that P's spike train can be driven by the train of one or a few active members of G*. These members would experience pre-then-post firing behavior to strengthen their synapses by STDP, while less active G* neurons would have few spikes and few spikes in the temporal window of P's spikes. Thus, the inactive members would not invoke much LTD, so the net result would be an increase in total synaptic weight. Hence, a lower bound also emerges.

Note that in both Song et al.'s STDP model and in the Oja rule (equation (10.9)), postsynaptic activity directly affects weight leakage, forming a negative feedback loop that stabilizes each neuron's vector of afferent weights. Furthermore, in both models the leakage associated with postsynaptic firing is larger than the weight gain stemming from presynaptic activity. This is achieved by $m_- > m_+$ in the STDP

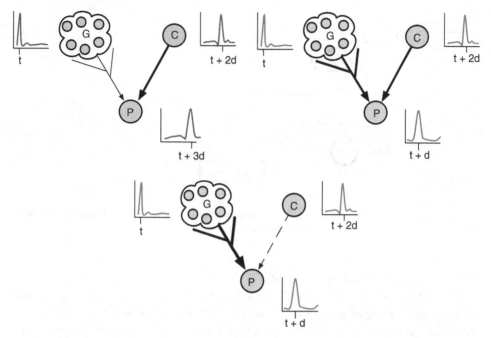

Figure 10.17
Emergence of cooperative postsynaptic control. (*Top left*) A group (G) of synchronously firing (i.e., cooperating) neurons helps stimulate postsynaptic neuron P, as does the single presynaptic neuron C. By
spiking just before P, neuron C achieves the greatest LTP (thick line), while connections from G receive
some, but less. (*Top right*) The slight increase in G's efferent connections enables it to stimulate P to fire
earlier than C. (*Bottom*) This produces strong LTP on G's efferents (thick lines) while invoking LTD on
C's efferent (dashed line). Thus, C's influence upon P diminishes, while G now controls it.

model. With the Oja rule, presynaptic activity (u_i) increases synaptic weight by
the factor v, while weight leakage involves a v^2 factor. Since the activation functions (Oja, 1982) are linear (not, for example, sigmoidal), output values often
exceed 1, and thus, $v^2 > v$ for strongly stimulated postsynaptic neurons. Consequently, a highly active postsynaptic neuron will produce weight increases only
on the synapses from its most active afferents; all others will decrease.

    In both cases a local learning algorithm drives the neural circuit to a *poised* state
in the sense that most presynaptic firing patterns elicit no significant response from
the postsynaptic neuron; but those patterns to which it has become *tuned* can consistently cause it to fire. The poised state is thus an informational state for the
brain, indicating the presence of a salient activity pattern but not responding (with
a potentially false positive) otherwise.

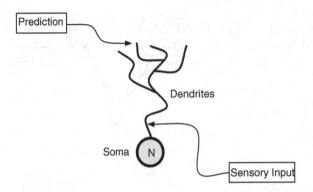

Figure 10.18
Connections for predictive learning in the brain. Sensory neurons provide proximal, bottom-up signals, while high-level, predictive inputs arrive via distal synapses.

10.5.3   Emergent Predictive Circuits

This STDP model has tight ties to the work of Artola et al. (1990), which characterizes important preconditions for LTP versus LTD, the combination of which plays in important role in a trial-and-error synaptic tuning mechanism that appears to underlie many predictive networks in the brain.

From the viewpoint of synaptic electrophysiology, the acquisition of declarative predictive models within the brain's hierarchical neural network has a very plausible explanation based on bimodal thresholding. As illustrated in figure 10.18, Artola et al. (1990) have shown that *weak* stimulation of neurons (in the visual cortex) leads to long-term depression (LTD) of the synapses that were active during this stimulation, while stronger stimulation incurs long-term potentiation (LTP) of the active synapses.

As shown in figure 10.18, a common connectivity pattern in the brain, particularly the cortex, involves a combination of bottom-up and top-down pathways, wherein lower-level sensory signals enter a neuron via proximal dendrites, while higher-level signals (carrying predictive information) synapse distally (Hawkins, 2004; Mountcastle, 1998). All other factors being equal, this gives the bottom-up signal greater influence than the top-down signal upon the activity of postsynaptic neuron N. In short, sensory reality dominates speculation. However, experience and learning can change this relation.

Three learning cases (summarized in figure 10.19) are worth considering with respect to neuron N, its low-level sensory inputs, S, and its high-level predictive inputs, P. First, if S is active but P is not, then the effects of S on N will produce a high enough firing rate in N to incite LTP of the S-to-N proximal synapse. Hence, N will learn to recognize certain low-level sensory patterns.

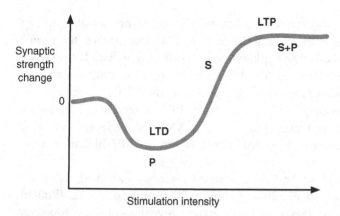

Figure 10.19
Changes in synaptic strength as a function of postsynaptic stimulation intensity. The three presynaptic scenarios are when only S is active, only P, and both S and P (S + P).

Second, if both P and S provide active inputs to N, then an even higher firing rate of N can be expected, so LTP of both the S-to-N and P-to-N synapses should ensue. In essence, the predictive and sensory patterns create a meeting point at N by tuning the synapses there to respond to the P-and-S conjunction. In fact, after repeated co-occurrences of S and P, the synapses in N may strengthen to the point of responding to the P-or-S disjunction as well, in effect saying that it *trusts* the prediction P even in the absence of immediate sensory confirmation.

In the third case, when only P is active, the distal contacts of the P axons may only suffice to weakly stimulate N, thus leading to LTD: a weakening of the P-to-N synapses. Hence, future signals from P will not suffice to fire N, and thus P's predictions will not propagate through N in the absence of verification from S. In short, the system learns that P is not a good predictor of S.

These examples of LTP and LTD seem to agree with Song et al.'s STDP model. In case 1, if S and N are both firing at a high frequency, with S spikes beginning slightly earlier and thus helping to initiate the N spikes, then LTP of the S-N synapse seems likely. In case 2, with S and P spiking frequently enough to excite N at a comparable frequency, there should be ample pre-before-post activity to produce LTP on both the S-N and P-N synapses. In case 3, assuming that presynaptic P produces spikes at a much higher frequency than postsynaptic N, then more decrements than increments of the synaptic weight (e.g., w_a in figure 10.16) can be expected as many presynaptic spikes go *unanswered* by N: the weight decreases that follow presynaptic spikes are not balanced (or exceeded) by weight increases that follow postsynaptic spikes.

Given this basic wiring pattern for the integration of sensory inputs and top-down expectations, along with an established LTP/LTD mechanism, the brain's ability to predict has an interesting explanation in terms of search and emergence. This combination of bottom-up and top-down processing as a route to emergent intelligence in a neural substrate was popularized in the 1970s by *adaptive resonance theory* (ART) (Grossberg, 1976a; Grossberg, 1976b), which formalizes the interplay between reality and expectations in an ANN. More recent results in computational neuroscience provide further inspiration for ART-like approaches to cognitive modeling.

Contemporary models of several diverse brain regions—including the thalamacortical loop (Rodriguez et al., 2004; Granger, 2006), hippocampus (Wallenstein et al., 1998), and neocortex (Hawkins, 2004)—reveal a common topology, which I call the *generic declarative prediction network* (GDPN) (Downing, 2009, 2013). Declarative prediction requires machinery that can associate two patterns, both of which have strong correlations with external stimuli. The perception of the first stimulus in the environment should then stimulate thoughts of the second.

Depicted in figure 10.20, the GDPN consists of several neurons connected as in figure 10.18, with a layer of detector neurons (A, B, and C) sandwiched between a layer of sensory inputs and one of top-down predictor neurons (W, X, Y, and Z). Once again, note the proximal influence of the sensory signals and the distal connection of the predictors. In contrast to procedural networks, declarative topologies tend to have considerable recurrence: combinations of bottom-up and top-down links.

Consider a situation in which stimulus A precedes stimulus B. The following series of events explains how the network learns to *predict* B when A occurs in future situations (Downing, 2013).

First, at time t_1, stimulus A has a strong effect upon neuron A, via its proximal synapse. Neuron A then fires and sends *bottom-up* signals to W, X, Y, and Z at time t_2. At this level, as in many areas of the brain, neurons fire randomly, with probabilities depending upon their electrochemical properties and those of their surroundings. Assume that neuron X happens to fire simultaneously with, or just after, neuron A, and that synapse S1 is modifiable. Then the A-X firing coincidence will lead to a strengthening of S1, via standard Hebbian means. In reality, several such high-level neurons may coincidentally coactivate with A and have their proximal synapses (from A) modified as well.

When X fires, it sends signals horizontally and to both higher and lower levels. These latter *top-down* signals have a high fan-out, impinging upon the distal dendrites of neurons A, B, and C. Since entering distally, along unrefined synapses, these signals have only weak effects upon their respective soma, so at time t_3 neurons B and C are receiving only mild stimulation. At this stage, one can metaphorically say that X is *waiting* for B and C (and thousands of other low-level neurons) to fire, and X *hedges its bets* by investing equally and weakly in each potential outcome. In short, X performs a parallel search for a postsynaptic target that is a *reliable* successor to stimulus A: one that tends to activate directly after neuron A does.

At time t_4, when event B occurs, neuron B will spike at a high frequency because of proximal stimulation from below. This will cause further bottom-up signaling, as when A fired, but the critical event for current purposes involves the LTP that occurs at synapse S2. Previously, stimulation from X alone was not sufficient to fire neuron B at a high frequency. Now, however, with help from stimulus B, both neurons, X and B, are very active, causing S2 to strengthen. Thus, in the future, the firing of X will send stronger signals across S2, possibly powerful enough to fire neuron B *without help* from stimulus B.

Through one or several A-then-B stimulation sequences, S1 and S2 can be modified to the point that an occurrence of stimulus A will fire neuron A, as before,

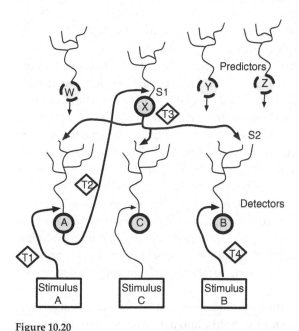

Figure 10.20
Generic declarative prediction network (GDPN). Neurons A, B, and C serve as low-level detectors for stimuli A, B, and C, while W–Z represent neurons at a higher level that serve as predictors of activity among the detectors. Only the axonal projections from X are shown, though W, Y, and Z have similar links to the lower level. The T1–T4 diamonds represent timesteps, while S1 and S2 denote important synapses.

but this will then directly cause X to fire, which in turn will fire neuron B. Thus, stimulus A will cause the brain to *predict* stimulus B.

Over time, neuron X ceases to hedge its bets and achieves a significant bias toward neuron B. This stems from both the strengthening of S2 and the weakening, via LTD, of X's synapses upon other detector-level neurons (that are not simultaneously activated by bottom-up signals); X's presynaptic spikes are not answered by postsynaptic spikes among these other detectors, thus causing LTD. So X becomes a dedicated prediction pathway between A and B. After the repeated presentation of many sequential patterns, LTP and LTD gradually convert a blanket of bet-hedging anticipatory links into a smaller population of dedicated connections between associated pattern-detecting neurons.

In the GDPN terms such as *detector* and *predictor* serve only as scaffolding to explain the slow emergence of a hierarchical network in which the neurons at higher levels represent more complex concepts, which are often detectors (and promoters) of lower-level activation patterns. These support prediction in the sense that the bottom-up formation of some lower-level patterns induces the higher level to activate further primitive patterns, thus *completing the picture* via top-down expectations.

The presence of a temporal ordering in this learning may seem arbitrary. For example, if a child learns the complex concept of a *red ball*, then does *red* originally predict *ball*, or vice versa? Rodriguez's and Granger's models of the thalamocortical loop (Rodriguez et al., 2004; Granger, 2006) provide one answer: rather than necessarily linking sequences of real-world events, GDPN networks can associate sequential neural states of perceptual processing, wherein the initial states tend to involve the most salient features (such as *red object*) with other aspects (such as roundness) registering in later steps. This work blurs the borders between prediction and conventional association, since any perception can now be interpreted as a time series of partial interpretations of sensory input, each predicting the next.

The GDPN is most clearly evident in the neocortex, where the individual neurons of figure 10.20 are replaced by cortical columns, each functioning as a processing module (Hawkins, 2004; Fuster, 2003). Bottom-up sensory interpretation involves cascades of neural firing from the back (sensory) areas of the brain to the front (executive) areas, while top-down, parallel search for reliable successors moves front to back.

This search is the GDPN's key contribution to emergent intelligence. Just as neurons migrate by filopodia searching for footholds, and axons find targets by successive growth and retraction, hierarchical neural layers tune themselves by flooding lower layers with exploratory signals and then allowing LTP and LTD to gradually convert mats of undifferentiated connections into finely specialized synaptic networks. This exploratory process is parallel and unsophisticated, but it produces networks capable of remembering, recognizing, and producing the intricate patterns of mental life.

10.6   Place Cells and Prediction in the Hippocampus

The GDPN is also based on the CA3 region of the hippocampus, a brain area exhibiting one of the highest densities of recurrent connections (Rolls and Treves, 1998; Kandel et al., 2000). Neuroscientists generally agree that recurrence is essential for the pattern storage and completion/retrieval that underlies associative learning (Rolls and Treves, 1998). When these associations include a temporal component, they become predictive; and indeed, the hippocampus is also considered a premier predictive area of the brain, particularly with respect to navigation (Burgess and O'Keefe, 2003; Gluck and Myers, 2000). What follows is a modified version of the account of place cell and prediction from Downing (2013).

This prediction is probably best exemplified by the well-documented phenomenon of *phase precession* in hippocampal place cells (Burgess and O'Keefe, 2003)—which also qualify as *context detectors* and predominantly lie in neighboring regions CA3 and CA1—whereby a neuron that codes for location X begins to fire (predictively) at locations prior to X along a familiar path. As detailed in figure 10.21, the formation of these predictive links between detector neurons coding for successive locations along an oft-traveled route is convincingly explained by Mehta (2001), who shows that standard spike-timing-dependent plasticity on the synapses between place cells can form asymmetric place fields such that a neuron is highly active prior to arrival at its place field but inactive immediately afterward.

Note that place cells are a particularly good example of declarative representation in that specific neurons fire when the animal resides in a particular location (X). Phase precession is an equally compelling example of declarative prediction, since the place cell fires on the approach to X. Furthermore, phase precession (and thus the predictions underlying it) may play a much deeper role in cognition, partly because of the dual roles of the hippocampus in both navigation and general memory formation.

As shown in figure 10.22, as a rodent moves along the corridor from locations A to G, STDP could easily lead to the formation of predictive connections between the detectors for each location, particularly when the intervals between arrival at successive locations are in the range of 0–50 msec (the time window for STDP).

Once formed, these synapses can be activated in sequence, with gaps much smaller than 50 msec. Lisman and Redish (2009) have shown that gamma waves (40–100 Hz) elicit this fast replay, with these high-frequency oscillations riding atop the slower (6–10 Hz) theta waves that characterize hippocampal activity. As shown at the top of figure 10.23, during one theta cycle, a good many *successive* context detectors can be activated, each by a gamma peak. At the peak of each theta cycle, the current location determines the active detector, but throughout the remainder of the cycle, succeeding detectors activate in a predictive manner.

This rapid sequencing brings several detector firings within the 0–50 msec window of STDP. So, for example. after cell A fires in response to current sensory input, cells B, C, D, and E fire in rapid succession via gamma stimulation. STDP then dictates that synapses from A to B, C, D, and E will all experience LTP, thus forming cell assemblies that manifest information chunking.

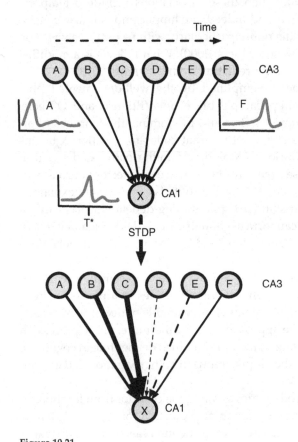

Figure 10.21
Emergence of phase precession in hippocampal place cells. (*Top*) Assume that a rodent moves through contexts A to F (detected in CA3) prior to and after sensing a key landmark, whose presence is signaled by the firing of place cell X in CA1. Then, firing times for the context detectors A–C should precede that of X, while D–F should fire after. (*Bottom*) These spike-timing differences produce graded LTP on the three left-hand connections and LTD on the right-hand ones. Hence, in the future, the presence of the left-hand contexts could fire neuron X prior to arrival at the landmark. Thickness of arrows indicates relative strength, with thin dashed lines being the weakest. Based on Mehta (2001).

The dual role of the hippocampus in both navigation and general memory formation raises the obvious question of whether this place cell chunking could manifest general information binding and integration. So the aggregated sequences could represent locations, steps in a procedure, or words in a memorable phrase or song melody. In all such cases, the same basic predictive machinery (grounded in the dynamics of STDP) combines with gamma-induced replay (and further STDP) to produce tightly linked neurons and neural firing patterns that may represent episodes or concepts.

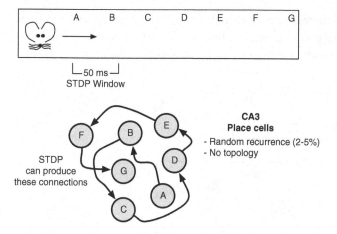

Figure 10.22
(*Top*) Corridor along which a mouse runs, with contexts (A–G) encountered approximately every 50 msec. (*Bottom*) A hypothetical connection pattern, formed via STP, among CA3 detectors for contexts A–G. From Downing (2013).

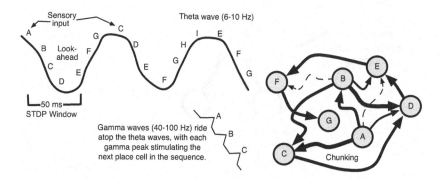

Figure 10.23
(*Left*) Gamma waves riding atop theta waves stimulate place cells in rapid succession. (*Right*) Rapid stimulation facilitates predictive chunking of temporally related contexts. Thicker lines denote stronger connections. From Downing (2013).

The leap from place cells to concepts is elaborated by Buzsáki (2006), who describes an interesting property of place cell learning: when mice move back and forth along a corridor, the same location (C in figure 10.24) binds to two different place cells (C_1 and C_2) depending upon the direction of travel. Hence, to the mouse, these are two different locations. However, in an open arena, without the constraints to movement imposed by corridor walls, the place cells tend to be omnidirectional: the same cell fires regardless of the angle of approach. Initially, they are unidirectional, as in the corridor, but with continued exploration of the arena, a unique place cell begins to represent the same location without directional bias.

Dragol et al. (2003) explain this situation with evidence of the continuous mapping and remapping of place cells to spatial fields (via LTP); repeated trials in an arena environment could easily stimulate this remapping. Thus, many approach

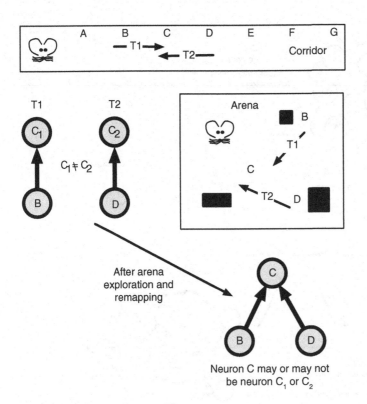

Figure 10.24
(*Left*) Unidirectional place cell formation in a corridor environment. (*Right*) Transition of unidirectional to ominidirectional place cells in an arena environment. T_1 and T_2 denote distinct temporal intervals, and B, C, and D, locations. From Downing, (2013).

episodes become bound to the same place cell, a process fitting of the term *generalization.*

Consider the neural network in figure 10.25, where two different contexts (B and D) originally predict context C, via independent place cells C_1 and C_2. In time interval T1, context B arises as a prelude to C, thus forming the five leftmost connections. During T2, context D arises and hooks up to C_2 via the six rightmost links. Now, before interval T3, assume that B-to-C_1 strengthen because of the repeated co-occurrence of contexts B and C while context D does not appear at all. This weakens synapses from the shared context (pentagons) to C_2 via STDP. Finally, at T3, assume that context D fires. Because of the weakened shared-context-to-C_2 links, C_2 may not fire, but the shared context between B and D now has enough strength to invoke C_1. The co-firing of D and C_1 can then strengthen several (previously weak) connections between them. Thus, in the future, D will easily trigger C_1. Furthermore, since B and D will both excite C_1, their shared context should be co-active with C_1 more than the unshared context of either. Links between the shared context and C_1 should then become particularly strong, thus representing the *essence* of the overlapping contexts. Buzsáki (2006) likens this generalization over experienced episodes to concept formation, wherein the invariants of many specific scenarios are eventually distilled into a general-purpose representation: a concept, which can be metaphorically described as a thought arrived at from many different paths of reasoning.

Since the hippocampus plays a key role in both navigation and memory consolidation (Andersen et al., 2007; McClelland et al., 1994), place cells and their interconnections (modifiable for both look-ahead and chunking) could provide the substrate for both spatial recognition and general concept formation. And in both cases, the predictive links formed by STDP provide a fundamental starting point. So if Buzsáki's analogy is correct, the hippocampus could be a critical junction between an advanced form of sensorimotor behavior (i.e., navigation) and some of the highest cognitive faculties: abstraction and concept formation.

10.7 Neural Support for Cognitive Incrementalism

As hinted earlier, the connection between physical and mental is believed to go much deeper than correlations between body movements and neural patterns. Cognitive incrementalism (Clark, 2001) posits sensorimotor activity as the foundation of higher-level cognitive activity (see chapter 1). This view is implicit in ALife research (a lot of which involves simple sensorimotor agents) that aspires to AI relevance, since the study and automation of advanced intelligence has little use for wall-following and floor-sweeping robots unless their underlying control mechanisms have implications for cognition.

Many good publications cover this topic in great depth (Clark, 1997; Clark, 2011; Pfeifer and Scheier, 1999), and a few include reasonable proposals for the neural basis (Deacon, 1998; Lakoff and Núñez, 2000; Llinás, 2001). Of particular note is the book by Pfeifer and Bongard (2007), which emphasizes the role of an efference copy of motor commands in forming a sensorimotor self-image. In a nutshell, the premotor cortex is believed to send commands not only to the motor cortex but also to higher-level cognitive areas, which include this information as part of the current body-world context.

Returning to Buzsáki's theory of concept formation via overlapping episodic memories (Buzsáki, 2006), the neural diagram of figure 10.26 illustrates a simple role of sensorimotor data in cognition. Imagine three episodes with overlapping

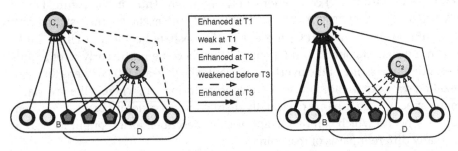

Figure 10.25
Emergence of a single detector for a general concept (*right*) from an earlier group of detectors (*left*). Pentagons represent neurons encoding shared aspects of contexts B and D; circles, unshared idiosyncratic details. Arrow thickness denotes relative synaptic strength. See text for explanation.

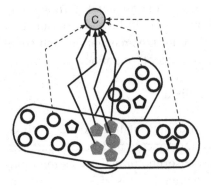

Figure 10.26
Three different neural groups representing three episodes, all of which contribute to the formation of a concept detector, C. Pentagons denote neurons with a sensory basis; circles reflect a motor grounding. Shaded shapes constitute the essence of the concept, with the majority being sensory, while most unshared neurons have a motor basis. Arrow thickness denotes relative synaptic strength.

sensory and motor components. The concept arising from them stems from the shared neurons, which, in this example, tend to be sensory, not motor. Though many of our procedural memories, such as the movements needed to slam a ping-pong ball, would seem to require very precise recordings of the motor actions, a declarative memory for a concept would seem less tied to motor specifics. For example, my memory of the beautiful view of a river valley from atop a steep hill has only vague connections to the actions that I used to get there, whether walking, running, cycling, or driving. My first visits to that spot might have had an episodic motor coupling, but after many visits using many forms of transport, the essential invariant (i.e., the concept) would probably involve the basic impression of being elevated above a wooded area with a river winding through it. A good many declarative concepts would seem to have a similar basis in sensory input. The sensory impression would prevail in later recall of the general location, with each motor efference copy simply being a different mental vehicle for helping to activate neuron (or neural population) C. In organisms with many muscles controlled by the cortex, these efference copies involve many neurons and thus have significant cumulative power to activate context detectors. Motor neuron activity thus has a bulk effect during the episodic memory formation, but most of this scaffolding has no presence in the final concept.

In general, the task of navigation could be a critical link between sensorimotor activity and cognition, largely because of its demands for recognizing and anticipating spatial locations. These predictions may constitute a key gateway between action and high-level reasoning. As discussed earlier, declarative concept formation may involve mechanisms for place-cell-mediated navigation, namely, STDP-based predictive linkage, gamma-cycle-driven replay of predictive sequences, and the consequent chunking, a key element of memory formation. So the predictive sequencing and aggregation underlying navigation may also manifest one of the most advanced cognitive faculties: abstraction.

10.8 Hebb Still Rules

Unsupervised learning is the most primitive form of adaptivity in ANNs, yet the most prevalent in the brain. Whether cooperative or competitive in nature, the process relies heavily on a fundamental feature of neural systems: Hebbian synaptic modification. This grounding in a simple local principle makes unsupervised learning very easy to implement in ANNs, but steering emergence toward meaningful global patterns requires careful modeling and parameter selection. Nature has had millions of years to evolve the proper components, which neuroscientists now labor to decipher, and Bio-AI researchers hope to use for computational purposes.

The emergent nature of unsupervised neural processing begins at the low level of STDP, continues through the dynamics of recurrent networks, and even appears in the interpretation of ambiguous figures such as the duck/rabbit and vase/kissing-couple staples of psychology textbooks. Behaviors at the level of networks of neurons have a clear *local search* nature. In the learning phase different combinations of synaptic weights are *tried* during the gradual tuning of a complex set of salient pattern detectors. This matrix of weights (W) is the focal, dynamic representation of learning. During recall, the important dynamic representation is the vector (V) of neural activation levels, which self-organizes to a form that is most compatible with W: a concise, imperfect record of the network's history.

Whether mice in boxes, ants on beaches, or teenagers in a new mall, the same basic principle holds: trial-and-error exploration combined with Hebbian synaptic tuning leads to representational refinement. Though the ant itself may remember nothing, lay down no mental trace, its pheromone trail serves the greater good of the colony: it constitutes an evolving structure, an extended memory, that enhances group performance. The mouse, on the other hand, internalizes much more of its environment by modifying the synapses onto place cells. Repeated wanderings, driven by little more than random inclinations, support information gathering. Over time, the emerging structures, whether networks of pheromones or interlinked CA3 cells, influence behavior, adding a purposeful or goal-directed appearance to movement. As for teenagers, whatever cannot be off-loaded to their extended memory of choice, the cell phone camera, probably ends up in CA3 as well.

One really can go a long way with the simple Hebbian rule of firing and wiring together. It appears to be the fundamental principle behind associational pattern formation, which in turn has been identified as one of the premier properties of animal intelligence (Hawkins, 2004). It is also one of the most obvious and impressive characters in the story of intelligence emerging.

11   Trial-and-Error Learning in Neural Networks

11.1   Credit Assignment

Imagine that you are the first-year coach of a professional basketball team. It is the first game of the season, and the score is tied with only seconds remaining. You call a time-out, consider your multitude of strategic options, and then decide to set up a play for the guy whom everyone just calls C. The play begins, C gets the ball, and though double-covered by two hard-nosed defenders, gets off a long shot. Swish! The ball goes through the net, you win the game, and C is carried off the court on the shoulders of his teammates. You have now learned some valuable information that will help throughout the season: C is a clutch performer.

Now consider a slightly different outcome. In the waning seconds C makes a nice pass to the guy whom everyone just calls B, who makes a shot to win the game. B acknowledges C's contribution by pointing and playfully saluting him before being whisked off the court on the shoulders of his teammates. This time B's stock rises in your eyes, but some of the credit is *passed back* to C as well. Though the pass from C to B was routine, a lesser player might have botched it in the heat of the moment. Both C and B will probably be mentioned in newspaper accounts of the game, and credit will be given where credit is due. In fact, basketball statistics include these *assists* as very tangible evidence of a player's worth.

But what happens in a third scenario, one in which C passes to B, who passes to A, who makes the winning shot? A gets the points and the shoulder ride to the locker room, B gets the assist, but does C get (or deserve) anything? It could be the case that before passing to B, C faked a pass or a dribble attack that froze A's defender in place. This made it easier for A to come free to get the ball and shoot the winning basket. Such a contribution by C would not show up on a statistics sheet, though you may notice it. You may even praise C more than B or A afterward in the locker room, since his fake-then-pass was obviously the key to the whole play. He *set up* a situation that then became routine for B and A.

Finally, imagine that C never makes the sports-TV highlight reels with spectacular last-second shots or assists. However, after many games, you realize that in a large majority of the close wins, C was on the court in the closing minutes. Once this majority achieves statistical significance in your own mind, you may begin to view C as a clutch player. In this case, getting credit to where it is due takes a long time, but gradually C's value rises.

In general, the evolving player values provide information that is essential for your strategic decision making throughout the season: the players with higher value are more likely to be played in the most difficult situations. And these values are updated game by game, play by play. Events that directly precede basketball rewards (such as points and wins) are obvious candidates for credit, but events more distant from the rewards require careful consideration as to their actual contribution to the outcome. Did the fourth pass in a series of eight really contribute to the score, or was it actually a mistaken move that done properly would have achieved a basket in five total passes? This bookkeeping — whether in the brain of the coach, the season's cumulative statistics, or some combination of the two — is the essence of the adaptive process known as *reinforcement learning* (RL).

RL has two key characteristics:

- An explicit search process in which alternative actions are attempted and local evaluations of their success/failure are recorded.

- A relatively infrequent and general feedback/reinforcement (as compared to the frequent and specific error signals of supervised learning) that must (somehow) translate into credit or blame to specific system components, whose temporal activity histories have varying degrees of overlap with the actual time period of the reinforcement. In short, the system must base the evaluations of actions on reinforcement signals that may not come until many steps into the future.

In addition to sports coaching, many other real-life situations involve RL. Usually, one does not receive constant feedback as to positive or negative value of actions, as is required for supervised learning procedures. Instead, one makes chains of decisions and performs many actions before encountering some value-laden consequence, such as a customer's praise for a beautifully renovated kitchen or a CEO's tongue-lashing for a poor quarterly result. In all such cases of positive and negative reinforcement, the task of assigning credit or blame where due, known as the *credit assignment problem*, is the main challenge. One general heuristic for *backing up* reinforcement is that the actions closest in time (and space) to the external feedback deserve a good deal of the credit/blame, with more distant actions getting less. These action value updates then affect decision making and behavior in efforts to improve performance, that is, to increase reward and decrease punishment.

RL is the epitome of trial-and-error search, with a proven ability to improve over time as a result of the bookkeeping embodied in credit assignment. Like a first-year coach at season start, an RL system begins with a repertoire of possible actions but little information as to the appropriate contexts in which to apply them. Through online (game time) experience, the system tunes these preconditions to craft an effective decision maker. But it usually requires many trials and a lot of error before noticeably intelligent behavior emerges.

Compared to other learning paradigms, RL has a unique combination of biological plausibility, practical utility, and emergence. Chapters 7 and 10 emphasized the emergent nature of unsupervised learning via local Hebbian synaptic modification, as so clearly exemplified by Kohonen networks. Furthermore, the search metaphor aptly describes movement in Hopfield's energy landscapes as well as in the weight vector spaces of supervised learning via backpropagation. Search and emergence are vital concepts in understanding the dynamics of neural networks, whether the transition to stable activation states or the adaptive tuning of synapses.

However, supervised learning algorithms achieve considerable problem-solving success but portray little emergence or biological realism, while unsupervised ANN methods often adhere closer to biology and local interaction protocols but yield fewer practical results. RL represents a happy medium between the two. When employed in ANNs, it has more biological plausibility than unsupervised techniques and a better chance of achieving problem-solving success in many real-life scenarios than either unsupervised or supervised approaches, since the former do not incorporate feedback, and the latter require a steady supply that the real world cannot usually provide.

In general, many AI applications have an RL flavor. In particular, AI systems that serve as controllers for multistep tasks, such as robot navigation or game playing, are rarely the beneficiaries of constant feedback. Instead, external reinforcements only occur when the agent's actions have an obvious good or bad consequence, such as winning a chess game or sending the robot crashing into a bookshelf. The agent must allocate reinforcement to those portions of the system that most significantly contributed to the overtly rewarded (or punished) result.

As shown in figure 11.1, the problem of using ANNs to do RL involves backing up a reinforcement signal to parts of the network that are spatially distant from those that actually produced the critical action (i.e., the output neurons) or that were not active at the exact time of the feedback signal.

This chapter examines solutions to this problem in the brain and in a few select ANN applications; but first, an overview of RL theory provides a helpful framework for evaluating RL in both natural and artificial neural networks.

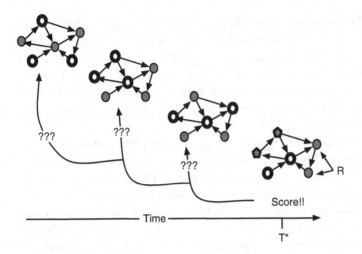

Figure 11.1
Credit assignment problem for neural networks: how to assign credit or blame to network components that were active (shaded nodes) but not at the same time T* as the reward/punishment, or that were active at T* but are spatially distant (pentagons) from components receiving explicit reward (R) at T*

11.2   Reinforcement Learning Theory

Much of the computational theory of reinforcement learning appears in Sutton and Barto's classic textbook (1998), which is mandatory reading for anyone interested in the subject. This section briefly summarizes some of the key points from that book.

Conventional RL is essentially *on-the-job-training*, in that the system uses a *control strategy* to work on a problem (e.g., search in a virtual environment) while simultaneously building a *policy*: knowledge about the most appropriate actions to take in different situations/states. A supplement to the policy is the *value function*, which provides evaluations of the expected future reinforcement for either a state or a state-action pair (SAP), depending upon the particular RL variant.

As the agent works and learns, it explores the space of possible states, actions, and rewards. In making action choices, the agent must strike a proper balance between exploration and exploitation: it can choose actions that have previously led to states that have yielded high future rewards (exploitation), or it can try actions that will take it to never visited or infrequently visited states whose evaluations are less certain (exploration).

Reinforcements in RL, whether rewards or punishments, occur rarely, often only at the end of a problem-solving attempt. The key feature of RL is its ability to *back up* the reinforcement to states along the path from start to finish. This is most easily visualized in maze-searching tasks (see figure 11.3), but the basic procedure applies to a wide range of tasks in which environmental feedback occurs only intermittently.

The three key components of an RL system are the agent, the environment, and the policy. A very important aspect of RL theory is a strict division between agent and environment such that actions are always determined by the agent and sent to the environment, and the environment determines future states and rewards. This means that if elements of the agent are components of the state or determinants of the reward, then they become part of the environment.

To illustrate this point, assume that a robot uses RL for a navigation task based on touch and sonar sensor inputs. The values of these sensors are normally considered part of the state for such a problem. If the robot runs into a wall, its own touch sensors would signal this condition, which is often associated with a negative reinforcement. Hence, the robot's body, which houses the sensor pool, would become part of the environment, while only the robot's controller (e.g., a neural network or finite-state machine) would constitute the agent.

After each round of action-choice, action-performance, and state-and-reinforcement updating, the learning module uses the new information to update the policy and value function. This insures that over time the agent's experience will lead to more effective decision making: actions will be chosen that most swiftly lead to rewards while avoiding punishments.

In *on-policy* RL the current policy governs control choices (i.e., exploitation dominates exploration), while in *off-policy* RL the controller runs independently of the policy, with action choices often made to more fully explore the space of possible states, actions, and rewards. In either case, the agent's experiences are used to update the policy, but in on-policy RL the agent's actions may stem from a relatively *immature* or *unrefined* policy base.

11.2.1   The State Space for RL

In conventional RL all possible states, Σ, are determined prior to any learning, with each state typically a point in a space whose dimensions are the relevant environmental factors and internal state variables of the agent. So for a maze-wandering robot, the dimensions might be discretized x and y coordinates along with the robot's energy level (and possibly the x and y components of its velocity vector).

A basic RL system would thus divide the world into MxNxE states, where M and N are the number of discrete bins into which, respectively, x and y coordinates fall, and E denotes the number of energy-level bins. The RL system would then work to determine the *values* of each such state, which is information only obtainable by problem-solving experience in RL: no instructor provides them.

A state's value is directly proportional to expected reward in both the near and more distant future. Value is a recursive relation: states with high value D are those that either have rewards directly associated with them or are direct predecessors (under the set of available actions) to states with high values.

Despite the diverse collection of tasks that are amenable to RL, basic maze-searching tasks are most frequently used for instructive purposes, because the value function is easily visualized as a discretized grid (the maze), with the number in each cell denoting the value of the state associated with that location. Figure 11.2 sketches one such scenario.

In tasks that are most amenable to RL, the agent has the opportunity to solve a problem many times. Hence, the agent can repeatedly return to a start state and work toward the goal. Each such attempt is called an *episode*, and each episode ends in a *terminal state*, which is frequently one in which a reinforcement (either reward or punishment) occurs. It is often the only state in which reinforcement occurs, although not necessarily. The arrival time at the terminal state is denoted T. Problems that do not break neatly into episodes are termed *continuing tasks*. They, too, can be solved with RL methods, but this normally requires reinforcements to occur throughout the task, not merely at the end. Figure 11.3 depicts a maze-searching task with several episodes, each concluded with a round of value updating for those states (shaded cells) involved in the episode. In this problem a reward is given when the agent reaches the upper right corner of the maze while a penalty is associated with the upper left corner of the maze.

The reward at time t is r_t, and the cumulative reward from time t until episode termination is R_t, where

$$R_t = r_t + r_{t+1} + r_{t+2} + \cdots + r_T. \tag{11.1}$$

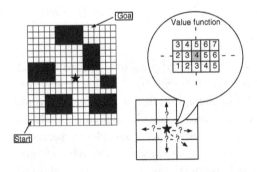

Figure 11.2
Classic RL problem: finding the best path from the start to the goal of a maze. (*Left*) An agent is currently at star-marked location. (*Right*) Through experience, the agent compiles a value function, which the agent's policy can consult to determine its next best move.

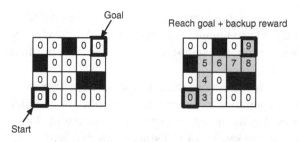

Figure 11.3
Example of the gradual updating of the value function (represented as a table) via the repeated combination of exploration and reinforcement backup. (*Top left*) Initially, the value table has no information. (*Top right*) After one random search (gray-shaded cells), which reaches the goal cell, a reward is received and backed up to all cells along the path, with the value diminishing (discounted) with distance from the goal. (*Bottom left*) Another random search finds a dead end, which incurs a penalty, which is also backed up and combined with previous values for the gray-shaded states. (*Bottom right*) After many episodes, the value table gives an accurate reflection of the expected future reinforcement from each cell. Filled black cells denote barriers.

In most RL scenarios a future reward is *discounted* by a certain fraction γ raised to the power of the number of timesteps into the future. This makes intuitive sense: reinforcements that might occur in the distant future have much less significance than those that could occur very soon. Formally, discounting requires an updated conception of R_t:

$$R_t = r_t + \gamma r_{t+1} + \gamma^2 r_{t+2} + \cdots + \gamma^{T-t} r_T. \tag{11.2}$$

or, more succinctly,

$$R_t = \sum_{k=0}^{T-t} \gamma^k r_{t+k}. \tag{11.3}$$

Reinforcement learning relies heavily on the ability to predict (immediate) future states and rewards given only the current state and action. Thus, previous states should have no effect upon the transition from the current state to the next state. This *path independence*—the current state, not the whole path of states leading to it, determines the future—has major implications for many machine-learning techniques and is commonly called the *Markov property*. Formally, the Markov property states that the exact probability of the next state-reward pair can be computed solely from the current state and action.

In RL one assumes that the policy has access to these probabilities, either from an external teacher or via its own experience in the problem domain. In the latter case, the Markov property insures that as the agent explores the space of states, actions, and rewards, it can accumulate statistical transition data that will eventually enable it to make accurate predictions of future state-reward probabilities based only on the current state and action.

The *value* of a state or of a state-action pair is directly linked to the expected reward tied to that state plus that attainable from the state; i.e., it is proportional to R_t. The *value function* embodies a mapping from states (or SAPs) to R_t. These mappings are always a function of the current policy, Π.

For states, RL characterizes the *state value function for policy* Π as

$$V^{\Pi}(s) = E_{\Pi}\{R_t | s_t = s\} = E_{\Pi}\{\sum_{k=0}^{\infty} \gamma^k r_{t+k} | s_t = s.\} \tag{11.4}$$

In short, the value of the current state at time t is the expected value of the cumulative reward, R_t, with discounting, γ^k. The summation on the right goes to infinity (instead of T) in order to cover both episodic and continuing tasks with one general expression.

Alternatively, some RL approaches, such as Q-Learning, utilize a value function for SAPs instead of just for states. RL expresses the *action value function for policy* Π as

$$Q^{\Pi}(s,a) = E_{\Pi}\{R_t | s_t = s, a_t = a\} = E_{\Pi}\{\sum_{k=0}^{\infty} \gamma^k r_{t+k} | s_t = s, a_t = a\}. \tag{11.5}$$

Once again, the evaluation reflects the expected cumulative reward.

To accurately calculate $V^{\Pi}(s)$ or $Q^{\Pi}(s,a)$ may require considerable computational resources for problems with large state spaces, and it is often intractable, since estimating the future reward from any given state may demand the exploration of a number of completion paths (from that state) that is exponential in the number of states. Thus, only for toy problems can perfect value functions be generated.

However, one can methodically estimate and update value functions based on a simple recursive relation:

The value of a state (or SAP), X, is estimated by considering discounted values of all immediate successor states (or SAPs) and the reinforcements achieved in the transition between X and the successor.

Thus, the values of successor states are *backed up* to those of their predecessors in a recursive process that gradually (through many episodes of search and backup) produces reasonably accurate value assessments for each state.

11.2.2 Three RL Methods and Their Backup Routines

The majority of RL methods employ the same basic concepts of agents, environments, policies, and value functions; and all utilize the basic framework of the Bellman equations (dynamic programming equations) (Bellman, 1957) in that the value of a state (or SAP) derives from the value(s) of its successor state(s) (or SAPs) and the immediate rewards achieved in transitioning to those successors.

The three main approaches described by Sutton and Barto (1998) and briefly reviewed here are dynamic programming (DP), Monte Carlo methods (MC), and temporal difference learning (TD). These vary across a few key features.

Dynamic programming (Bellman, 1957) is a brute force approach to RL that involves multiple rounds of complete value-function updates. A DP algorithm never actually attempts to solve the puzzle by search, i.e., by beginning at a start state and working toward the end state. Rather, it implicitly begins at the goal state and spreads reinforcement information backward to all possible predecessor states, which in turn propagate the information further backward. DP also assumes a complete environmental model (i.e., a complete set of accurate state transition probabilities and rewards associated with those transitions), with which it can precisely predict all possible transitions and reinforcements. If the state space is of reasonable size, DP can find optimal solutions, and if the space is too large, DP can still find reasonable approximations to optimal policies.

Monte Carlo methods do not require complete environmental models and do not perform complete value-function updates. Rather, they use experience in solving the problem to build approximations for transition probabilities and rewards, and they only update the value function for those states that are encountered during an episode of problem solving. These updates occur at the end of each episode.

Temporal difference methods are very local in nature. They update the value of a state immediately after transitioning from it and observing both the value of its successor state and the immediate reinforcement. Furthermore, they do not depend upon complete environmental information. TD algorithms include a step parameter, k, which indicates the number of successor states in an episode that will be used to update the current state's value.

Figure 11.4 illustrates the differences in backup procedures for these three RL methods. Notice that in DP the update scheme has considerable breadth but a restricted depth (of 1, although that can be parameterized as well). Breadth stems from the complete environmental model, which allows the algorithm to foresee all possible successor states and rewards based on all possible actions. In DP, *every*

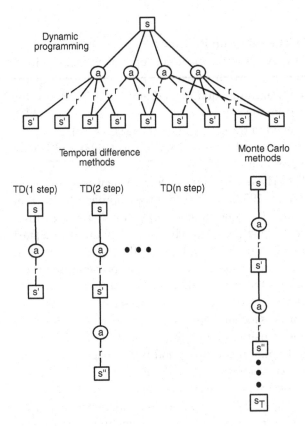

Figure 11.4
Backup schemes for three primary RL approaches. (*Top*) In dynamic programming a state, s, updates its value based upon all possible actions and all possible immediate successor states, along with the rewards attained in those transitions. (*Bottom left*) In temporal difference learning, value updates depend only upon the values of the next k states visited after state s during a task episode, where k is the step parameter. (*Bottom right*) In Monte Carlo methods no value updates occur until the end of an episode, when all states along the chain receive value information backed up from all successors between themselves and the terminal state, s_T.

state in the model updates its value based on this broad lookahead, in every round of the algorithm.

The other two methods lack a complete model and must experiment by selecting actions, accepting their consequences, and selecting more actions on the path to a terminal state. Whereas the TD methods vary as to the number of backup steps (i.e., the number of sequential successor states whose values are used to update the current state's value), the MC approach waits until the end of an episode and then gives each state in the episode a complete linear backup from s_T.

To illustrate how these backup schemes differentially affect the status of an update function, Figure 11.5 shows a simple maze search problem in which the initial status of the update function is the same for each method. However, after one round of updating, the differences are clear.

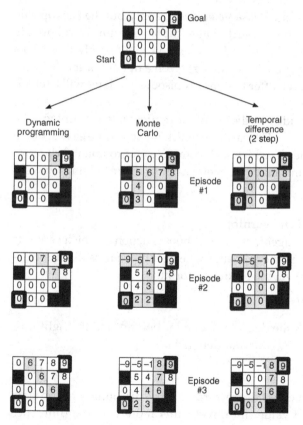

Figure 11.5
Three successive backup rounds for the three different RL approaches

The DP column of figure 11.5 displays three rounds of backup, beginning at the goal state and fanning back to predecessors. On each round, every state gets updated based on all possible actions (left, right, up, or down movement) and the values of all immediate successor states. However, most cells have successors with values equal to 0, so only those cells closest to the goal experience a change of value. After a few rounds, the wave of change gradually propagates back toward the start state. So despite the goal-to-start movement of this wave, the individual updates occur in a forward, breadth-first manner (as shown in figure 11.4).

In the Monte Carlo column, each row denotes a different problem-solving episode, with the path of that episode shown as shaded cells. Upon episode completion, all states along the path receive updates based upon all their successors, both immediate and distant. In episode 2 the agent reaches the upper left corner and attempts to move off the board, thus incurring a severe penalty, whose negative effect propagates back along the episode's path.

In the temporal difference column, the same episodes apply, but the backup rule only considers two steps into the future, such that s_t, the state at time t, will update its value based upon the values $V(s_{t+1})$ and $V(s_{t+2})$. As shown, only those cells within a few steps of the goal (or penalized corner) receive updates after the three episodes in this example. However, after two more episodes, updates will back up all the way to the start state.

This section only gives some idea of the scope of reinforcement learning. The remainder of this chapter focuses on TD learning, which is usually the easiest of the three to apply to realistic problems, which lack complete environmental descriptions, have huge state spaces, and may not even admit problem solving by well-defined episodes.

11.2.3   Temporal Difference (TD) Learning

In typical TD versions of RL, the agent, in state s, chooses action a, which converts the system to state s' and incurs reinforcement r. The agent then uses this information to immediately update $V(s)$ via the following formula:

$$V(s) \leftarrow V(s) + \alpha[r + \gamma V(s') - V(s)], \tag{11.6}$$

where α is the learning rate, γ is the discount factor (as before), and the rightmost bracketed term is known as the *TD error* and denoted δ:

$$\delta = r + \gamma V(s') - V(s). \tag{11.7}$$

TD error is the difference between (1) the reward plus the discounted future state value and (2) the current state value. Intuitively, this represents the difference between r (the actual reward at time $t+1$) plus $V(s')$ (the prediction of future rewards at time $t+1$), and $V(s)$ (the prediction of future rewards at time t). Thus,

TD learning refines the prediction of the future given by $V(s)$ to incorporate one new fact r and a prediction $V(s')$ that is further into the future but presumably closer to the goal and thus a more accurate prediction than $V(s)$.

Q-Learning   Q-learning is an off-policy form of TD learning invented by Watkins and Dayan (1992). It focuses on the evaluation of state-action pairs; $Q(s, a)$ denotes the value of choosing action a while in state s. Temporal differencing arises in updating $Q(s, a)$ for the current state, s_t, and most recent action, a_t, by using the difference between the current value of $Q(s_t, a_t)$ and the sum of the reward, r_{t+1}, received after executing a_t in s_t and the discounted value of the resulting new state, s_{t+1}, where this value is greedily based on the best possible action that can be taken from s_{t+1}, or $\max_a Q(s_{t+1}, a)$. Hence, the complete update equation is

$$Q(s_t, a_t) \leftarrow Q(s_t, a_t) + \alpha[r_{t+1} + \gamma \max_a Q(s_{t+1}, a) - Q(s_t, a_t)]. \tag{11.8}$$

As before, γ is the discount rate and α is the step size or learning rate. Once again, the bracketed expression is δ, the TD error. The logic behind this update rule is straightforward: if performing a in s leads to positive (negative) rewards and good (bad) next states, then $Q(s, a)$ will increase (decrease), with the degree of change governed by α and γ.

Eligibility Tracing   As shown in figure 11.5, backups in Monte Carlo methods and in TD learning are similar, particularly with n-step temporal differencing. Essentially, the step size is a parameter that determines the location of the RL algorithm on a spectrum from pure, single-step TD to MC.

   A slightly different approach to parameterizing this relation between TD and MC employs *eligibility traces*. In RL these are implemented as simple real-valued flags, attached to each state, s, that indirectly indicate the elapsed time since s was last encountered during problem-solving search. As this time increases, the eligibility decreases, indicating that s is *less deserving* of an update to $V(s)$, or $Q(s, a)$, based on any recent rewards or penalties. Conversely, states with high eligibility should reap the full benefits (or suffer the full punishment) of a recent reinforcement. In short, eligibility traces handle the *temporal credit assignment problem*.

   Formally, the eligibility trace for state s at time t is denoted $e_t(s)$ and updated as follows:

$$e_t(s) = \begin{cases} \gamma \lambda e_{t-1}(s) & \text{if } s \neq s_t \\ \gamma \lambda e_{t-1}(s) + 1 & \text{if } s = s_t, \end{cases} \tag{11.9}$$

where s_t is the state encountered at time t, γ is the discount factor, and λ is the *trace decay* factor. Together, these two factors (both in the range [0, 1]) determine the rate at which an eligibility trace decreases with each passing time step after a state has been visited, i.e., after $s = s_t$. Notice that when $s = s_t$, $e_t(s)$ gets incremented by 1, but on all other steps it will decay as long as $\gamma\lambda < 1$.

In TD learning with eligibility traces, often denoted TD(λ), the basic sequence of events is repeated for each step of an episode (Sutton and Barto, 1998, 174):

1. $a \leftarrow$ the action with highest probability in the policy component $\Pi(s)$.

2. Performing action a from state s moves the system to state s' and achieves the immediate reinforcement r.

3. $\delta \leftarrow r + \gamma V(s') - V(s)$.

4. $e(s) \leftarrow e(s) + 1$.

5. $\forall s \in S$.

    a. $V(s) \leftarrow V(s) + \alpha \delta e(s)$.

    b. $e(s) \leftarrow \gamma \lambda e(s)$.

At step 3 of this algorithm, the TD error, δ, is computed based on the values of the current and best-successor states, along with the reward, just as before. At step 5a, that error, whether positive or negative, is then applied to every state of the search space, as the $V(s)$ value of each state gets updated, and by an amount directly proportional to the eligibility trace of s. $e(s)$, At step 5(a), α is a learning rate. The eligibilities are updated at step 4 (for the current state) and then decayed for all states at step 5b.

In practice, TD(λ) may use various tricks, such as a cache containing only those states in the current episode, to avoid updating the value of every state after each action choice, but in theory, each state gets an updated value on every step of the algorithm, with the size of each update diminishing quickly for $\lambda < 1$.

For Q-learning, the accommodation of eligibility traces works similarly, with the key difference being in the computation of δ, the TD error:

$$\delta = r_{t+1} + \gamma \max_a Q(s_{t+1}, a) - Q(s_t, a_t). \tag{11.10}$$

The Q values for each state-action pair (SAP) are then updated, using eligibility traces and the TD error, as follows:

$$Q(s,a) \leftarrow Q(s,a) + \alpha \delta e(s,a). \tag{11.11}$$

Note that now the eligibility trace is attached to the SAP, not to the individual states, but the basic update is similar to that of step 5a in the algorithm.

The net effect of eligibility traces is best illustrated by the backups of figure 11.6. Compared to the two-step TD, the TD(λ) enables backup of the TD error through the entire state chain, with the value attenuated by $\lambda\gamma = (0.9)(0.9) = 0.81$ between each pair of states.

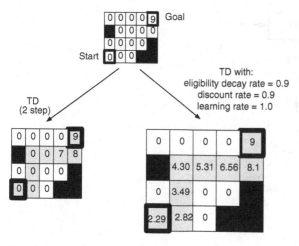

Figure 11.6
Two temporal-difference approaches: TD with two-step backups (from figure 11.5) and TD($\lambda = 0.9$), which uses eligibility traces. In the grid at the right, the search stumbles upon the upper right corner, which terminates the episode and begins the updating with a TD error, $\delta = 8.1$. Each state along the path gets a backed-up version of 8.1. Because of differences in eligibility, each previous state gets the fraction 0.81 ($= \gamma\lambda$) of its successor's update, as dictated by step 5a of the TD algorithm.

11.2.4   Actor-Critic Methods

A popular variant of TD learning, known as the actor-critic method, is characterized by a complete separation between the value function and the policy. The *actor* module contains the policy, $\Pi(s)$, whereas the *critic* manages the value function, $V(s)$. Many TD models fit into the actor-critic framework, but let us focus on TD(λ) and the use of eligibility traces to update both $\Pi(s)$ and $V(s)$.

In this approach, the critic module manages $V(s)$ and computes the TD error (which is based on $V(s)$ for the current and successor state). The actor then utilizes the TD error for its own update. This is shown in figure 11.7.

TD error (δ) is computed as in equation (11.7), and eligibility traces updated as in equation (11.9). In addition, the critic modifies its value function as in step 5a of the TD(λ) algorithm:

$$V(s) \leftarrow V(s) + \alpha\delta e(s). \tag{11.12}$$

Now, however, the policy updates rely directly upon the TD error and thus only indirectly upon $V(s)$, whereas in the standard generalized policy iterator (Sutton and Barto, 1998), the policy directly accesses the value function to compute changes.

The probabilities of choosing each action are modified in a simple two-step process. First, the current probabilities are modified by the TD error, attenuated by the

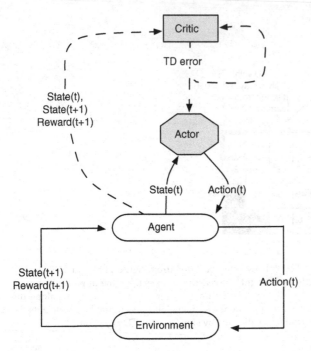

Figure 11.7
The actor-critic paradigm. Solid lines show a behavioral sequence wherein the actor selects an action based on the policy and performs it in the environment, which then returns a new state and (possibly null) reward. Dotted lines show the learning process, whereby the states at two successive time points along with the reinforcement are sent to the critic, which uses that information to compute the TD error, which is then used (by the critic) to update $V(s_t)$ and sent to the actor to update the policy.

eligibility trace:

$$\Pi(s,a) \leftarrow \Pi(s,a) + \alpha \delta e(s,a). \tag{11.13}$$

Then all modified probabilities (which may not sum to 1 after the update) are normalized (to sum to 1):

$$\Pi(s,a) \leftarrow \frac{\Pi(s,a)}{\sum_{a' \in \Pi(s)} \Pi(s,a')}. \tag{11.14}$$

In considering examples of reinforcement learning in artificial neural networks and the brain, the TD(λ) actor-critic model moves to center stage for these reasons:

- Temporal differencing is the most feasible approach to state and policy updating because of

☐ the enormous search space size of typical ANN problem domains, which make comprehensive updates very costly;

☐ the difficulty of wiring neural circuitry to handle DP or MC approaches.

• Eligibility traces are a simple bookkeeping mechanism requiring no extra caches (of, for example, all states in the current episode) or (temporary) rewiring of the network.

• The actor-critic separation of $V(s)$ from $\Pi(s)$ maps nicely to modular neural networks, whereas a tightly integrated combination of the two does not.

Note that although RL learning is typically characterized as an adaptive paradigm in which feedback only occurs intermittently (such as at the end of a task), TD methods move RL much closer to supervised learning in that the TD error signal can be computed at each timestep and immediately employed to modify both state values and policy components (which are just weights in an ANN). The key difference between TD and supervised learning is that in the latter the error signal is generated by an external observer, who presumably knows the correct responses to each input. In TD the error signal is self-generated, based on deviations in the system's own predictions (of distances from intermediate to goal states).

11.3 Reinforcement Learning and Neural Networks

A wide variety of natural and artificial systems combine RL and ANNs, but in two fundamentally different ways.

First, AI systems frequently employ ANNs to serve as actors or critics to perform Q-learning. These ANNs are typically trained by methods other than reinforcement learning, such as evolutionary algorithms or backpropagation. They constitute approximations to complete value functions and policies, which are infeasible to exhaustively compute in large problem domains. Thus, the ANN makes RL fully realizable in situations other than the toy problems of AI and RL textbooks.

Alternatively, neural networks can be trained using RL principles. As mentioned earlier, intermittent feedback and credit assignment issues are ubiquitous in real-life situations, whereas the constant feedback required by supervised methods is a rarity. Hence, networks deployed as the controllers of various agents (including living organisms) can benefit from RL.

This section includes a few examples of each such RL-ANN combination.

11.3.1 Reinforcement Learning via Artificial Neural Networks

A classic combination of ANN and RL was introduced by Ackley and Littman (1992). As shown in figure 11.8, they used an evolutionary algorithm to evolve

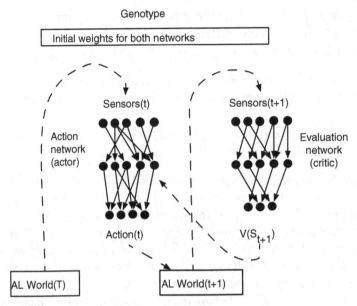

Figure 11.8
A pair of evolved neural networks used by Ackley and Littman (1992) to perform reinforcement learning in an artificial life (AL) World. The current state of the AL world is input to the actor network, which chooses an action that (potentially) causes a change in the world state. This new state, as perceived by the agent's sensors at time $t+1$, is fed into the critic, which outputs an evaluation of that state. Whether positive or negative, the evaluation is used to modify the actor's weights so that similar actions are chosen or avoided, respectively, in similar future situations.

the initial weights of an interacting pair of neural networks that corresponds to an actor and critic. The weights of the critic network remain fixed throughout the agent's lifetime, whereas the actor weights are modified by a variant of backpropagation, wherein the output from the critic network (an evaluation of the next state of the system) acts as an error signal for supervised learning.

By analyzing the population variance of each gene in the genome, Ackley and Littman could assess the selection pressure upon different parts of the genome, where low variance indicates high selection pressure. This revealed that selection pressure for a good critic network was very high during the early generations, when the critic was essential for proper learning in the actor. Over time, however, as the actor's circuits began to evolve proper weights (and thus were not dependent upon backpropagation learning to achieve good behavior), the selection pressure on the critic diminished, as reflected by increased variance among the genes coding for critic weights.

Along with providing interesting evidence for the Baldwin effect (Baldwin, 1896; Weber and Depew, 2003), an intriguing theory of learning's ability to influence evolution, these simulations showed that an evolutionary algorithm could tune an actor-critic network for RL. Although their networks did not compute TD error or use eligibility traces, the essence of value functions and policies was embodied in the network pair.

11.3.2   TD-Gammon

Perhaps the most famous combination of RL and ANN is Tesauro's backgammon player (Tesauro, 1995), known as *TD-Gammon*, which is considered one of the best players in the world (among humans or machines). This work serves as a strong advertisement for RL in general, and for the use of ANNs in complex domains for which RL has no chance of generating accurate evaluations for every state in the (immense) search space.

TD-Gammon's ANN functions as a critic to assess $V(s_t)$ for the state of the backgammon board at any time t. The ANN uses backpropagation to learn the proper mapping from s_t to $V(s_t)$, with both the TD error and eligibility traces playing key roles in the training process, which originally comprised hundreds of thousands of backgammon games that TD-Gammon played against itself.

Figure 11.9 outlines the basic algorithm used by TD-Gammon. The state of any backgammon board is encoded by 198 input neurons, while between 40 and 160 hidden nodes have been used in various versions of the system. At any time point t in the game, the following algorithm applies:

1. Encode the current state of the backgammon board, s_t, and run it through the critic ANN, yielding $V(s_t)$.

2. Roll the two dice.

3. Consider all possible moves from s_t given that roll of the dice. Each such move constitutes an *action* in the search process, with action i producing the next state s_{t+1}^i and possibly achieving a reward, r^i. The only reinforcement given by TD-Gammon is a positive reward for a win; all else yields $r^i = 0$.

4. Choose that move, a^*, such that

$$a^* = \operatorname*{argmax}_{a^i} r^i + V(s_{t+1}^i). \tag{11.15}$$

Here, $V(s_{t+1}^i)$ is computed for each possible future state s_{t+1}^i by running it through the critic ANN, as illustrated by the miniature versions of the ANN in the search tree at the bottom right of figure 11.9.

5. Compute the TD error: $\delta_t = r^* + V(s_{t+1}^*) - V(s_t)$.

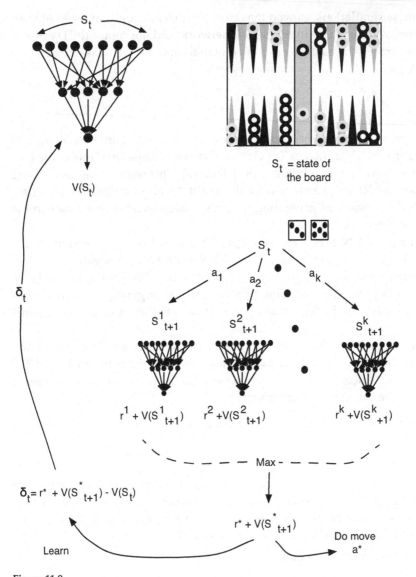

Figure 11.9
The general operation of TD-Gammon. (*Top right*) Board configurations are converted into 196-element vector states. (*Top left*) When fed into the critic ANN, vector states produce outputs corresponding to $V(s_t)$. (*Bottom right*) Given the current board state and a roll of the dice, many actions are possible, with each leading to a new state at time t+1. Each such state can be evaluated using the same critic ANN, and that action producing the maximum combination of reward and state evaluation is chosen. (*Bottom left*) The reinforcement combined with the evaluation differences of the two successive states yields a TD error term, which provides the basis for an immediate update of the critic ANN.

6. Use δ_t as the error signal for gradient descent updating of the ANN, where each ANN weight, w_j, changes as follows:

$$w_j \leftarrow w_j + \alpha \delta_t e_j, \tag{11.16}$$

where α is the learning rate and e_j is the eligibility trace associated with w_j. The eligibility trace decays using the standard formula $e_j \leftarrow e_j \gamma \lambda$, but it is incremented via

$$e_j \leftarrow e_j + \frac{\partial V(s_t)}{\partial w_j}. \tag{11.17}$$

In other words, the more w_j contributes to $V(s_t)$, the greater the increase to e_j.

Initially, the critic has random weights and gives meaningless evaluations, but even poor backgammon play eventually leads to a win, particularly when the system plays against itself. Upon achieving the winning state, s_T, and its associated positive reinforcement, the immediate predecessor state, s_{T-1}, will receive a boost in its evaluation via a large TD error. The critic ANN will then use this error to increase its output response to s_{T-1}. Through the course of thousands of games, the good evaluations of next-to-winning states will get backed up to earlier and earlier game states, and the ANN will begin to produce realistic evaluations for many game states, from late, intermediate, and even early stages of play.

The system truly *bootstraps* its way to success. TD-Gammon does so well that human players have begun to adopt some of its rather unusual strategies.

11.3.3 Artificial Neural Networks That Adapt using Reinforcement Learning
Returning to the diagram of figure 11.1, the credit assignment problem for artificial neural networks appears less daunting now that we have a mechanistic understanding of RL.

Figure 11.10 revisits the problem armed with two key concepts from TD learning: eligibility traces and TD error. Here, each synapse in the network includes an eligibility trace e, whose value is reset to 1 whenever both ends of the synapse are active within a relatively narrow time frame (in typical Hebbian form). In all other situations, the trace decays.

The network can presumably calculate the TD error, δ_t, at each timestep and then use it to update the value function. In addition, it uses the combination of eligibility and TD error to modify the weights of individual synapses, and it is these changes that embody policy updating, since the strengths of connections between nodes represent the likelihood of each node's exciting its downstream neighbors.

Figure 11.11 gives a broader view of a general RL network, which can house the rudiments of both an actor and critic. First, assume that neurons in the upper layer (A) have additional efferent connections (not shown) to an action-producing area, such as premotor or motor regions of the brain. Thus, A constitutes an actor circuit, wherein sequences of patterns in A map to action sequences; and a strong tendency for pattern P1 to activate pattern P2 entails that the network sanctions the action of the latter as a proper follow-up to the action of the former. In general, it helps to just think of these patterns as states, with varying levels of importance (value) to the system; many of these states can lead to actions via their connections to other parts of the system.

The value of any pattern P1 is reflected in the strengths of connections emanating from its constituent nodes. These can be divided into three classes:

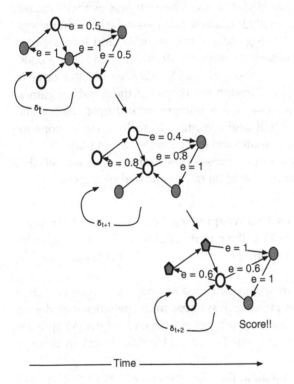

Time ⟶

Figure 11.10
Credit assignment problem with eligibility traces and error terms of TD learning. Notice that the long feedback links of figure 11.1 are replaced by local feedback loops, since TD error can be computed at each timestep, including the one at which the score/reward occurs, so there is constant feedback, and eligibility traces e keep track of the elapsed time since a synapse was last active, with greater delays indicating less receptivity to the latest error feedback.

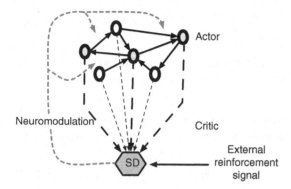

Figure 11.11
Neural network that performs reinforcement learning. SD denotes a saliency detector, while the neuromodulator is a diffuse signal sent to all synapses in the actor network. SD and its incoming and outgoing connections constitute the critic. Shaded nodes in the actor network represent neurons that recently fired, and thick dashed lines entering SD denote recently active synapses.

1. Connections among the neurons of P1
2. Connections from the neurons of P1 to the neurons of other patterns, such as P2
3. Connections from the neurons of P1 to the saliency detection level (SD)

When class 1 links have high synaptic strength, then P1 has a more likely chance of becoming active: when a subset of P1 neurons fires, the network has a tendency to complete the pattern. High weights among the class 2 links bias the network such that P1 often activates P2 as its successor. Finally, the class 3, P1-to-SD, connections provide a more explicit representation of P1's value: strong links mean that SD receives considerable stimulation when P1 is active, and this in turn can result in large amounts of neuromodulator being sent back to A.

In this abstract model the critic resides in SD, its neuromodulator-producing efferents, and all its afferent connections, whose strengths reflect the values of the actor layer patterns in which they participate. SD can trigger on a high-valued pattern or on an external reinforcement. As described later in a popular interpretation of the basal ganglia, SD can also trigger on the comparative values of patterns, thus performing a crude form of temporal differencing, although this requires a more elaborate connection topology between the upper and lower networks.

The critic's neuromodulatory signals, which can help strengthen or weaken synapses, can have at least three important effects corresponding to the three classes of synapses. For example, consider a simple case in which P1 is active, followed closely by P2, whose activity coincides with a positive external reinforcement (thus guaranteeing that SD fires and broadcasts neuromodulator). This causes P2's inherent value to increase in three ways:

- P2's class 1 synapses strengthen, thus increasing the likelihood of pattern P2 forming in the future. In geometric terms, the basin of attraction for P2 becomes wider and/or deeper.
- The P1-to-P2 class 2 synapses strengthen, thus increasing the likelihood that P2 is the preferred successor to P1. In geometric terms, the barrier between the basins for P1 and P2 is reduced.
- The P2-to-SD class 3 synapses strengthen, thus raising the explicit value of P2, i.e., the value upon which the critic bases decisions, in, for example, temporal difference calculations.

In all three updates, the eligibility traces (not shown in figure 11.11) play a vital role in distributing credit/blame to the proper synapses. Hence, the credit assignment sequence of figure 11.10 arises from the neural architecture of figure 11.11. A more advanced version might include a multilayered actor circuit such that pattern P1 in layer 1 could activate P2 in layer 2, which would then stimulate its corresponding action. By expressing the patterns in different layers, it becomes easier to segregate intralayer pattern completion from interlayer sequential pattern activation (see chapter 7).

11.3.4   The Emergence of RL Mechanisms

The abstract architecture of figure 11.11 has several different manifestations in natural (Barto, 1995; Houk et al., 1995; Prescott et al., 2003) and artificial (Niv et al., 2002; Soltoggio et al., 2008; Izhikevich, 2007) neural networks. These topologies may arise from evolutionary and developmental processes as fully formed anatomical structures, or they may slowly emerge during lifetime learning. Contemporary neuroscience cannot fully explain origins of the former type, but several interesting studies in computational neuroscience shed light on the latter while giving a few hints as to the former.

The most impressive example of emergent RL comes from Izhikevich (2007), whose spiking ANN (see chapter 6) is elegantly extended for handling eligibility traces, wherein each synapse has its own trace/plasticity variable, c, which is updated at each timestep of the simulation via the following:

$$\frac{dc}{dt} = \frac{-c}{\tau_c} + \delta(t - t_s)\text{STDP}(\tau), \tag{11.18}$$

where τ_c is the decay time constant for plasticity, t_s is the time of the most recent spike of the pre- or postsynaptic neuron, and $\delta(t - t_s)$ is the Dirac function, which only returns a 1 if t and t_s are equal; otherwise, it returns 0. Hence, the second term only comes into play on a timestep when one of the two ends of the synapse spikes. Finally, τ is the difference between the most recent pre- and postsynaptic spikes, and $STDP(\tau)$ computes the change in synaptic strength for τ using a conventional STDP curve (Fregnac, 2003), which gives weight increases for pre-before-post firing, and decreases for post-before-pre spikes.

Izhikevich then updates each synaptic weight (w) using c and d (the global concentration of neuromodulator, e.g., dopamine):

$$\triangle w = cd. \tag{11.19}$$

On any timestep devoid of pre- and postsynaptic activity, the eligibility simply decays. Spikes, however, produce an STDP-induced change in c, provided that $|\tau|$ lies within the typical STDP time window (about 25 ms in Izhikevich's model). This update to c is then passed on to w, but only in the presence of a sizable neuromodulator concentration (d). In the model, d decays on each timestep and is only fortified when a salient event occurs, such as when the agent acquires food.

As long as τ_c is not set too low, the eligibility trace enables a combination of pre- and postsynaptic spikes to affect w even though the reinforcement signal (i.e., jolt to d) comes several seconds later. In fact, even if there is a small degree of intervening spiking of the pre- or postsynaptic neuron (i.e., noise), this will not dilute c as long as the noisy spikes are not correlated within the STDP time window. Hence, a correlated firing sequence of the pre- and postsynaptic neuron can be *remembered* for several seconds by c, even in the face of noise, until neuromodulator arrives to *seal the deal* and modify w.

Using this simple scheme, Izhikevich achieves the simulated emergence of operant conditioning in a population of 1,000 neurons. Even more impressive, classical conditioning arises from the same group of neurons; in fact, what emerges is a global interaction topology whose functionality closely resembles temporal differencing.

Figure 11.12 illustrates several interesting aspects of these operant and classical conditioning simulations. Izhikevich divided a network of 1,000 neurons into many overlapping subsets, five of which are shown in the figure:

- S neurons represent an input stimulus.
- In the operant conditioning task, groups A and B represent alternative actions in response to stimulus S. They become involved in a spike-producing competition.
- In classical conditioning, group A neurons encode the unconditioned stimulus, while C neurons represent the conditioned stimulus.

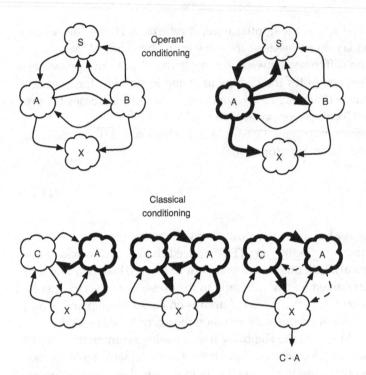

Figure 11.12
(*Top*) Operant conditioning wherein repeated dopamine is given in response to A but not B. This strengthens A's intra- and intergroup connections, making it the more likely response to S in the future. (*Bottom*) Presentation of conditioned stimulus C prior to unconditioned stimulus A (which originally triggers saliency detector X) leads to increase of C's intra- and intergroup connections, since C is active just prior to the dopamine signal from X. Over time, C becomes strong enough to fire X on its own, leaving A to fire after X, which reduces the A-to-X synapses (bottom right, dotted lines) by the long-term depression (LTD) component of STDP. The average strength of intragroup synapses is denoted by the thickness of each group's outline; arrow thickness reflects the intergroup synaptic strength. Though neuron groups may share neurons, they are drawn disjointly for ease of readability.

- Neurons in X are presumed to project to the ventral tegmental area (VTA), which secretes dopamine when stimulated. Hence, X is akin to SD in figure 11.11.

To show operant conditioning (i.e., the ability to learn to perform actions that bring reward), the network receives repeated rounds of random stimulation of the S neurons. This produces activation throughout the network, and if more A than B neurons produce spikes, then dopamine (*d*) is boosted by what can be viewed as the external stimulation of area X. This produces long-term potentiation (LTP) on all synapses with pre-then-post synaptic activity within the STDP window. Since more A than B neurons were active, the S-to-A, A-to-A, and A-to-X connections will exhibit more LTP than the efferents and afferents of B, in particular, the S-to-B and B-to-X links. Over many trials, this produces a strong preference to perform action A in situation S; and by increasing the A-to-X connection weights, it allows A to directly stimulate dopamine production.

Thus, the network learns two things: to prefer action A over B in response to stimulus S, and to predict reward (by inducing dopamine production) when area A activates.

Switching to a classical conditioning context but retaining the results of the operant simulation, area A is now viewed as an unconditioned stimulus, US: when active, it immediately invokes an expectation of reward via its strong links to X and dopamine production. In the classic case of Pavlov's dog, US is the sight of food, which invokes salivation, a salient indicator of future pleasure.

In his classical conditioning simulations, Izhikevich introduces another group of neurons (C) to represent a conditional stimulus (e.g., the bell in the Pavlovian experiments). He randomly stimulates these neurons on average 1 second prior to stimulating the A neurons, which excites the X neurons, leading to dopamine production and synaptic modification. Now, connections involving both C and A are fortified; and in particular, the C-to-X connections will, on average, experience more LTP than LTD (since the neurons of C begin firing before those of X). Through repeated trials, C becomes a direct indicator of future reward, and as the C-to-X bond strengthens, X neurons will begin to spike before those of A (since stimulation of C precedes that of A). As this trend continues, the A-to-X synapses will experience LTD, yielding the somewhat counterintuitive scenario in which the US no longer invokes the reward expectation.

Interestingly, this is exactly the same phenomenon observed in the brain's basal ganglia: dopamine is only secreted when reinforcement is *not* expected (Fellous and Suri, 2003; Schultz, 1998), i.e., not predicted by a prior context. Once a conditioned stimulus arises to give advanced warning of a reward, the unconditioned stimulus loses its ability to stimulate dopamine production. As described in great detail in Downing (2009), the basal ganglia enable these predictions to back up further and further in time—a phenomenon that also emerges in Izhikevich's models, where a second conditioned stimulus, C2, prior to C, takes over the role of exciting X, while C and A lose that ability because of LTD. Neuroscientists attribute the brain's main RL capabilities to the basal ganglia, and Izhikevich shows that this same reward-predicting behavior can emerge from nothing more than a collection of spiking neurons whose synapses adapt via STDP and eligibility traces.

Other studies (Soltoggio and Steil, 2012) show similar emergence using only simple Hebbian learning among rate-coding neurons: synaptic tuning stems from the comparative activation levels of neurons, not their actual spike times. A related set of simulations (Niv et al., 2002) involving an abstract foraging task show that when given a fixed ANN topology but parameterized Hebbian learning rules, evolution favors parameters that embody temporal differencing. On this same foraging task, other simulations (Soltoggio et al., 2008) show that when given the freedom to configure ANN topologies, evolution chooses combinations of normal neuron-to-neuron connections plus neurons that broadcast neuromodulatory signals. Together, these results indicate a potential bias of both evolution and learning toward the emergence of the rudiments of RL; and these may have been evolutionary precursors to more anatomically grounded RL machinery believed to reside in the basal ganglia.

11.3.5   Reinforcement Learning in the Basal Ganglia

The basal ganglia (BG) of the mammalian brain exhibit several characteristics of a reinforcement learning system, thus leading many prominent researchers to posit RL as a central functionality of this region (Doya, 1999; Houk et al., 1995; O'Reilly and Munakata, 2000). RL systems learn associations between environmental (and bodily) states and various rewards or punishments (i.e., reinforcements) that those states may incur, either immediately or at some time in the future. Thus, the system learns to predict the reinforcement from the state. Naturally, RL provides a survival advantage, since it enables organisms to behave proactively instead of merely reactively.

As shown in figure 11.13, the basal ganglia are large midbrain structures that receive convergent inputs from many cortical areas onto the striatum (consisting of caudate nucleus and putamen) and the subthalamic nucleus (STN). The striatal cells appear to function as a layer of competitive context detectors (Houk, 1995), since each neuron receives inputs from circa 10,000 cortical neurons; their electrochemical properties are such that they only fire if many of those inputs are active; and they have intralayer inhibitory connections.

Strong evidence (Strick, 2004; Graybiel and Saka, 2004) indicates that the basal ganglia are arranged in parallel loops wherein a striatal cell's inputs come from a region of a particular cortex, such as the motor cortex (MC). Their outputs to the substantia nigra pars compacta (SNc), substantia nigra pars reticulata (SNr), and entopeduncular nucleus (EP) are eventually channeled back to the MC in the form of both action potentials (via the thalamus) and the neuromodulator dopamine. A great majority of these loops appear to involve the prefrontal cortex (PFC) (Houk, 1995; Kandel et al., 2000; Strick, 2004), thus indicating BG contributions to attention, possibly as the mechanism for gating new patterns into working memory (Graybiel and Saka, 2004; O'Reilly, 1996).

Accounts of BG functional topology vary considerably (Houk, 1995; Houk et al., 1995; Prescott et al., 2003; Graybiel and Saka, 2004; Granger, 2006), but several similarities do exist. First, the striatum appears to consist of two main neuron

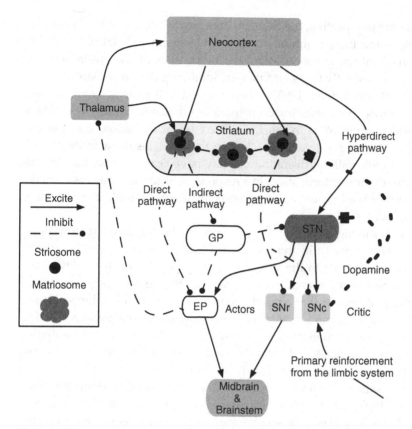

Figure 11.13
Functional topology of the basal ganglia and their main inputs. Based on Houk (1995), Houk et al. (1995), Prescott et al. (2003), Graybiel and Saka (2004). The actor denotes the direct outputs of the BG: EP and SNr, while the critic consists of the diffuse neuromodulatory output from SNc. Matriosomes are primarily gateways to the actor circuit, while striosomes have direct pathway links to both actors and critics.

types: striosomes and matriosomes, with the former surrounded by the latter. Several prominent researchers (Barto, 1995; Houk et al., 1995) characterize the BG as a combination of actor and critic, with the matriosomes and pallidal neurons (EP and SNr) as the actor's input and output ports, respectively, while the striosomes and SNc compose the critic. Although this characterization is not completely consistent with other sources, such as Joel et al. (2002), the matriosomes and striosomes are often characterized as respectively supporting action selection and state assessment (via dopamine signaling from SNc). See Houk et al. (1995) and Graybiel and Saka (2004) for overviews of the empirical data and theoretical models.

From an abstract perspective, the BG map contexts to actions. When a context-detecting matriosome fires, it inhibits a few downstream pallidal (GP and EP) neurons. In stark contrast to the striatum, the EP consists of low-fan-in neurons, most of which are constantly firing and thereby inhibiting their downstream counterparts in the thalamus (Houk, 1995). When a striatal cell inhibits a pallidal neuron, this momentarily disinhibits the corresponding thalamic neuron, which then excites a cortical neuron, often in the PFC. The cortical excitation links back to the thalamus, creating a positive feedback loop that sustains the activity of both neurons, even though pallidal disinhibition may have ceased. Thus, the striatal-pallidal actor circuit momentarily gates in a response whose trace may reside in the working memory of the PFC for many seconds or minutes (Houk, 1995; O'Reilly and Munakata, 2000).

Since the PFC is the highest level of motor control (Fuster, 2003), its firing patterns often influence activity in the premotor (PMC) and motor (MC) cortices, while the MC sends signals to the muscles via the spinal cord. In addition, the sustained PFC activity provides further context for the next round(s) of striatal firing and pallidal inhibition that embody context detection and action selection, respectively. Via this recurrent looping, the basal ganglia execute high-level action sequences. The situation-action rules housed within the BG may constitute significant portions of commonsense understanding of body-environmental interactions, whether consciously or only subconsciously accessible.

The BG learn salient contexts via dopamine (DA) signals from the SNc, which influence the synaptic plasticity of regions onto which they impinge (Kandel et al., 2000). DA acts as a second messenger that strengthens and prolongs the response elicited by the primary messenger. For example, when a striatal neuron, S, is fired via converging inputs from the cortex, the primary messenger is the neurotransmitter from the axons of the cortical neurons (C) that recently fired. The immediate response of those S' dendrites (D) connected to the active axons is to transmit a synaptic potential (SP) toward S's cell body. The summation of these D inputs will lead to S's production of a new action potential (AP). If dopamine enters these dendrites shortly after SP transmission, a series of chemical (and sometimes physical) changes occur which make those dendrites more likely to generate an SP (and a stronger one) the next time its upstream axon(s) produce neurotransmitters. Since the chemicals involved in this strengthening process are conserved, those dendrites that did not receive neurotransmitter may become less likely to produce an SP later on, even when neurotransmitters reach them. Thus, in the future, when the C neurons fire, the likelihood of S's firing will have increased, whereas other cortical firing patterns will have less chance of stimulating S. In short, S has become a detector for the context represented by C. Without the dopamine infusion, S develops no bias toward C and may later fire on many diverse cortical patterns.

In unfamiliar situations the SNc fires upon receiving stimulation from various limbic structures, such as the amygdala, the seat of emotions (LeDoux, 2002), which triggers on painful or pleasurable experiences. The ensuing dopamine signal encourages the striatum to remember the context that elicited those emotions—the stronger the emotion, the greater the learning bias. Because of the biochemical temporal dynamics (Houk et al., 1995), the striatal neurons that become biased (i.e., learn a context) are those that fired approximately 100 ms *prior* to the emotional response. Hence, the BG learn a context (C) that *predicts* the reinforcing situation (R).

Since dopamine signaling is diffuse, the matriosomes and striosomes in a striatal module are both stimulated to learn. Hence, the critic not only learns to predict important states but assists in the learning of proper situation-action pairs by the actor circuit.

Again, descriptions of the critical topological elements differ—see Joel et al. (2002) for a review—but many experts name two paths from the cortex to SNc (Graybiel and Saka, 2004; Prescott et al., 2003). The first, often called the *hyperdirect pathway*, bypasses the striatum and directly excites the subthalamic nucleus (STN), which in turn excites SNc. The second, termed the *direct pathway*, involves a strong inhibitory link from striatum to SNc. The hyperdirect pathway is quick but excites SNc for only a short period. The direct pathway is slower, but it inhibits SNc for a much longer period.

This timing difference between excitation and inhibition enables these predictions (of reinforcement based on context) to regress backward in time such that very early clues can prepare an organism for impending pleasure or pain (Downing, 2009). As pointed out by Joel et al. (2002), physiological evidence indicates that the excitatory and inhibitory signals to SNc cannot both come from the striatum but more likely from the prefontal cortex (via the hyperdirect pathway) and the striatum, respectively.

11.3.6 An Actor-Critic Model of the Basal Ganglia

Houk et al. (1995) analyze the basal ganglia's anatomy with respect to the formal model of a TD(λ) actor-critic. The result is a compelling description of the basal ganglia as an RL system.

As shown in figure 11.13, there are many pathways from the striatum to the SNc (which produces dopamine). There is still controversy as to the exact topology of this circuitry, but many of the models agree on the presence of both a fast excitatory connection from the input layers (such as the striatum and cortex) to both the

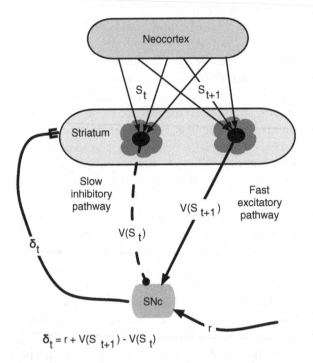

Figure 11.14
Proposed implementation of a TD critic by the SNc of the basal ganglia. Striatal detectors are denoted by black ovals.

SNc and the action/actor nucleii (such as the EP), and a slower but longer-lasting inhibitory connection.

Thus, s_t, the inhibitory signal from a context detected at time t, could reach SNc simultaneously with s_{t+1}, the excitatory signal from a context detected later, at time $t + 1$. The reward for time $t + 1$ would also reach SNc at a similar time point.

Figure 11.14 diagrams the basic relations between the three signals converging upon SNc. In this model it is assumed that the strength of the signals from the striatum reflect the *value* of the contexts that excited the given striatal detectors. Thus, the signal from the striosome for s_t to SNc represents $V(s_t)$.

Since the $V(s_t)$ signal is inhibitory, while the other two are excitatory, the SNc can be viewed as additively combining these factors to produce a term similar to the TD error:

$$\delta_t = r + V(s_{t+1}) - V(s_t). \tag{11.20}$$

From the earlier formalism for TD learning, remember that δ_t denotes the difference between the prediction (of future reinforcement) associated with one problem-solving state and the combination of immediate reward and prediction associated with its successor state. Thus, it reflects the error in the original prediction. A higher error indicates more *surprise* at time $t+1$ based on the system's expectations at time t. In the brain, dopamine signaling correlates with this surprise factor manifest in the mismatch between expectations and reality, or between expectations and successor expectations (Fellous and Suri, 2003; Schultz, 1998).

The magnitude of δ_t is reflected in the amount of dopamine broadcast by SNC, and this will in turn affect the degree to which synapses onto recently active striosomes are strengthened. And stronger synapses onto the striosome for s_t will insure that the signal from that striosome to SNc will be stronger in the future, thus representing an increase in the $V(s_t)$ signal.

Thus, a high value of δ_t leads to an increase of $V(s_t)$, just as in TD learning. In contrast, when $V(s_t) \approx r + V(s_{t+1})$, then $\delta_t \approx 0$, no dopamine is released, and the synapses onto the s_t striosome will not change, and thus neither will the strength of the $V(s_t)$ signal.

Furthermore, as mentioned earlier, there are chemical processes at work in the striosomal synapses that manifest eligibility traces, such that only synapses that were first activated within a small time window of the δ_t signal—and preferably about 50 ms prior—become stronger. Thus, the basal ganglia implement TD(λ) learning.

To summarize the argument so far, signals from the striosomal detectors to SNC can represent $V(s)$ for any context, s, and dopamine signals can control modifications of these signals based on a relation, detected in SNc, that is akin to TD error. This explains the critic component of the model.

In the actor circuitry, dopamine signals also have a strong effect upon synaptic change of the matriosomal neurons of the striatum, so the equivalent of a TD error also influences changes to the actor.

Assume that any active chain of neurons in the actor circuit corresponds to a situation-action pair (SAP). That SAP then leads to a new state, s^*, when its action is performed by the agent in the environment. If the TD error associated with s^*, δ_{s*}, is high, then the ensuing strong dopamine signal will lead to strengthening of the synapses in the SAP circuit. Hence, the given action will be more likely to be performed in the given situation, if it results in a *surprisingly good* state, i.e., one with a high TD error. Clearly then, the TD error governs modifications to both the actor and critic circuits.

In summary, the basal ganglia appear to be a prime example of reinforcement learning in a biological neural network. The following factors combine to support this argument:

- The separation of actor and critic circuits (in an area of the brain known for serial circuits with little overlap)

- A clear reinforcement signal to the SNc from the limbic system

- The differential time delays in excitatory versus inhibitory signaling from the striatum to the SNc
- The integration of inputs in the soma of SNc neurons resembling the calculation of a TD error term
- The correlation of this TD error with dopamine transmission, which extends throughout the basal ganglia and cortex
- The effects of dopamine upon synaptic strength that mirror the effects that TD error has on the value function and policy in TD learning
- The biochemistry of synaptic modification, which involves time delays and attenuations that resemble eligibility traces

So while a neuroscientist may point to the basal ganglia and say that they perform RL, largely because of the dopamine transmission, an AI researcher may take the argument much further and pinpoint a host of mechanisms that parallel the formal descriptions of RL provided by Sutton, Barto, Watkins, and others. Granted, some of these apparent brain-RL similarities rely heavily on speculation as to the exact nature of BG circuitry. Still, they provide a rich framework for explaining how the knowledge of complex sequences can emerge from local, Hebbian-style learning supplemented with a very general and diffuse feedback signal—one that has no instructional content as to how the myriad neurons should coordinate their activities but only conveys a simple message that something salient has occurred. Coordination emerges from local synaptic modifications carried out repeatedly over many problem-solving episodes, as the network forms an elaborate tapestry of mental and physical capacities from the simple products of trial-and-error search.

11.4   Experience Needed

As an algorithm, trial-and-error search is hardly a marvel of engineering; more likely, it plays the role of *last resort* when no truly insightful solutions exist. However, the complexity of many problems, particularly those requiring sequences of actions, can easily defy all rational problem solvers. In these cases, the best hope is random exploration with bookkeeping (i.e., memory) that gradually bootstraps itself to informed exploitation. On-the-job training is never easy, but the experience gained can produce a much more robust problem solver than one based on downloaded heuristics and expertise. By analogy, many companies are just as likely to promote a long-time, nuts-and-bolts employee to a top position as to hire an outsider with a Harvard M.B.A. Experience matters, immensely.

At first glance, the RL algorithm appears hopelessly resource-consuming, particularly with respect to dynamic programming approaches. But Monte Carlo methods simplify the paradigm considerably, and then TD techniques introduce a bootstrapping process that, as Tesauro shows, brings complex problems within the reach of RL, and additionally motivates realistic interpretations of brain circuits as reinforcement learners. The common occurrence of RL scenarios in real life makes these results all the more significant for AI.

For AI systems, the trick is to handle this bookkeeping in a reliable and efficient manner. Reliability involves recording the proper information, while efficiency reflects an accelerated transition from exploration to well-informed exploitation. An important lesson from Tesauro's work is that in complex domains with enormous state spaces, RL can still achieve this efficiency despite the inability to generate accurate evaluations for every state in the space. Instead, the ANN can serve as a stellar substitute for a complete lookup table by learning a *general evaluation function* that is just as applicable to completely new scenarios as to previously experienced conditions.

In effect, temporal differencing allows the agent to achieve the high-frequency feedback of supervised learning, since the evaluation of each new state can be compared to that of its predecessor to generate a TD error term, which directly governs weight modifications in the ANN. The agent therefore exploits its own predictions to generate a steady stream of learning stimuli.

In general, many ANN applications to cognitively challenging tasks (particularly games and puzzles) rely on an actor-critic separation in which the ANN constitutes the critic, by mapping states to evaluations, but the actor relies on more conventional code, such as A* or minimax search, which uses the critic's evaluations plus its own model of the task (e.g., transition tables giving the likelihood of moving from one state to another when performing a particular action) to select promising actions. So the ANN does not handle action selection or the interaction between criticism and action. The ANN has deep, implicit *understanding* of problem states and their salience, but still, claims of a neural network *playing* game X hint of exaggeration. In essence, standard search tools generate options using brute force or simple heuristics, and the ANN evaluates them. This is similar to problem solving with evolutionary algorithms, where relatively *dumb* genetic operators produce new variants, while significant task-specific information resides in the evaluation/fitness function.

As mentioned in the discussion of brain evolution, the key to exploiting emergent natural processes may lie in an understanding of the biases that have emerged to bridle the variation-generating processes. As discussed earlier and in many other neuroscience contexts, the simple constraints imposed by STDP-based

Hebbian learning and neuromodulation seem to have profound consequences upon the neural interactions that become canalized during an agent's lifetime. The biases that they impose may well have shaped the entire evolution of intelligent behavior. Izhikevich starts with these low-level mechanisms and produces the rudiments of RL, whereas others assume the basic RL machinery and teach or evolve ANNs to fulfill roles of critics or actors. From many different angles of inquiry, scientists are gradually gaining an appreciation for the powers of trial-and-error learning, its potential for AI, and its role in the saga of intelligence emerging.

12   Evolving Artificial Neural Networks

Evolving Artificial Neural Networks (EANNs) are the hallmark synthetic version of emergent intelligence. Though they vary along several dimensions, they often exhibit emergence across all three primary adaptive levels: evolution, development, and learning. They are also the default representation in evolutionary robotics (ER), the field that most tangibly illustrates the true potential of biologically inspired routes to AI.

In their seminal ER textbook Nolfi and Floreano (2000) cite several reasons for the popularity of ANNs and EANNs for robotic control:

- There is normally a smooth mapping between ANN parameters and the behaviors they produce. This enhances adaptive search.

- It is relatively straightforward to implement any or all of the three adaptive levels in EANNs, thus maximizing agent flexibility.

- Many aspects of ANNs can be evolved, including connections, weights on connections, neural time constants, learning rules, and neural layers.

- ANNs can easily link sensors to actuators in both continuous and discrete domains.

- Since artificial neurons typically sum many inputs to determine their output, they are robust to noise in several of those inputs.

- EANNs permit the investigation of biological questions about the evolution of intelligent behavior.

Not coincidentally, Nolfi and Floreano are two of the most influential pioneers and proponents of EANNs, which allow ER researchers to explore their own search space of possibilities for the emergence of intelligence. Multiple combinations of adaptive processes in a wide variety of task situations (using a diverse repertoire of robot morphologies) can all be examined within the same basic search and representational framework.

EANNs also play a key role in trial-and-error search (a topic of prime relevance for ER as well). As detailed in chapter 11, the computational complexities of reinforcement learning (RL) can be tamed by ANNs trained as search-state evaluation functions (critics), as in Tesauro's backgammon player (Tesauro, 1995). But the actor half of the actor-critic framework typically relies on tried-and-true, general-purpose GOFAI search algorithms such as A* and minimax, which use a few heuristics but have no deep insights into the domain. Sophisticated actors are difficult to hand design for complex, theory-weak domains, particularly those where the agent must perform many actions between feedback events. This paucity of corrective information also marginalizes backpropagation-trained ANNs.

Hence, to take full advantage of the neural network's ability to achieve complex mappings between situations and appropriate actions, a different approach to ANN parameter tuning is needed, and evolutionary algorithms have impressively answered the call. Now evolution comes up with parameter vectors (of synaptic weights, activation thresholds, time constants, and even topological configurations) that enable ANNs to fulfill starring actor roles in many practical (and often entertaining) AI systems. In these, one can honestly say that the ANN embodies the brunt of the system's intelligence.

12.1   Classifying EANNs

Several publications (Yao, 1999; Nolfi and Floreano, 2000; Floreano and Mattiussi, 2008; Floreano et al., 2008; Gauci and Stanley, 2010) are wholly or partly devoted to illustrating and classifying the breadth of EANN research. Those publications clearly motivate and inform this chapter; its fundamental EANN characteristics mirror many of those found there.

For our purposes, four (predominantly orthogonal) axes define the space of EANNs: genotype, phenotype, genotype-phenotype mapping, and adaptivity. The first, the distinction between data-based and program-based genotypes, is a clear separator of EANN systems. The second, phenotype, refers to those aspects of the ANN that are under evolutionary control. The third, genotype-phenotype mapping, denotes the direct or indirect nature of the conversion process from genotype to ANN. Finally, adaptivity involves the combination of evolution, development, and learning employed by the system.

Typically, EANN genotypes are data-oriented: they encode key parameters for either the ANN itself or for a generator function that will produce the ANN. However, some of most groundbreaking work in evolutionary computation has roots in a genetic programming approach to EANNs (discussed below).

As phenotypes, ANNs can offer up many features to evolutionary control, such as the number and sizes of neural layers; properties of synapses such as their weights and learning rules; the connection topology; and individual neuron parameters such as time constants, activation functions, and thresholds. All these options have been explored, often extensively, in the broad body of EANN research.

A diverse collection of genotype-phenotype mappings graces the EANN literature. These span a spectrum from the (simplistic but popular) direct encoding of synaptic weights for a fixed topology to highly sophisticated schemes in which (data- and program-oriented) genotypes specify developmental recipes for both coarse and fine-grained details of network topologies.

Finally, many subsets of the POE trilogy of adaptation—phylogeny, ontogeny, and epigenesis—appear in EANNs. By definition, an EANN needs to evolve, but no other adaptivity can be presumed. Many classic and contemporary EANN applications (P systems) simply evolve fixed weights to solve complex input-output mapping problems. More ambitious PE approaches evolve initial weights that a learning algorithm fine-tunes during the course of an agent's lifetime. Others—PO systems—evolve a developmental algorithm that produces both the topology and a set of fixed weights. The most challenging systems embody the entirety of POE by, for example, evolving developmental routines that generate topologies and learning schemes.

These four axes of classification will not be the basis for a thorough survey of the EANN field but merely a collection of characteristics that assist in the comparison of several systems in the next few sections of this chapter. I begin with a look at several original/classic forays into the field, followed by a slightly more methodical focus on one of the four axes, the genotype-phenotype mapping, and some of the more popular, successful or unique systems, grouped by their use of direct versus indirect encodings.

12.2 Early EANN Systems

Originally, EANNs were designed to *replace* learning with evolution, not to integrate the two mechanisms. In the late 1980s, AI researchers began to recognize that despite its major advance over the delta rule and perceptrons, backpropagation for multilayered, feedforward networks still suffers from many drawbacks:

- Like other gradient descent methods, it often gets stuck at local optima.
- Extensions of the algorithm to recurrent networks are complex and inefficient.
- The supervised learning that it embodies puts strong demands on the problem scenario: for each input, target outputs must be provided.

Many AI situations, especially those involving autonomous agents, can benefit from neural networks but cannot guarantee the presence of an omniscient teacher to provide target outputs for each input vector, particularly when an agent such as a robot samples its environment and acts many times per second. In addition, many of these problem-solving scenarios require ANNs with memory, which often entails recurrence. Robotic applications are similar to many game-playing situations in which feedback comes only at the end of the task; an instructor cannot evaluate and correct each individual move. Such tasks are amenable to reinforcement learning, while basic pattern clustering aspects of the problem can call for unsupervised learning. But supervised learning of correct moves is usually infeasible. Yet even when data exists for supervised learning, and the network includes no feedback, backpropagation search can still stagnate at local optima.

As a remedy, Montana and Davis (1989) proposed the evolution of ANN weight vectors for a standard supervised learning task: classification of patterns in sonar data. As shown in figure 12.1, their genotype was a simple list of real numbers, each representing one weight of the ANN. These were mutated by the addition of small increments (in the range −1 to 1). Crossover points were restricted to chromosomal locations that were boundaries between the afferent weight sets of individual neurons (i.e., the weights on all incoming links to a neuron). These sets of input weights are common building blocks in EANNs, so their crossover operator was biased to treat these as atomic units.

The Montana and Davis system outperformed standard backpropagation on the supervised learning task, but it was beaten by newer optimized versions. They therefore concluded that the approach was useful for supervised learning but probably best for unsupervised situations.

In backpropagation learning the data set is split into two parts: a training set s_1 and a test set s_2. Members of s_1 are then sent through the network many hundreds or thousands of times in order to gradually adjust the weights to minimize output error. When training is complete, the members of s_2 are run through the network *just once*, and the total error is recorded. Low error indicates that the network has generalized from its experience with s_1, i.e., that it has learned a general concept and not simply memorized the training cases.

With EANNs that evolve weights, a similar separation of the data set occurs. The weights are essentially *guessed* during the initialization of the first generation, filtered by selection, and refined by genetic operators over many generations of networks. They are not learned in any one generation. Each ANN only processes the training set one time, without modifying its weights; and fitness is based on the accuracy of the network during that single pass. Fitness tends to gradually increase across the generations, and at the end of the evolutionary run, the best individual can be assessed for generality by invoking the test set.

In general, the evolution of ANN weights works equally well for supervised feedforward, unsupervised, recurrent, and even spiking neural networks. As long

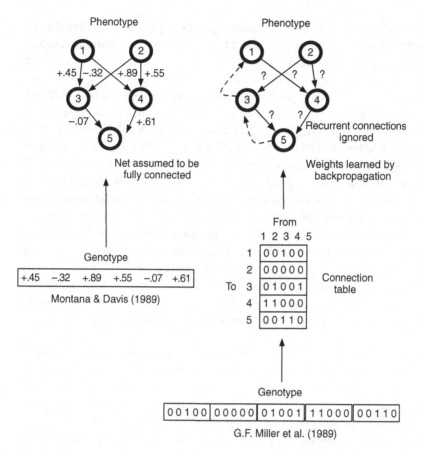

Figure 12.1
Genotype-phenotype conversions in the EANNs of Montana and Davis (1989) and G. F. Miller et al. (1989)

as the genotype representation supports a broad search of weight space, and the fitness function appropriately dispenses credit, the EANN can converge on an intelligent solution. However, no network style is immune to the curse of dimensionality: as networks get larger, and the number of connection weights tends to increase as the square of the number of nodes, performance suffers, often dramatically.

Another early EANN system, by G. F. Miller et al. (1989), uses evolution to determine the ANN's connection topology. As shown in figure 12.1, the genotype is a binary array wherein rows are laid side by side to form one long bit vector. The array, a, represents the complete network topology in a very straightforward manner: for any neurons n_i and n_j, $a_{ij} = 1$ if and only if n_i receives input from n_j;

otherwise $a_{ij} = 0$. The Miller et al. system uses a genetic algorithm to evolve the network topology, but the weights are initialized randomly and then learned via backpropagation. Their genetic operators include bit-flipping mutation and crossover, with crossover points restricted to those between each node's input weight set. Their system solves many simple problems, such as pattern copying and XOR, but there is no clear indication that it can scale up to more complicated problems.

G.F. Miller et al. (1989) were the first to combine evolution and learning for ANN phenotypes, and Chalmers (1990) followed shortly thereafter by evolving learning rules for ANNs. However, the combination of evolution and learning dates all the way back to 1975 and the original genetic algorithm (GA) of Holland (1992), which included an elegant learning mechanism known as the *bucket brigade* algorithm. Holland's GA encodes phenotypes that denote problem-solving rules run by a simple rule-based system (RBS). The combination of Holland's GA and RBS is known as a *classifier system*. It uses the bucket brigade to perform credit assignment along a temporal sequence of rule invocations, wherein credit is awarded not only to rules whose application directly leads to a solution but also to early rules that help *set up* a solution. These credits affect the prioritization of rules during problem solving, and thus the bucket brigade embodies adaptive lifetime performance, i.e., learning.

Kitano (1990) contributed the first combination of evolution and development for ANNs. His phenotype is a neural network generated from a connection table, identical in form to those of G.F. Miller et. al. However, whereas the latter use genotypes that directly correspond to these tables, Kitano uses context-free grammar rules as the basis for *growing* these tables, as shown in figure 12.2.

Kitano employs a bit-flipping mutation operator and totally unrestricted crossover. In addition, Kitano uses a mutation rate that is specific for each newly generated child genotype. The rate is inversely proportional to the Hamming distance between the parents. Hence, the children of similar parents are more likely to be mutated. This makes intuitive sense relative to the goal of exploration: if the parents are very different, then crossover alone will probably create unique children, but if the parents are similar, then supplementary mutation is needed to find child genotypes that are not (near) clones of either parent.

Once evolved and developed, Kitano's networks receive random initial weights that are then tuned via backpropagation. Hence, Kitano was the first to design a complete POE system. Also, at first glance, Kitano appears to have solved the scaling problem in that large phenotypes can be generated from short rule sets. His results are better than those of direct-coded genetic algorithms, but the test conditions are somewhat suspect (Floreano and Mattiussi, 2008), and no demonstration with large ANNs is given. Hence, Kitano's work primarily serves as a proof of principle.

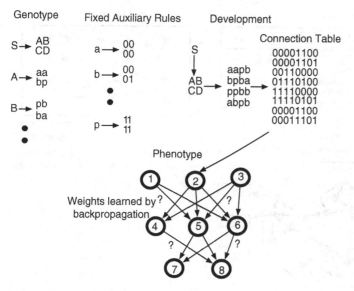

Figure 12.2
Kitano's (1990) encoding of neural networks as context-free grammars. (*Top left*) The genotype encodes the higher levels of rewrite rules, while the lowest (auxiliary) level rules are fixed. (*Top right*) Development is a series of rewrites. First, S is rewritten as a four-letter 2 × 2 array. Then each of these four capital letters is rewritten as a 2 × 2 array of lowercase letters. Finally, the lowercase letters are expanded into 2 × 2 binary arrays as dictated by the auxiliary rules. (*Bottom*) The ANN connection topology, as specified by the connection table. For example, the upper row of the connection table specifies the downstream neighbors of neuron 1: neurons 5 and 6. Recurrent connections, for example, those in the bottom row of the table, are ignored.

The PolyWorld system of Yaeger (1993) was the first to employ a complete POE approach in an ALife environment, one in which a population of agents move about, eat, mate, and fight. As shown in figure 12.3, each agent consists of a genotype that encodes many properties of both its body and neural network brain. Hence, a significant developmental process is required to convert the genotypic parameters into complete ANNs. Synaptic weights are randomly initialized and then modified by unsupervised Hebbian learning as the agent acts in the environment. The behaviors that eventually emerge in PolyWorld include edge running, flocking, and cannibalism.

One of the prime showcases for EANNS, the field of evolutionary robotics, made formidable progress in the mid 1990s (as fully documented by Nolfi and Floreano, 2000). Most of this early work uses small wheeled robots (either real or simulated) with infrared sensors, light sensors, or low-resolution cameras. The typical controller is an EANN for which only the weights are evolved. However, experiments

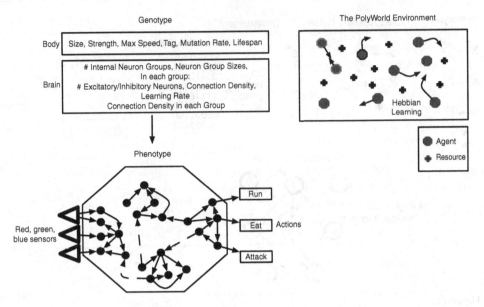

Figure 12.3
PolyWorld system (Yaeger, 1993). (*Left*) Conversion of genotypes to phenotypes. (*Right*) Artificial world in which agents live.

were performed with many different phenotypic factors encoded in the genome. For example, in work by Beer and Gallagher (1992) evolution controls weights along with the firing thresholds and time constants of their CTRNN neurons (see chapter 6), which control a six-legged robot. Some elaborate PO systems (Husbands et al., 1994; Nolfi et al., 1994) employ evolved developmental regimes in which neurons grow axonal projections to their targets, while one of the early PE systems (Urzelai and Floreano, 2001) evolves separate learning rules (applied to each incoming synapse) for each neuron. In these systems the evolved behaviors range from simple reactivity to minimally cognitive behaviors requiring memory of the recent past.

Considered a true classic of evolutionary computation and ALife is the groundbreaking work of Karl Sims (1994). His virtual creatures have a three-dimensional body plan (morphology) and an ANN controller, both designed by evolution and run in a physical ALife simulator that was well ahead of its time in the early 1990s. Sims evolved the weights and node functions for a unique variant of ANN in which neurons do more than simply sum their weighted inputs and apply a common activation function. His nodes can include many different functions, most of which do not sum weighted inputs but, for example, multiply them or compute their minimum or maximum. These are the forerunners of compositional pattern-producing networks (CPPNs) (Stanley, 2007), described later in this chapter.

Sims employed a graph-based representation for genotypes, both for the neural networks and the morphology. The former is a complete neural network and thus is equivalent to the phenotype, whereas the latter serves as a recursive generator program for segmented bodies. Complex mutation and crossover operators insure the production of valid genotypes and continuous variance in the population, and a wide variety of competitive formats facilitate selection such that skillful (though often oddly shaped) creatures evolve in a few dozen generations (with a population of size 300). The videos of these evolved creatures are a staple in many ALife lectures, and this work has been a huge source of inspiration for many in the field. Nearly a decade passed before 3-D simulators of equal power became common enough that other ALife researchers could begin to produce similar results (Bongard and Pfeifer, 2001; Bongard, 2002), many of which show the intricate relations in the coevolution of body and brain.

Reflecting on his work, Sims (1994) expresses an important underlying motivation for the ALife route to AI:

> Perhaps the techniques presented here should be considered as an approach toward creating artificial intelligence. When a genetic language allows virtual entities to evolve with increasing complexity, it is common for the resulting system to be difficult to understand in detail. In many cases it would also be difficult to design a similar system using traditional methods. Techniques such as these have the potential of surpassing those limits that are often imposed when human understanding and design is required. The examples presented here suggest that it might be easier to evolve virtual entities exhibiting intelligent behavior than it would be for humans to design and build them. (38)

In addition to the potential creative advantage of evolutionary over human-designed artifacts, Sims points to an important drawback of bio-inspired approaches to problem solving: the solutions can be very hard for humans to understand. This is particularly acute with EANNs, which do not readily divulge the logic of their controlling, decision-making activity. Tossing out a heap of synapses, and letting evolution (or backpropagation, for that matter) tune them, is a far cry from structured and systematic design. Nature and engineering work in very different ways, and the early days of EANN research showed few signs of common ground. Simulating nature's design process produced results that were often improvements over contemporary methods, and in ALife circles, frequently entertaining as digital life-forms, but they had little in common with engineered artifacts or problem solvers.

12.3   Direct Encodings

Most EANN systems that have achieved legitimate engineering success employ a direct encoding of the ANN in the genotype: each connection weight, neuronal behavior parameter, or synaptic learning factor maps to an individual gene. For all but the simplest problems, this entails a large genome and thus a large (and often very rugged) search landscape. Finding optima on these landscapes involves tuning a daunting set of ANN parameters, which, by hand or even with automated search tools, is no easy chore. However, direct encodings provide such a straightforward and reliable basis for EANN applications and research that they are hard to displace as the methodology of choice, despite the intimidating scaling issues.

In the mid 1990s, in the lab of Risto Miikkulainen at the University of Texas, a line of direct-encoding EANN research began that has substantial biological roots, fruitful engineering ties, and some important indications of common ground between the two. Paramount among these similarities is that both engineers and nature begin with relatively simple designs and gradually complexify. Miikkulainen and coworkers achieved this in two fundamental ways: by evolving individual neurons and their incoming and outgoing weights as a package, and by evolving complete networks whose initial sizes were restricted but could gradually increase.

12.3.1   Competing Conventions Problem (CCP)

One of the early motivations for Miikkulainen's work was the competing conventions problem (CCP). As originally recognized by Radcliffe (1990) and elaborated by Whitley (1995), the symmetry of neural network solutions allows the same functional unit to appear in different locations of different neural networks. Then crossover of these networks produces functional redundancy (and the concomitant loss of other functions). As a simple example, consider two networks for performing exclusive or (XOR), as shown in figure 12.4.

This is a permutation problem, since N behaviors can be distributed $N!$ ways across N internal nodes of an ANN. Crossover among any two ANNs that do not employ the exact same permutation can potentially lead to functional redundancy and loss. As described by Whitley (1995), prior to 1995 several solutions had been proposed for this problem, but none were convincing enough to make EANNs a more efficient approach than backpropagation for the general case of supervised learning in standard feedforward ANNs.

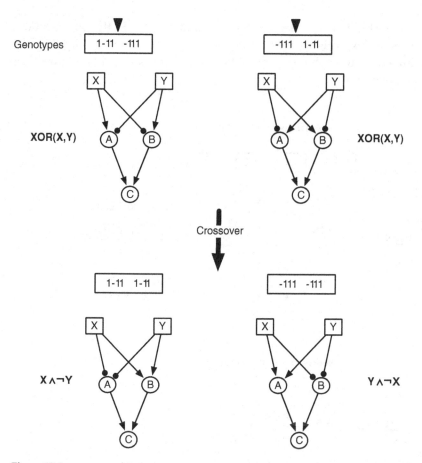

Figure 12.4
Competing conventions problem. Each segment of genes encodes the input weights followed by the output weight for neurons A and B. Appropriate combinations of weights and firing thresholds enable nodes A and B to perform an AND operation, while node C acts as an OR. Both upper ANNs compute XOR, but with internal nodes A and B performing opposite roles in each ANN. After crossover, the children each get a pair of functionally equivalent nodes and no longer compute XOR. Triangles above the genotypes denote crossover points. All nodes have a firing threshold of 1, and the outputs of all nodes are presumed to be either 1 or 0. Connections ending in arrows are excitatory and have a weight of +1; those ending in circles are inhibitory, with a weight of −1.

12.3.2   SANE

In the late 1990s, Moriarty and Miikkulainen (1997) devised the SANE system, which effectively evolves individual neurons, thus circumventing CCP entirely. In SANE, the ANN's size is predetermined, with a fixed number of input, hidden, and output nodes. Networks are strictly feedforward, but the three layers are not necessarily fully connected to one another. Each EA genotype encodes functional parameters for one hidden layer neuron. For any hidden node, H, this specification consists of pairs (index, weight), where the index selects an input or output node, and the weight codes the value for the arc between that node and H. Figure 12.5 illustrates this encoding.

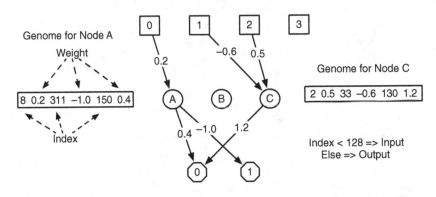

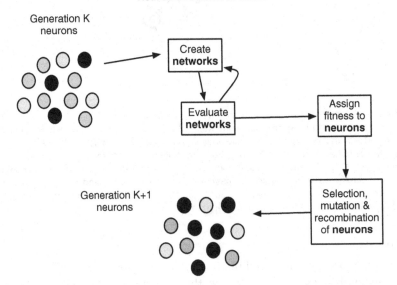

Figure 12.5
SANE system (Moriarty and Miikkulainen, 1997) for designing neural networks by evolving populations of individual hidden layer neurons. (*Top*) Genotypic encoding of (index, weight) pairs for the hidden layer neurons. (*Bottom*) SANE's evolutionary process, wherein a neuron's fitness is the sum of the fitnesses of the best five ANNs in which it participates.

Individuals' specifications are then evaluated in terms of their ability to cooperate with other specifications to form complete ANNs. Essentially, each individual participates in many random groupings with other individuals, with the fitness of each individual being the sum fitness of the best five groups in which it participates. Figure 12.5 (bottom) summarizes this process.

Extensive testing of SANE in domains such as game tree search and robot control verifies that it

- finds good solutions faster than standard whole-network evolutionary approaches,
- maintains population diversity over the entire simulation (since individuals cooperate, not compete, with other individuals),
- quickly adapts to changing environments.

These results made SANE the first system to convincingly show that the evolution of ANN topology has benefits over merely evolving the weights of a fixed network. In many EANN applications, neighboring layers are assumed to be fully connected. In the course of evolution some weights tend toward zero and thus implicitly modify the topology. So in theory there is no need to explicitly evolve the topology, since weight changes produce the same effect. However, SANE clearly outperformed an equivalent but fully connected EANN on a host of problems. Intuitively, it makes more sense to only generate necessary connections than to overproduce and slowly prune. So SANE benefits from that strategy, although brain development appears to follow the generate-and-prune approach (Striedter, 2005).

The SANE brain difference lies in the representational form: SANE encodes ANN connections directly, so each new link requires another gene, which essentially expands the search space. SANE simplifies search by keeping chromosome size low compared to that of a fully connected network. However, real brains are only generatively encoded by DNA; the generate-and-prune approach does not complicate evolutionary search. On the contrary, it can simplify evolution, since concocting a developmental strategy where neurons simply grow connections that are later tuned by learning is easier than crafting one full of special cases (for when to forge connections or when not to).

SANE skirts CCP by never explicitly linking neurons together in the genome. Although several similarly functioning neurons may exist in the population, functional heterogeneity is maintained naturally, since collections of *diverse* cooperating neurons are needed to solve problems, and all neurons cooperating in successful ANNs are rewarded with higher fitness. A neuron and its incoming and outgoing synaptic weights are packaged as units of selection, which struggle individually to survive, but their ability to cooperate determines their fitness.

What typically emerges from a SANE run are multiple subpopulations, each characterized by a distinct weight package. Combining the packages from each subpopulation produces a complete, high-fitness neural network. However, during fitness testing, there is no guarantee that neurons from each of these emerging groups will be selected for combination. The enforced subpopulations (ESP) extension to SANE (Gomez and Miikkulainen, 1997) remedies this problem by creating an explicit subpopulation for each neuron in the network. Basically, it shortcuts one level of emergence to facilitate evolutionary emergence of proper connections and weights between the prototypes of each subpopulation. A useful side effect is support for recurrent links between the hidden-layer neurons, which enables ESP to evolve ANNs with memory. The authors took advantage of this memory to evolve controllers for a difficult predator-prey task, while the explicit subpopulations enabled it to double the speed of SANE in finding solutions to the classic pole-balancing problem.

12.3.3   NEAT

At the heart of CCP lies an inability to detect functional similarities among subcomponents, thus leading to redundancies and omissions during crossover. A similar problem arises when chromosomes are allowed to change size during evolution. This is a desirable property in terms of complexification: genomes begin small/short and gradually grow as evolution essentially finds good regions of a coarse-grained search space and then increases the resolution of search in those areas.

Discussed earlier in the context of brain evolution, gradual complexification via duplication and differentiation can also aid evolving ANNs, though this can introduce genetic complications. As shown in figure 12.6 (*top*), an XOR network (phenotype) represented by a direct encoding of its weights can tolerate duplication (without altering the function that it computes) but not differentiation, whereas in figure 12.6 (*bottom*) the network cannot even handle duplication without losing its functionality. Ideally, a phenotype should remain functionally equivalent after genetic duplication and then gain extra functionality after differentiation without losing earlier-acquired abilities. Although duplication and differentiation in the generative encodings of biology tend to complexify this way, direct encodings of complete EANNs do not.

Duplication and differentiation also create problems for crossover, as shown in figure 12.7. When genome length differs between chromosomes, crossover can produce children with duplicates or omissions of important genes (or their derivatives), just as in CCP. The problem lies in the alignment of genes on the crossing chromosomes. Just as CCP stems from a failure (of genetic operators) to recognize functionally equivalent genes at different chromosomal locations, figure 12.7 (*top*)

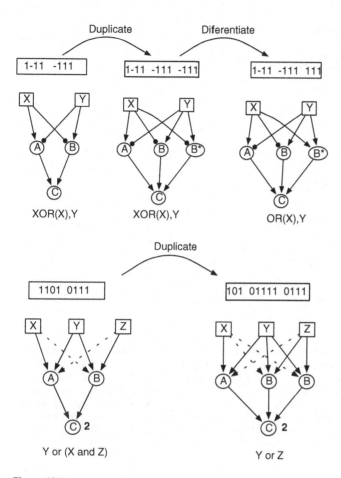

Figure 12.6
Effects of duplication and differentiation upon a direct-encoded ANN. (*Top*) Although duplication of node B and its connections retains the original XOR functionality, differentiation via a sign change to the X-B* connection changes the function to an OR. (*Bottom*) Duplication of an interior node and its connections alters functionality. Connections ending in arrows are excitatory, with a weight of +1 and those ending in circles are inhibitory, with a weight of −1 dotted links have a weight of 0. All nodes have a simple step function for activation and a firing threshold of 1 unless otherwise indicated by an integer beside the node.

shows recombination that produces redundancies and omissions due to an inability to detect equivalent genes between chromosomes and align them accordingly. This would seem to reduce the effectiveness of crossover by producing children that lack certain core genes.

To partly solve this problem, Stanley and Miikkulainen (2002) use an abstraction of the biological process of *synapsis*, wherein identical genes align during

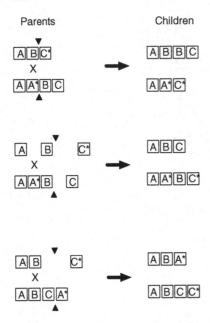

Figure 12.7
Crossover problems created by duplication and differentiation, but remedied by proper alignment of
genes prior to crossover. (*Top*) From the original genome ABC, differentiation (of gene C) produces
ABC*, while duplication and differentiation (of A) produce AA*BC. Crossover of these two descen-
dant chromosomes produces children with duplicates or omissions of the three original genes or their
mutated versions. (*Middle*) By aligning identical genes between the parents, crossover never produces
exact duplicates in the children and can reduce omissions. (*Bottom*) Ordering of the genes based on
when they arise can still produce omissions; here A* presumably arose after A, B, and C but before C*.

recombination. In their NeuroEvolution of Augmenting Topologies (NEAT) system, each gene includes a historical marker that denotes its relative time point
of origin, producing an alignment as shown in figure 12.7 (*bottom*). This allows
genomes to align properly, with identical genes always appearing in corresponding spots. Hence, crossover itself never causes gene duplication and stands a better chance of combining phenotypic building blocks. The latter fact stems from the
simple observation that two genes coding for different traits will never align and
thus never be prevented from appearing in the same child.

The actual genotype representation in NEAT codes for the number and type
of network nodes, along with their connections and final weights; no learning is
involved. Networks are initialized to have three layers: input, hidden, and output,
although the hidden layer can effectively have many topologies, some of which
give it a multilayered look. A typical genotype and corresponding phenotype

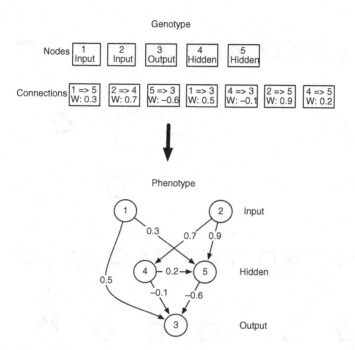

Figure 12.8
Genotype and corresponding phenotype in NEAT. Each rectangle denotes a gene. Historical markings on both node and connection genes are not included.

appear in figure 12.8. Genomes are mutated by adding a new node or adding a new connection between existing nodes.

12.3.4 Speciation for Complexification

In evolutionary systems innovative phenotypes with useful new traits often arise. However, unless those traits integrate well with preexisting traits to form a *better* organism, the new phenotype may lose in competition with its contemporaries. Figure 12.9 illustrates this problem for EANNs: the goal is to evolve a network that detects objects with a fine mixture of black and white coloring.

Network A represents a useful first attempt: it fires on any scene with considerable black-white content regardless of the distribution. For example, a large black block beside a comparable white block would suffice, since each block would independently trigger one of the two intermediate nodes.

Through mutation, new nodes (pentagons) are added, each of which receives inputs from a few neighboring input nodes. These detect local black-white mixtures. Network B contains one such pentagon, but note that the firing threshold

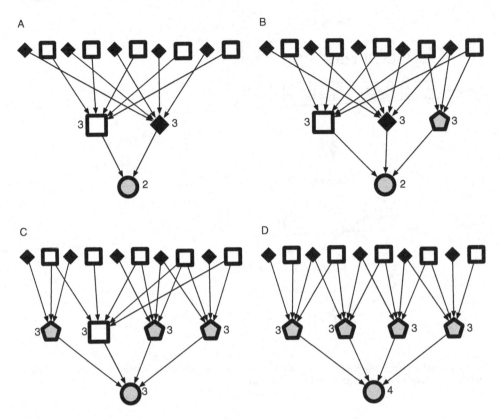

Figure 12.9
Gradual complexification (from A through D) of an ANN for detecting fine mixtures of black and white.
White squares denote white detectors; black diamonds, black detectors; gray pentagons, detectors of
local black-white mixtures. Circles represent output neurons; the upper layer of squares and diamonds
in each network, input neurons that topologically map to a visual field (e.g., black diamonds on the left
detect patches of black in the left visual field). All connections have a weight of 1, all nodes have step
function activations, and firing thresholds appear beside each node.

of its output neuron retains the value of 2; it has not properly co-evolved with
the intermediate layer. Hence, it will fire on the white-block, black-block scene but
also on scenes dominated by black or white, and having one black-white patch on
the right. Thus, the output node would fire on many scenes lacking the desired
global black-white mixture, such as that of a white ghost carrying a small black
lantern. To enforce a condition of well-mixed blacks and whites, many more mutations are needed. In network C, two additional pentagons appear, one of which
replaces the black-detecting hidden node, and the output neuron's firing threshold

has increased, making it slightly harder for some undesirable patterns to elicit a response. Finally, with network D, an effective solution emerges, one that only triggers on multiple occurrences of black-white patches.

Evolution's problem stems from the fact that the intermediate networks could easily lose in competition with the simple yet reasonably effective network A. The coarse solution often dominates fine-grained attempts because more detail often means more error possibilities. The intermediate networks can never get a foothold in the population, since they are outperformed by their ancestors long before they can fine-tune their specializations. Hence, all of the stepping stones to a good solution wash away too early in the process.

Evolution rarely gives new designs an opportunity to work out the bugs, and standard evolutionary algorithms are equally impatient. However, artificial evolution can be supplemented with explicit niche mechanisms such that new designs have a fighting chance against established solutions. For example, fitness sharing (Goldberg, 1989) divides an individual's raw fitness by the number of individuals that it resembles, where resemblance can entail similar fitness values, genotypes, or phenotypes. Clearly, the latter two conditions seem more reasonable, since two individuals of like fitness need not be similar solutions. Fitness sharing prevents a population from converging, since as more individuals become similar, they must share fitness among a larger and larger group, thus reducing each agent's worth. This allows unique new solutions to get a foothold in the population, because they have some promising properties, and they need not share fitness with many others.

Unfortunately, judging similarity among graphlike structures such as neural networks is an NP-complete problem (Garey and Johnson, 1979), so proper fitness sharing would appear intractable at the phenotype level. However, NEAT's historical markers enable a simple similarity check among pairs of individuals: compare their marking lists. An individual must then share its fitness with all others having similar genotypes.[1] So NEAT exploits its history-based similarity metric and fitness sharing to nurture, not filter, innovation. In this way, neural networks can gradually complexify over the evolutionary generations. This may be the only manner in which a relatively unintelligent search process (e.g., evolution) can solve complex problems. To begin with large genomes creates an intractable search space. By beginning small, the EA can find useful general solutions that can then improve by elaboration as new genes arise to code for special accessory traits. At the end of an evolutionary run, the final genome may be large, but the entire space that it represents never needs to be searched. Only subspaces of proven utility are investigated.

Gradual complexification in NEAT leads to impressive performance results. On the classic pole-balancing control problem, it finds solutions 25 times faster than cellular encoding (Gruau, 1994) and five times faster than ESP (Gomez and Miikkulainen, 1997). In addition, NEAT is the adaptive mechanism in an exciting video game, NERO (Stanley et al., 2005), in which teams of warriors are trained via an interactive evolutionary algorithm in which the human user adjusts fitness parameters according to the desired type of fighting unit; evolution determines the neural network controllers of the individual fighters; and group-level fighting strategies emerge on the battlefield.

In summary, NEAT supports gradual complexification of neural networks via the accretion of genetic material. It does so by including a simple yet powerful piece of information in each gene: a historical marking. This concept can supplement many EAs, whether they evolve neural networks or other structures.

12.3.5   Cartesian Genetic Programming

Despite many of the conceptual advances and biological inspirations in NEAT, it was recently defeated quite handily by a popular version of GP known as Cartesian genetic programming (CGP) (J. Miller et al., 2000), which was originally designed for evolving digital circuits but has been extended in several different ways (Khan et al., 2011; Khan et al., 2013) to produce neural networks. On several benchmarks , two different direct-encoded CGP EANNs soundly bested SANE, NEAT, and ESP on single- and double-pole balancing tasks (Khan et al., 2013; Turner and Miller, 2013). Other CGP EANNs achieve very impressive results in breast cancer diagnosis, with over 95 percent accuracy on a popular benchmark data set.

CGP uses a more *complete* direct encoding than SANE and NEAT in that the user decides ahead of time upon the number N of neurons and the maximum number M of incoming connections (per neuron); the fixed-size chromosome is then dimensioned to handle all N neurons and MN connections. This representation is not fully complete, however, since one neuron may receive several afferents from the same neuron and none from several others. This allows evolution to still control a good deal of the topology, which Miikkulainen and Stanley have shown to be an advantage over strictly complete encodings for each weight.

Figure 12.10 shows the essence of the CGP genotype and genotype-phenotype mapping. The top of the figure shows the coarse chromosomal structure, which consists of N groups, one for each noninput neuron. The group begins with the function housed within that neuron (e.g., sigmoidal, linear, hyperbolic tangent) and is followed by M triples consisting of an index, weight, and switch (i, w, s). The index identifies an afferent neuron, the weight gives the strength on that afferent's incoming connection, and the (binary) switch determines whether or not

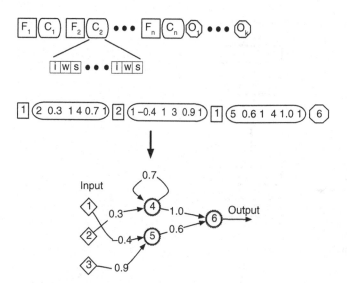

Figure 12.10
CGP direct encoding for ANNs. (*Top*) The genome consists of one group per neuron that determines the activation functions (squares), afferent connections (ovals), and indices of all output neurons (octagons). (*Middle*) Simple genome that codes for neurons 4–6 of a six-neuron network, where neurons 1–3 are inputs and the genome choses neuron 6 as the output. For illustrative purposes, all connections have a switch value of 1. (*Bottom*) The resulting network, including a recurrent link for neuron 4.

the given connection will appear in the network. The switch allows evolution to quickly remove connections rather than waiting until a weight evolves to zero. It also enables search on a neutral landscape, since when the switch is off, afferents and weights can change without affecting fitness.

12.4   Indirect Encodings

Many of the problems discussed earlier, such as competing conventions and road-blocks to complexification (scaling difficulties), stem from direct genetic encodings, which most EANN applications employ despite their lack of biological realism. Generative indirect encodings can often remedy the situation, although (unsurprisingly) *expanded indirect encodings* (see chapter 8) typically cannot.

Consider the expanded indirect genomes of figure 12.11 (*bottom*). Let $a_1 a_2 ... a_n$ be the direct representation, where it is assumed that a_i is 1 or -1 for all i. To compute the corresponding indirect representation, $b_1 b_2 ... b_n$, where b_i is 0 or 1, use the following algorithm:

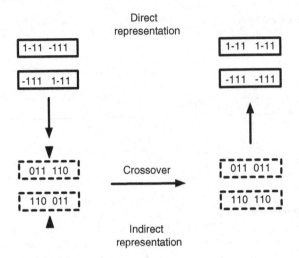

Figure 12.11
CCP using an expanded indirect representation. The genomes (*top left*) are direct encodings of XOR networks from figure 12.4, which are then converted to indirect representations before crossover. The children (*bottom right*), when converted back into the direct representation, exhibit the duplications and omissions characteristic of CCP.

1. $a_0 \leftarrow 1$.
2. For $i = 1$ to n, do:

    a. If $a_i = a_{i-1}$, then $b_i \leftarrow 0$;

    b. else $b_i \leftarrow 1$.

In short, b_i indicates whether a_i is different from a_{i-1} (assuming that $a_0 = 1$). Of course, not all indirect encodings produce these same duplications and omissions for these particular XOR networks. The key point is that going to an indirect representation does not solve the problem. These codes have the same length as the direct encoding and can often be divided relatively cleanly into modules (that then get swapped during crossover). The modularity is only relatively clean, since interactions between the modules can affect the final phenotypes. For example, a 0 as the first bit of a module can represent 1 or −1, depending upon the last trait encoded by the preceding module.

However, in moving to a generative encoding, the phenotypic modules may no longer have corresponding segments in the genotype. Once again, consider the two weight vectors (i.e., phenotypes) for the XOR networks in figure 12.4: [1 −1 1 −1 1 1] and [−1 1 1 1 −1 1]. Now examine the two context-free grammars of figure 12.12, each of which generates one of these vectors. First note

$$S \rightarrow AB \qquad\qquad\qquad\qquad S \rightarrow BA$$
$$A \rightarrow 1B \qquad\qquad\qquad\qquad A \rightarrow 1B$$
$$B \rightarrow -1A \qquad\qquad\qquad\qquad B \rightarrow -1A$$
$$A \rightarrow 1 \qquad\qquad\qquad\qquad A \rightarrow 1$$
$$B \rightarrow 1 \qquad\qquad\qquad\qquad B \rightarrow 1$$

$$S \rightarrow AB \rightarrow 1B-1A \rightarrow 1-1A-11B \rightarrow 1-11-111 \qquad S \rightarrow BA \rightarrow -1A1B \rightarrow -11B1-1A \rightarrow -1111-11$$

Figure 12.12
(*Top*) Context-free grammars for producing simple ANN weight vectors containing only −1's and 1's. (*Bottom*) Derivation of the XOR weight vectors of figure 12.4 using the respective grammar.

that the grammars are identical except for the first rule of each. Assuming that the rule lists constitute a genotype and that crossover can only occur between (not within) individual rules, then the near equality of the grammars insures that child genomes will be identical to one of their parents - and thus be capable of producing one of the XOR phenotypes. Hence, CCP vanishes for this generative representation.

Essentially, the genome lacks modules that correspond to the phenotypic modules [1 −1 1] and [−1 1 1], thus discouraging CCP. At first glance, the grammar variables A and B might appear to fit the bill, but they do not work independently. Rather, they repeatedly spawn one another in a doubly recursive procedure to create alternating lists of 1's and −1's punctuated by a single 1 (the weight on the connection from a hidden node to the output node). This intertwined activity of A and B makes it impossible to segregate the genome into units that are susceptible to duplication and omission (of phenotypic components) following crossover. Although this is not a general property of all grammars, it is common for generative representations to include recursion and an intricate interaction scheme among the variables. Hence, individual genotypic variables often have pleiotropic effects: one may help determine several phenotypic traits. So it seems reasonable to assume that many generative representations have a general immunity to CCP.

They also readily support complexification by duplication and differentiation. Making a second copy of a rule does not change the set of possible products of a grammar, while mutating that copy tends to increase (and should never decrease) the set of possible derivations (assuming that the grammar is context-free). In neither case does the resulting grammar lose any of its original functionality. This amenability to complexification changes a bit when grammars are context-sensitive, as illustrated with the G2L system, discussed later in this chapter.

The most frequently cited advantage of generative encodings is their ability to *scale*: small genotypes can produce large phenotypes. Since evolutionary search occurs in genotype space (with fitness determined in phenotype space), anything that reduces genome size should reduce search complexity. With direct encodings, the genotype and phenotype have the same size, so large phenotypes entail large genotypic search spaces and longer odds of finding good solutions (e.g., ANN parameter vectors).

Generative genomes cannot always produce every point in phenotype space, and they are often biased to form those with a recursive or symmetric structure. For example, a generator for ANN weight vectors might tend to produce sequences of equal or gradually increasing/decreasing values. Finding a one-off weight vector with little correlation between neighboring values is a big challenge for a generative encoding, but for a direct encoding, this is typically no harder than finding a set of equal weights, since direct encodings enable traits to evolve more or less independently.

Thus, the theoretical advantage of generative representations seems highly problem-dependent. However, many ANNs exhibit weight correlations and are thus amenable to the improvements that generative EANNs offer; and the topological (hence correlated) nature of many biological neural networks increases the plausibility of a strong generative contribution to their design.

12.4.1   Cellular Encoding

One of the first generative encodings to get beyond the proof-of-principle stage was cellular encoding (CE) (Gruau and Whitley, 1993; Gruau, 1994), an approach that also revolutionized genetic programming. John Koza's genetic programs (Koza, 1994; Koza et al., 1999) have been enlisted in the design of everything from electrical circuits to antennas to automatic controllers, some of which have been patented and many of which infringe upon existing patents (Koza, 2003). It turns out that a critical feature in the genetic program (GP) used for these human-competitive achievements comes directly from Gruau's work with the evolution of neural networks.

Traditionally, genetic programs use genotypes that very closely resemble phenotypes. In many cases, the genotype merely needs to be *packaged* so that it actually compiles and runs. However, with a developmental approach to GP, the genotypes do maintain the same basic syntax—they are still nearly executable computer programs—but they now encode a *recipe* for growing the phenotype through a multistep process. This is analogous to (though not nearly as complex as) the growth of a human body from DNA code.

In cellular encoding, each GP tree is interpreted as a set of growth instructions, with individual GP functions performing simple operations on the growing ANN, such as node copying, weight assignment to an arc, and so on. In essence, the GP genome governs *duplication and differentiation* activity of the growing phenotype, with the key decision being how the new child node is integrated into the subnetwork surrounding its parent.

Figure 12.13 shows the basic components of the CE developmental process. The ANN begins as an *embryo* consisting of the known inputs and outputs plus a single internal node. Pointers from the ANN to the GP tree (dotted arrows) denote *read heads*, which indicate the GP command that will be used to modify that ANN node. After a command executes, it sprouts new read heads for each of its children in the GP tree while relinquishing its own pointer.

Development begins to unfold on the left side of the second row of figure 12.13, where the topmost S command produces a serial expansion of the original node (N0), making the new node, N1, a child of N0. N0 and N1 then each receive a read head for the next round. These heads point to the two children of the root S command. Next (second row, right), the level-2 P node dictates a parallel expansion of N0, with copies of N0's input and output arcs attached to the new node, N2.

In row 3 (left), node N1 performs the S command, creating node N3 in series with N1, and then (right) node N0 performs a second parallel split by executing the P command, thus producing N4 and a three-node top layer.

Then in row 4 (left), node N2 executes an E (End) statement, terminating activity for that node and removing its read head, and then (right) node N1 executes a P statement, creating a new parallel node, N5, which receives copies of all of N1's input and output arcs.

Finally, in row 5 (left), nodes N3, N0, and N4 (in that order) execute E commands, terminating their activity and deleting their read heads, and then (right) node N1 executes the A command, which increments the threshold of its transfer function (thicker circle) and then terminates activity. N5 executes an E command and also terminates to complete development.

Not shown are commands such as those for modifying ANN arc weights or for making recursive calls to the GP tree (i.e., moving a read head back to the root), which CE also includes. The use of recursion allows CE to create networks with repetitive structure, which facilitates the control of multilegged agents, one of several tasks to which this approach has been successfully applied.

12.4.2   G2L

Inspired by both CE and L systems (Lindenmayer and Prusinkiewicz, 1989), Boers and Sprinkhuisen-Kuyper (2001) designed a generative encoding, G2L, for producing network topologies. G2L consists of an L system grammar whose rewrite rules are applied in parallel to a developing string. The final, expanded string serves as a partial direct ANN encoding, consisting of node symbols followed by zero or more integers denoting offsets to neighbors. This syntax resembles that of the CGP

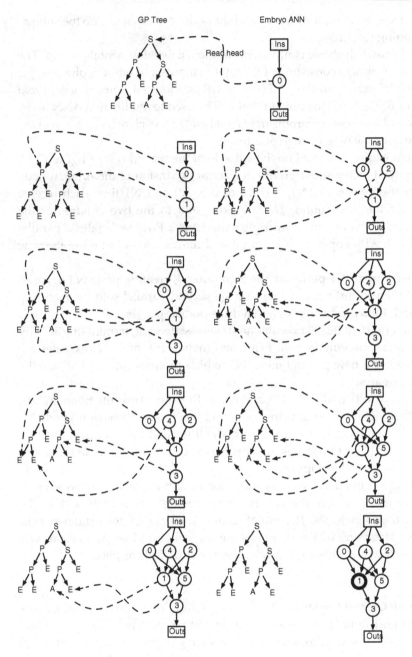

Figure 12.13
Generating a neural network from a cellular encoding (Gruau and Whitley, 1993). Development proceeds left to right, top to bottom.

ANN genome, although the latter gives absolute addresses of afferents to a unit, while G2L employs relative addresses to efferents (i.e., downstream neighbors).

Boers and Sprinkhuisen-Kuyper (2001) provide a detailed account of G2L, which includes a description of their complex binary chromosome for encoding grammar rules. The following example begins at the rule level and follows the gist of G2L, but it uses a slightly different L system (parameterized instead of context-sensitive) for clarity of illustration. Thus, each nonterminal symbol can have any number of parameters associated with it. These behave like function arguments in rules where the symbol is an antecedent.

Consider the rewrite rules in the left-hand column of figure 12.14 (similar to those generated by G2L). These generate three-layered networks where input nodes connect to alternating hidden nodes, such as the ANN on the bottom of the figure or the upper left network of figure 12.9.

Rule 1 takes a start symbol (S) with two parameters, n and h, the number of input and hidden nodes, respectively; it rewrites S as parameterized A and B symbols along with an unparameterized C. These represent the three layers of the ANN. Rule 2 applies to A symbols whose n value exceeds 0. It replaces A with a group-generating term (G) and a new A with a reduced first parameter, while the second

$S(n,h) \rightarrow A(n,h)B(h)C$	$S(8,2)$
$A(n,h): n > 0 \rightarrow G(h,n)A(n-h,h)$	$\Rightarrow A(8,2)B(2)C$
$A(n,h): n \leq 0 \rightarrow \emptyset$	$\Rightarrow G(2,8)A(6,2)b2B(1)c$
$G(c,n) \rightarrow anG(c-1,n)$	$\Rightarrow a8G(1,8)G(2,6)A(4,2)b2b1B(0)c$
$G(0,n) \rightarrow \emptyset$	$\Rightarrow a8a8G(0,8)a6G(1,6)G(2,4)A(2,2)b2b1c$
$B(c) \rightarrow bcB(c-1)$	$\Rightarrow a8a8a6a6G(0,6)a4G(1,4)G(2,2)A(0,2)b2b1c$
$B(0) \rightarrow \emptyset$	$\Rightarrow a8a8a6a6a4a4G(0,4)a2G(1,2)b2b1c$
$C \rightarrow c$	$\Rightarrow a8a8a6a6a4a4a2a2G(0,2)b2b1c$
	$\Rightarrow a8a8a6a6a4a4a2a2b2b1c$

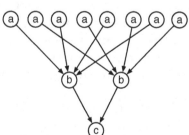

Figure 12.14
G2L example. (*Top left*) Context-free grammar with parameterized symbols. (*Top right*) Sequence of parallel symbol replacements producing a direct encoding for an ANN topology. (*Bottom*) Corresponding network.

parameter, which keeps track of the number of hidden nodes, remains unchanged. Rule 3 represents a base case for A: when its *n* value goes below 1, replace it with the empty string, ∅.

Rule 4 handles the production of each group of input neurons, where each member of a group connects to a different hidden neuron. It replaces G with *an* plus a reduced G. Here, the symbols *a*, *b*, and *c* denote neurons in the input, hidden, and output layers, respectively. So rule 4 produces an input layer neuron followed by an integer, which serves as a *connection offset* at which we find the neuron to which neuron *a* will send its output. Rule 5 provides the base case for the group generator.

Rule 6 produces the neurons of the hidden layer, all of which send their signals to the single neuron in the output layer. Hence, the connection offset of each hidden neuron is 1 less than that of its left neighbor. Rule 7 is the hidden-layer base case, and rule 8 generates the single output layer node.

The right-hand column of figure 12.14 shows the use of these eight rules to derive a direct encoding for an ANN with eight input, two hidden, and one output nodes. Note that this direct encoding resembles those used by CGP: a node followed by its connections. For example, the first segment, *a8a8*, directs the translator to

1. create a node labeled *a*,
2. make a connection from it to the node that is eight spots after it in the complete node list (i.e., the ninth node),
3. create another node labeled *a*,
4. connect it to the tenth node.

Though not shown in this example, many offsets can follow a single node, indicating many outgoing connections from it.

The G2L framework provides a nice tool for illustrating general properties of generative encodings. For example, the potential large-scale effects of point mutations become quite obvious. Figure 12.15 shows the result of a simple modification to rule 4 of the grammar: it produces a vastly different direct encoding, which yields a completely new network. This should not be surprising, since these grammars function as little computer programs, and everyone knows the effects (often catastrophic) of small changes to code.

However, most small (one symbol) changes to the direct encoding *a8a8a6a6a4a4a2a2b2b1c* (top right of figure 12.14) translate into small phenotypic changes, such as the redirection of a single connection, as shown on the left of figure 12.16. But, since the connection offsets constitute relative addresses, the whole topology can change with the addition or deletion of a single node symbol (see figure 12.16). Note that an absolute location address, as employed by CGP

$S(n,h) \rightarrow A(n,h)B(h)C$ $S(8,2)$

$A(n,h): n > 0 \rightarrow G(h,n)A(n\text{-}h,h)$ $\Rightarrow A(8,2)B(2)C$

$A(n,h): n \leq 0 \rightarrow \emptyset$ $\Rightarrow G(2,8)A(6,2)b2B(1)c$

$G(c,n) \rightarrow \mathbf{ac}G(c\text{-}1,n)$ $\Rightarrow a2G(1,8)G(2,6)A(4,2)b2b1B(0)c$

$G(0,n) \rightarrow \emptyset$ $\Rightarrow a2a1G(0,8)a2G(1,6)G(2,4)A(2,2)b2b1c$

$B(c) \rightarrow bcB(c\text{-}1)$ $\Rightarrow a2a1a2a1G(0,6)a2G(1,4)G(2,2)A(0,2)b2b1c$

$B(0) \rightarrow \emptyset$ $\Rightarrow a2a1a2a1a2a1G(0,4)a2G(1,2)b2b1c$

$C \rightarrow c$ $\Rightarrow a2a1a2a1a2a1a2a1G(0,2)b2b1c$

                                                              $\Rightarrow a2a1a2a1a2a1a2a1b2b1c$

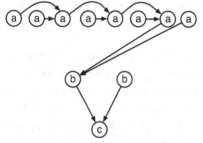

Figure 12.15
Same example as in figure 12.14 but with *an* replaced by *ac* in the fourth rule of the grammar

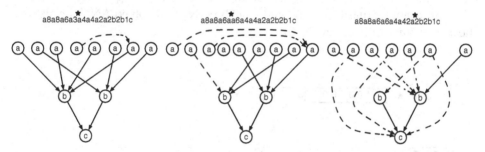

Figure 12.16
Effects of point mutations (at starred locations) to the direct encoding produced by G2L in figure 12.14. New connections are drawn with dashed lines. (*Left*) Modified connection offset. (*Middle*) Added node. (*Right*) Deleted node.

(J. Miller et al., 2000), ameliorates this problem by greatly reducing the scope of a point mutation.[2]

12.4.3   HyperNEAT

Ken Stanley and coworkers have extensively explored generative encodings via their HyperNEAT system (Stanley et al., 2009). Though NEAT employs a direct encoding for ANNs, its basic machinery can produce many types of networks, including those that can serve as generators for ANNs. Borrowing an idea from Karl Sims, Stanley's group looked to computational networks in which nodes can house many different functions. Their version, called compositional pattern-producing networks (CPPNs), typically incorporates continuous functions such as sine, cosine, and a Gaussian to produce patterns with considerable correlation between neighboring values: similar inputs to the CPPN give similar outputs. In HyperNEAT, the basic NEAT evolutionary engine (using historical markings and gradual complexification) produces CPPNs, which then serve as generators for ANN weight vectors. In short, it uses direct-encoded CPPNs to act as generative-encodings for ANNs.

As shown in figure 12.17, HyperNEAT uses evolved CPPNs as simple mappings from the location descriptors of neurons to weights: the layer numbers and intralayer indices of two neurons go in, and a weight comes out. This can apply to any pair of neurons as well as to recurrent links from a neuron to itself. The use of continuous functions in the CPPNs tends to produce well-correlated weight vectors. For example, if neurons 1, 2, and 3 have locations (4, 7), (4, 8), and (5, 3), respectively, then a CPPN fed with (4, 7, 5, 3) will probably produce an output similar to when fed with (4, 8, 5, 3). Thus, the weights on the neighboring 1-to-3 and 2-to-3 connections will correlate well, and, in general, the weight vector will display significant regularity, as emerged in a HyperNEAT-evolved ANN for playing checkers (Gauci and Stanley, 2010).

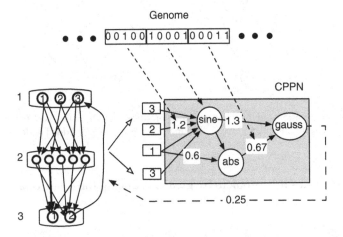

Figure 12.17
HyperNEAT, which uses an evolved, direct-encoded CPPN to generate synaptic weights. The layer and position indices for two neurons, (3,2) and (1,3), are input, and the CPPN produces the weight value for the connection between the two neurons. Only some CPPN weights are shown.

The inspiration for HyperNEAT comes from two interdependent features of the brain: numerous neurons and prevalent topological maps, which preserve spatial relations between neural layers and either other neural regions or sensory or motor spaces. By exploiting CPPNs to generate well-correlated neural mappings, Hyper-NEAT scales to large ANNs (with millions of connections). These reap benefits on tasks that require topology, such as visual discrimination, where HyperNEAT soundly defeats a direct-encoded genome (Stanley et al., 2009).

The CPPN adds a second level of adaptivity (ontogeny) to NEAT, yielding a PO system. More recent work by their group (Risi and Stanley, 2010) incorporates learning (epigenesis) as well, to achieve full POE status. Learning seems essential for a few types of scenarios. First, if the optimal ANN for a task is large and lacks symmetries and other regularities, then direct encoding has problems (because of ANN size), while generative encoding can produce large, correlated networks but not the intricate detail of an unstructured weight vector. Only learning can tune a large network to a needle-in-the-haystack collection of synaptic strengths. Second, many ANNs must perform in dynamic environments where the mapping from ANN inputs to outputs must change in the course of deployment. Artificial evolution and development typically provide one ANN for the lifetime of the agent; adaptations at smaller time scales require learning.

Motivated by the dynamic environment problem, Risi and Stanley (2010) developed Adaptive HyperNEAT, which employs CPPNs to generate both synaptic weights and learning parameters. They experiment with three different approaches to CPPN-based learning:

- *Plain Hebbian model* As in HyperNEAT, the CPPN receives the locations of two neurons, but now it outputs the learning rate on the synapse between them. A standard Hebbian learning rule is used for all synapses.

- *Hebbian ABC model* The CPPN again receives only two neuron locations, but now it outputs five items: the initial weight, learning rate (η), and parameters A, B and C for the following (standard) generic learning rule:

$$\Delta w = \eta(Auv + Bu + Cv) \tag{12.1}$$

where Δw is the weight change, and u and v are the outputs of the pre- and postsynaptic neurons, respectively. Each synapse thus receives a specially tailored version of the rule, which it uses throughout the simulation.

- *Iterated model* This CPPN receives two neuron locations plus their current activation levels (u and v), along with the current weight (w) on their synapse. It then produces Δw. Thus, the CPPN embodies a very adaptive learning rule, and calls to the CPPN occur throughout the running of the ANN. Initial weights are determined by calling the CPPN with the last three inputs set to zero.

A comparison of these models on T-maze navigation tasks shows that the plain Hebbian approach generally fails, while the iterated version clearly dominates the ABC model in mazes where the intermediate signals have a deceptive, nonlinear relation to the final rewards. As expected, the iterated model pays a much higher computational cost for its superior adaptivity.

SANE, NEAT, and their successors are just a few of the many hundreds of EANN systems in the literature. However, Miikkulainen' and Stanley's work has strongly influenced the EANN community, to the point where some conference sessions are solely, though coincidentally, devoted to extensions of one of their methods. Miikkulainen is also well known for his exceptional EANN tutorial given at many major conferences on evolutionary computation. In general, no review of EANNs is complete without touching upon his and Stanley's contributions.

From the prospective of sophisticated intelligence and its emergence, a few features of their work stand out. First, both focus on small units from which complexity gradually arises. In SANE each neuron and its connection strengths must compete based on its ability to cooperate, and over many generations useful combinations arise. In NEAT small initial networks slowly accumulate additional neurons and synapses to zoom in on particularly promising areas of design space. The complexification strategy differs from duplication and differentiation in that new nodes are not copies of existing nodes (since they do not inherit the internal nor local topological properties from another node), and new connections often cause some existing connections to deactivate. However, as shown in figure 12.6, the utility of duplication and differentiation for direct encodings of ANNs seems very limited.

Second, they both stretch direct encodings to include topology: evolution decides when to add neurons and precisely how they should interconnect. This saves evolutionary time—in contrast to beginning with large layers of fully interconnected neurons and letting evolution drive some weights to zero—while still allowing the search process to find intricate, unstructured, needle-in-the-haystack parameter vectors (which are often elusive for generative encodings). Thus, at the evolutionary time scale, these systems are extremely adaptive.

Both ESP and NEAT utilize speciation to preserve diversity. In NEAT this shelters innovative new topologies from selective dominance by established phenotypes, while in ESP it prevents competition between neural packages that actually need to cooperate in order to solve a problem. In both cases, speciation proves critical to success.

Finally, although SANE and ESP adapt only phylogenically, NEAT spans the entire POE spectrum, first as a direct encoding for topology and weights, and then as the direct encoding for CPPNs that generate both fixed weights (PO) and initial weights plus learning behavior (POE). Sims, Stanley, and their co-workers have gotten a lot of adaptive mileage out of this versatile tool.

12.5   Evolvability of EANNs

Despite the impressive results of several decades of EANN research, the field has yet to produce anything *brainlike*: a large neural network that performs a truly complex task. Part of this inability to scale up stems from the strong reliance on the direct representation, an extreme oversimplification of biology but possibly a necessary abstraction at the moment, as researchers grapple with a host of issues at the crossroads of robotics, dynamic systems theory, evolutionary computation, and machine learning. Just as evolution works best when complexity increases gradually, so too does Bio-AI research.

The beauty of HyperNEAT and its successors lies in the ability of relatively simple CPPNs to produce complex, geometrically structured ANNs; and the surface similarities between these and biological neural networks, along with the promise of scalability, certainly justify further research in this direction. However, other EANN work has more explicitly incorporated principles and processes that were touched upon earlier with respect to the evolution of the brain: duplication and differentiation, modularity, weak linkage, and exploratory growth. These projects, though often exploratory in their own right, have also yielded impressive results and interesting insights.

The pillars of evolvability are rarely the main focus when one sets out to solve problems with EANNs, but they have slowly crept into many experiments. Few (if any) projects have incorporated all four of the primary mechanisms, but many utilize a subset. These mechanisms, their relation to EANNs, and some of the systems that use them are explored in the following sections.

12.5.1   Duplication and Differentiation
First, it is important to distinguish between copying and modification at the genotype versus phenotype level. For example, in cellular encoding (Gruau, 1994), the P (parallel) and S (serial) commands both duplicate a neuron, and P copies all input and output connections as well. Other developmental operators may then differentially modify the parent and its clone. This is clearly an important feature of CE and many other rewrite systems. Note that this is duplication and differentiation of the growing phenotype, not of the genotype; and it is a common feature of generative representations.

Many EAs and EANNs have chromosomes that can grow in the course of evolution, and this has obvious advantages in terms of gradual complexification, as illustrated quite early by the SAGA system of Harvey (1993) and more recently by NEAT. However, growth alone cannot guarantee one of the prime advantages of growth by duplication: neutral complexification. By adding copies of genes, evolution creates a relatively safe haven (by preventing trait omission) for exploration via differentiation.

In the direct encodings for ANNs (such as those used by SAGA and NEAT) and other phenotypes, any form of genomic growth (whether a duplication or a simple random extension) is hard not to notice in the phenotype, since these representations, by definition, have a 1-to-1 mapping to traits: add a gene and one adds a trait. As shown in figure 12.6, this structural change need not affect the overall function of the phenotype, but in many cases it will.

One interesting exception is the evolutionary robotics work of Calabretta et al. (2000) in which portions (modules) of direct-encoding genomes can duplicate to form competing control units for a robot. Initially, the competition between two identical modules has no significance, but after gradual divergence the paired subcircuits specialize to different aspects of the task and outperform both non-modular networks and those in which all pairs of competing modules are present at birth (though with different connection weights). The advantage of gradually sprouting new modules instead of having all present innately appears to lie in the freedom to duplicate only after the parent version has reached some level of behavioral sophistication. In the innate case, the early competition between multiple, unrefined modules confounds evolution, thus slowing its progress.

Calabretta et al.'s networks have a human-designed framework tailored for neutral complexification, including a binary *duplicate* gene associated specifically with a hand-crafted group of neurons and connections. Whereas the 1-to-1 aspect of most direct ANN encodings renders neutrality less likely—genetic changes yield different phenotypes—many generative representations have natural possibilities for hiding a genetic duplication and taking full advantage of search on a neutral landscape. For example, consider the simple grammar below:

$$S \to aA \quad A \to aA \quad A \to B \quad B \to bB \quad B \to b.$$

This generates strings containing one or more a's followed by one or more b's, such as *aaaabbbbbb* (which could easily be translated into some simple form of ANN). The second rule may then duplicate to yield

$$S \to aA \quad A \to aA \quad A \to aA \quad A \to B \quad B \to bB \quad B \to b.$$

Naturally, the effects of this change depend upon the rule interpreter, but redundancy in the rule base will not necessarily affect the phenotypes. In fact, it certainly does not alter the space of producible phenotypes. However, it could affect the probability distribution: if rules are chosen randomly in cases where several apply to the current string, then extra copies mean an increased probability of being chosen. Hence, the new grammar might, on average, produce strings with more a's than b's. Only when the copied rule mutates, for example to A $\rightarrow$ BA, yielding

$$S \rightarrow aA \quad A \rightarrow aA \quad A \rightarrow BA \quad A \rightarrow B \quad B \rightarrow bB \quad B \rightarrow b,$$

does the phenotype space actually broaden (quite dramatically) to any string of a's and b's that begins with an a and ends with a b, such as *abbbaaababab*.

Few EA publications highlight their use of duplication and differentiation, but it appears in many implementations, particularly in genetic programming, where duplication of subtrees is a common genetic operation (Koza et al., 1999). A very versatile new representation, analog genetic encoding (AGE) (Mattiussi and Floreano, 2007; Mattiussi et al., 2008), incorporates key mechanisms from genetic regulatory networks (GRNs) and supports the evolutionary design of many types of analog devices, including ANNs. AGE uses duplication and differentiation of genes representing GRN rules to aid complexification and to achieve performance comparable to NEAT on a double-pole-balancing task (Floreano et al., 2008). Another, more abstract, GRN-based EANN (Reisinger and Miikkulainen, 2007) outperforms NEAT on a board game, and it too exploits duplication and differentiation.

Bongard and Pfeifer describe one of the most elaborate and impressive uses of this bio-inspired mechanism (Bongard and Pfeifer, 2001; Bongard, 2002). In their artificial ontogeny (AO) system, evolved GRN rules govern a developmental process that produces the simulated bodies and ANN brains of complex multicellular creatures, whose fitness stems from their performance on a physical task (such as block pushing). In contrast to Sims (1994), Bongard and Pfeifer put very few constraints (such as symmetry) on the evolved bodies and controllers, which grow together during a sophisticated morphogenesis that begins with a single cell. Their GRNs produce chemicals that can diffuse within and between cells and stimulate different styles of neural network growth, such as the serial and parallel duplication operations of cellular encoding (Gruau and Whitley, 1993). The details of the system are well beyond the scope of this book, but the results leave little doubt that artificial developmental systems can evolve not only target patterns (Federici and Downing, 2006; J. Miller and Banzhaf, 2003) but fully functional structures (J. Miller and Banzhaf, 2003).

Bongard and Pfeifer's work also accentuates the potential utility of duplication and differentiation in Bio-AI. In a personal communication Bongard elaborated upon their results (Bongard and Pfeifer, 2001) with recalled sightings of telltale footprints in the genetic record. Apparently the successful genomes of the final generation contained abundant copies and modifications of ancestral GRN rules. And this signature was much more prominent in the presence of selection pressure than in neutral runs. To Bongard and Pfeifer, this was a strong indication that duplication and differentiation were important factors (not just artifacts of their model) in producing highly fit individuals.

12.5.2   Weak Linkage

In AI systems weak linkage refers to an ease of interaction among components. Simple interaction protocols entail more potential configurations of parts or mechanisms and thus increased adaptivity on both evolutionary and lifetime time scales.

In EANN work, common indicators of weak linkage are tags, which provide an indirect yet flexible means of controlling interactions. For example, directly encoded neurons may have evolved input and output tags such that when the output tag of neuron X matches the input tag of neuron Y, then a connection between X and Y forms; the weight may then stem from the actual matching degree. A small mutation to Y's input tag can then significantly impact its potential afferents.

Simulated GRNs are another common medium for weak linkage. Consider this simple set of GRN rules:

$$A \rightarrow B \quad C \rightarrow D \quad E \rightarrow A$$

where A–E are bit strings of some fixed length L. In order for these rules to produce the autocatalytic loop $A \rightarrow B \rightarrow D \rightarrow A$, the product B's string needs to match that of C; and D's string needs to match E's. This representation supports (but does not guarantee) weak linkage, depending upon L and the conditions for matching. If L is large and perfect equality between bit strings is required, then few interactions will occur. But more flexible matching criteria increase the odds of one rule's product (consequent) matching another rule's regulator (antecedent), thus expanding the space of possible interaction topologies and presumably the breadth of higher-level behaviors.

Figure 12.18 gives an overview of four GRN genotypes that are representative of those found in the ALife literature. The upper left network shows GRN rules spread randomly throughout the genome. A short starting sequence (S, e.g., the bit pattern 1111) reveals these locations to a decoding procedure, and other short sequences (not shown) may demarcate the regulatory and product/structural regions within the gene. AGE (Mattiussi and Floreano, 2007; Mattiussi et al., 2008)

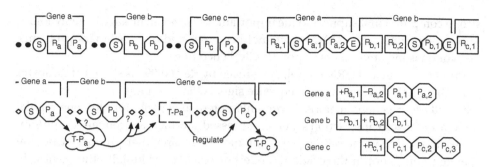

Figure 12.18
Four artificial GRN genotypes. In all cases, the matching of protein(s) to a gene's regulator(s) will promote or inhibit creation of that gene's product(s). Circles labeled S and E represent start and end markers, respectively; black dots, noncoding material; diamonds, material that can participate in many different regulator sequences; rectangles, regulatory sequences; octagons, products.

employs this simple encoding in the following manner: every GRN rule represents a neuron, and the presence and weights of connections between two neurons depend upon relations between their regulatory and product regions. For example, if a genome has five rules, then five neurons are created; to determine the connection from neuron A to B, the product of rule A is compared to the regulator of rule B. A match-evaluating function, M_{age}, determines whether to produce the A-to-B link and its weight. The properties of M_{age} determine the degree of weak linkage that AGE can achieve. Note that AGE is somewhat of a hybrid between a direct and generative encoding, since each of the N rules corresponds to a neuron, but interactions between $2N$ gene products and regulators can account for anywhere from 0 to N^2 connections and weights.

The upper right network of figure 12.18 is the cornerstone of Eggenberger's pioneering work (Eggenberger, 1997b,a) into GRN-based EvoDevo systems. This representation lacks noncoding regions; everything prior to a start sequence encodes regulatory factors, and product sequences continue until a special end substring appears. Eggenberger employs a complex match evaluator, M_{egg}—based on string similarity, an evolved affinity parameter (housed within the product sequence), and the local concentration of regulatory factor—to determine the product level. Each cell within the developing multicellular system has a chemical environment and a copy of the GRN genome. Products may remain within a cell, attach to its exterior as receptors, or travel extracellularly as signals. The emerging signal-receptor relations between cells (i.e., neurons) then govern connectivity and weights in the ANN. Again, the matching function, M_{egg}, controls weak linkage, but compared to AGE, Eggenberger's systems exhibit an extreme degree of ontogenetic flexibility because of the complexity of the growth process. This gives strong computational support to the important role of weak linkage in evolvability. Though Eggenberger's EANNs have not been assessed on benchmark problems, his work has inspired others to pursue GRN encodings.

Another often-cited GRN model, introduced by Reil (1999), is shown at the lower left of figure 12.18. Here, a start sequence signals the beginning of a gene product, while all other genetic material represents potential contributors to gene regulation. In an intricate interaction scheme, expressed products are converted into simulated transcription factors (e.g., T-P_a), which then can either produce structural change (though not in the original model) or promote or inhibit other genes. To perform this regulatory role, the transcription factor must match *any subsequence* residing between the end of one gene's product region and the next start sequence. The match must be a perfect one, which defeats weak linkage, but the potentially long gap between gene products (and corresponding lengthy source of possible regulatory matches) not only saves it, but puts weak linkage under direct evolutionary control. Essentially, the weakness of linkage is proportional to the distance between genomic start signals. Bongard and Pfeifer (2001) based their (more complex) GRN upon this model, although in retrospect, theirs more closely resembles that of AGE, developed several years after their own.

The fourth model (figure 12.18, lower right) comes from Reisinger and Miikkulainen (2007), one of the most successful GRN-based EANNs to date, and one motivated by three key contributors to evolvability: duplication and differentiation, weak linkage, and modularity. The genome is a simple list of regulator-product relations, with variable sizes for each sublist. Mutation can alter the number of rules as well as the lengths of their regulator and product lists. Each unique product translates into a neuron, and the presence and strength of a connection from neuron A to neuron B derives from the ability of symbol A to match each of the regulators in all the rules that produce B. Regulators marked as negative have an inhibitory effect upon their products. The matching criteria also evolve as properties of each regulator in each rule. So once again, evolution controls weak linkage.

Reisinger and Miikkulainen use this generative representation to create critics (i.e., board state evaluators) for the game of Nothello (a slightly simplified version of Othello), where it clearly outperforms both NEAT and a different direct-encoded EANN, which uses a fixed, fully connected topology. The authors attribute this success to increased evolvability, which they document with a comparison of mutation landscapes: whereas the fitness of a NEAT genome drops dramatically with increasing mutational severity, the fitness of their generative genomes decreases very gradually.

They argue that weak linkage (which NEAT lacks) contributes substantially to that difference. They observed much denser connection patterns in generative versus NEAT networks, which seems advantageous for the Nothello domain. With just a few mutations, evolution can alter the linkage level and concomitant synaptic density, whereas direct encodings require much more genomic change for such an extensive phenotypic shift.

This and other work by the same group (Reisinger et al., 2005) highlight their commitment to biological principles of evolvability (Gerhart and Kirschner, 2007). As evidenced in figure 12.18, their representation abstracts away significant detail from natural genetic regulatory systems (and more so than many other artificial GRNs), but it preserves the vital essence. And they achieve very high performance, even compared to systems with more of an engineering than a biological focus. The key mechanism that requires careful attention is weak linkage, and it clearly takes center stage in their work.

12.5.3 Modularity

Whereas weak linkage promotes component interaction, modularity involves a significant constraint on those couplings: the number of local couplings far exceeds the number of global. Although its contributions to evolvability have been hard to confirm in computational studies, modularity is a very popular theme in the EANN literature and an often-praised feature of systems, where the concept pertains to the genotype, phenotype (i.e, ANN), or genotype-phenotype mapping.

Some EANN studies pursue the question of whether modularity is, in fact, useful, and for which environments and tasks. These typically have strong modular biases in the representation(s) and then test whether a selective advantage occurs. Other studies investigate whether (and under what conditions) modularity can emerge through one or more of the three adaptive levels (POE). These are usually very special scenarios designed to favor modularity (when and if it emerges), but for the most part not biased to produce it.

Generally speaking, the artificial evolutionary advantages of genomic modularity is well accepted. If several genes constitute a unit of some sort, then it makes sense that they reside near one another on the chromosome, thus avoiding separation during crossover, as formalized by schema theory (Holland, 1992). In EAs without crossover, the relative chromosomal locations of genes should have little performance impact.

With fixed-topology, direct-encoded EANNs, it is common practice (Nolfi and Floreano, 2000) to group together the genes coding for all a neuron's incoming weights, and to forbid the choice of crossover points within these weight packages. This recognizes a neuron's incoming weight vector as performing an important detector function, and any hard-won combinations deserve protection from fracture. In general, direct-encoded genetic algorithms need not restrict crossover in this way, since the functionality encoded by the kth gene on two different chromosomes is typically the same, e.g., the setting of one of 30 dials controlling a factory process. But because of competing conventions, these functions differ for corresponding genes (or gene groups) in ANN encodings. Hence, recombining the kth neuron weight vectors of two chromosomes is often similar to mixing apples and oranges and yields a worthless hybrid.

Modularity of ANN phenotypes seems to carry an advantage when performing several different tasks. By dedicating distinct portions of a network to each job, the system avoids cross-functional conflicts when an agent must task-switch or when behaviors for only one of the tasks must undergo adaptation. Several EANN results indicate a relation between decomposable tasks and modular phenotypes.

As mentioned earlier, the evolutionary robotics work of Calabretta et al. (2000) shows clear advantages of modularity (of genotype, phenotype, and mapping) for tasks that require functional specialization. Similarly, in an extension of the AGE system, Dürr et al. (2010) enforce modular genotypes, phenotypes, and a modular mapping. This significantly improves performance over nonmodular AGE on a T-maze task, which the modular ANN tackles by separating collision avoidance and navigation into the predefined subnetworks.

Frequently, EANN research into phenotypic modularity involves strong biases in the genome or developmental process such that modular phenotypes easily arise. When the fitness function also cleanly divides into distinct functionalities, then, naturally, the modular phenotypes prevail in the population. Thus, these studies can exaggerate the inevitability of emerging and dominating modular solutions. However, a few studies have shown emergent modularity from unbiased representations.

In a seminal paper Kashtan and Alon (2005) employ an abstract retina simulation to show an increase in evolved phenotypic ANN modularity when the fitness function switches periodically (on an evolutionary time scale) between two separate pattern recognition tasks that share subtasks. Essentially, the evolved modular subunits allow rapid reconfiguration for intergenerational task switching.

Motivated by the general idea that modularity should have a selective advantage in noisy environments (because of the stability that encapsulation provides), Hoverstad (2011) investigates the effects of noise in the genotype-phenotype mapping upon ANN modularity. Using a variant of Kashtan and Alon's abstract retina, he convincingly evolves modular phenotypes, but only in the presence of a noisy mapping.

Another characteristic of mappings is their own modularity: the degree to which genes or gene groups that code for particular traits interact primarily among themselves. Direct encodings have trivially modular mappings, since each gene maps to a single trait, with no interference among genes. In generative representations, pleiotropy normally abounds, thus destroying most obvious forms of modularity. However, gene activity can still be restricted to certain portions of the phenotype, thus taking advantage of scaling while providing some degree of encapsulation.

Using the same GRN-based developmental approach to evolving morphologies and neural networks as discussed earlier (Bongard and Pfeifer, 2001), Bongard (2002) has investigated the modularity of the genotype-phenotype mapping. Expectedly, his generative representation houses considerable pleiotropy, but detailed lesion studies of the highest-fitness individuals reveal a modularization of GRN interactions: highly pleiotropic genes affect many aspects of neurological or morphological development, but not both. Though not investigated in his work, this segregation may result from duplication and differentiation in the same way that different copies of Hox packages control body and brain development: a useful general gene complex arises and then specializes for different design tasks.

AGE also employs a GRN, but one that codes directly for neurons and generatively for connections. Thus, the ANN's weight vector is susceptible to pleiotropic effects, wherein one mutation affects many traits. However, the authors found that these effects were mitigated by the segregations imposed by their modular mapping: though their genome remained generative, it had only a very limited ability to produce cross-modular connections. This modular AGE work shows that generative representations can take advantage of hardwired modularity, and Bongard shows that evolution can gravitate toward modular mappings. But can a developmental EANN find modular networks without any user-imposed representational biases?

In a detailed study of HyperNEAT applied to Kashtan and Alon's retina task, Clune et al. (2010) achieve superior performance with hand-crafted modular networks, which are quite easy to generate using CPPNs. However, HyperNEAT cannot find these networks during evolutionary search. It can only find modular networks for simplified versions of the task. Since Kashtan-Alon and Hoverstad use direct encodings, while HyperNEAT is generative, could this indicate, somewhat paradoxically, that a modular phenotype poses a greater challenge for a developmental than a direct encoding?

More generally, regardless of the representational form, does evolution fail to discover optimal, modular networks because initially selection favors ad hoc, nonmodular ones and then has trouble transitioning from these local optima? In the black-and-white ANNs of figure 12.9, note that the optimal network D is also the most modular, yet evolution could have a hard time getting there once nonmodular solutions arise, such as in network A. Many modularity-enhancing tweaks to this nonmodular network will also decrease fitness.

This brings up the interesting question of whether neural network modularity has a widespread selective advantage because of various functional factors, such as specialization for divide-and-conquer problem solving, or whether modularity is merely a by-product of biological development. In their most recent work Clune et al. (2013) explore the hypothesis that phenotype modularity emerges as a side effect of a selective advantage for reduced network connections. They evolve direct-encoded ANNs to perform Kashton and Alon's artificial retina task and find vastly improved performance and high modularity when the fitness function includes a preference for low wiring costs. However, no modularity arises when fitness stems solely from task performance.

In a different network domain, that of digital circuits, Miller et al. (2000) use CGP and find that the best evolved circuits are nonmodular and use 20 percent *less* material (i.e., units and connections) than do conventional modular, human-designed (and patented) circuits for the same task.

In general, the modularity of phenotypes is far from a ubiquitous virtue. As pointed out by Hoverstad (2011), a software engineer strives to write modular code, but compiler optimization often moves copies of subroutines inline to save function calls. This has the interesting effect of removing a module while making other components more isolated (i.e., they become bigger modules with fewer external dependencies). Just as a good cook often works on different parts of the recipe in separate bowls but then mixes everything together in the end, certain aspects of brain and ANN building may benefit from early isolation but later integration. As described by Fuster (2003), the brain's maturation process supports this hypothesis, since specialized areas for vision, audition, and touch mature much earlier than high-level areas such as the prefrontal cortex, where signals from multiple modalities combine.

From an engineering perspective, whenever main tasks share a subtask, it makes sense to develop one module, accessed by several subnetworks, rather than evolving (or learning) the same thing twice. But requiring each large module to access the shared subunit does decrease its own modularity (and thus heightens its susceptibility to deleterious mutations of the shared subunit). So encapsulation creates safe environments for adaptivity, but shared encapsulations demand more distant communication and can increase vulnerability to widespread perturbation. The proper balance of this trade-off in a neural system would seem very problem-dependent, and information-theoretic analyses by computational neuroscientists (see chapter 13) provide interesting general insights.

No matter how one slices the modularity question (into isolated issues), it is very difficult. In many cases, modularity studies in computational systems do little more than confirm the obvious: modular phenotypes work best on (some) modular tasks. But evolution frequently finds very efficient nonmodular solutions that surprise engineers, many of whom solve problems in a top-down, decompositional manner. What artificial evolution definitively tells us about modularity is that its importance merits guarded (not blind) acceptance. These experiments also point out another key disconnect between human and biological design processes: whereas we can build systems piece by piece, module by module, and eventually assemble the nonautonomous components into a functioning whole, evolution traffics in autonomous, working systems. Nature cannot test components in isolation, only as pieces of a complete organism. These vastly different routes to success make modularity a no-brainer in one case and a brain teaser in the other, especially when it comes to the evolution of brains. Thus, it makes sense for Bio-AI systems, particularly EANNs, to support modularity but certainly not to impose it.

12.5.4   Exploratory Growth

Duplication and differentiation, along with weak linkage, are biological mechanisms that have rather neatly fit into various Bio-AI systems. The utility of modularity (especially phenotypic) is less obvious, since the field's own research has revealed both the difficulty of evolving modular designs and the surprising virtues of many nonintuitive, nonmodular solutions. However, the mechanism of exploratory growth has an even tougher road to acceptance.

Though this may be the only feasible manner in which the organs and subsystems of an organism can hook up, the constraints of computational design have little if any aspect of physical space (except with hardware implementations of neural networks). If the tags of two components indicate that they should interact, then the growth algorithm connects them. There seems little need for a time-consuming process in which the components *find one another*. Thus, a lot of the work into growing axonal connections or gradually duplicating neurons in ALife systems (Astor and Adami, 2000; Husbands et al., 1994; Nolfi et al., 1994) may occupy a researchers' no-man's land: too abstract to answer detailed questions in neurodevelopment—perhaps best handled by the systems in (van Ooyen, 2003)— while rather superfluous and computationally expensive for AI.

For example, the work of Nolfi et al. (1994) exhibits a complex developmental process in which both internal ANNs and external environmental signals influence neural activation, which in turn affects network connectivity. When neurons achieve a threshold (also evolved) activation level, they begin to grow axonal extensions; when these simulated growth cones intersect other neurons, a connection forms. Thus, the topology gradually adapts to the impinging external factors. Not surprisingly, their results show that ANNs developing in a particular environment perform better in it than those grown in a different setting. Of course, biologists like Hubel (1995) already knew this, and typical ANN problem solving either has no significant sensitivity to the temporal aspects of training case presentation or adapts during learning by synaptic change, not growth. So the relevance of the work seems questionable.

Similarly, Astor and Adami (2000) simulate a complex environment, complete with artificial chemicals that diffuse between cells and trigger various activities, including neuronal duplication and axonal/dendritic growth. The authors hand-design a genome capable of growing an ANN that performs classical conditioning, but they give no evidence that evolution can find such elaborate recipes within their detailed representational framework. The research is exploratory in its own right and very impressive in terms of the computational abstractions of biological processes, but it surely lands in this no-man's land as well.

Another very ambitious project (Khan et al., 2011) uses a creative extension of Cartesian genetic programming to support a host of biological processes, including neuronal birth and death; dendritic and axonal growth, shrinkage, and interactions; and dynamic modification of the location and strengths of synapses. The authors evolve networks to navigate in the Wumpus world, and they also perform coevolutionary simulations wherein both agent and Wumpus have CGP-generated ANN controllers. The complexity of the networks and tasks makes the results extremely difficult to analyze, so few conclusions can be drawn from the work other than that the complexity of the neural simulations does not parlay into visibly sophisticated behavior of the agent or Wumpus.

Downing (2007a) has an intermediate abstraction level of neuron groups that serves as the atomic unit during development (in much the same way as in Poly-World (Yaeger, 1993)). Evolved tags associated with each group determine a host of properties, including learning mechanisms and rates, sensitivity to neuromodulators, probabilities of connecting to other groups, the extent to which single connections will branch (thus innervating neighboring neurons), and the duration of the group's developmental phase. Despite all this detail, the model circumvents the simulation of axonal growth via algorithms (based loosely on neural Darwinism (Edelman, 1987) and Deacon's law (Deacon, 1998)) that iteratively introduce connections and then use a group's total connectivity (along with that of competing groups) to compute a biasing factor that affects the connections generated in the next developmental round. The system evolves ANNs that produce simple motion in a simulated starfish, but the complexity of the model dwarfs that of the task.

These and other elaborate ALife systems should serve as flashing caution signs along the low road to AI. Many aspects of neurobiology have been left out of EANNs, but usually justifiably so. The computational costs of detailed simulations often fail to pan out. So while throwing a million simulated neurons at a problem may help, adding a simple detail such as multiple synapses between a pair of neurons seems inefficient (since the sum total of all synapses between two units is effectively summarized by a single connection weight). Whether a synapse forms proximally or distally on a dendrite tree makes a difference in a brain, but a simple difference in weights (high for proximal and lower for distal) may be all that a simulation requires.

Despite these apparent drawbacks, several of these approaches address important issues regarding the sequential nature of the three POE processes and its true significance for neural network assembly and tuning. Examining the complete series of events that produces a mature brain, the genetic component (P) clearly falls into place first: recombination during meiosis forms the unique new DNA profiles that ride inside the gametes, two of which join to form a zygote. From there, however, the stories of the O and the E are less straightforward.

A common lockstep scenario for brain formation and maturation consists of two clearly distinct phases: prenatal development wherein neurons are produced and linked together and postnatal learning, wherein synaptic strengths are modified to enhance behavioral control. Though convenient for computational models and general explanations, this oversimplifies temporal relations whose details may prove useful for EANN research. For example, many studies, summarized by Sanes et al. (2011), find high levels of long-term potentiation (LTP) and long-term depression (LTD)—both forms of synaptic tuning—during prenatal development. In fact, the rates of LTP and LTD (i.e., learning rates) are actually very high during development and much lower during adult life. In addition, Shors (2009) reveals that neurogenesis occurs throughout life, particularly in the dentate gyrus of the hippocampus, but those neurons only hook up to other neurons (and ultimately survive) if the organism subsequently performs cognitively challenging tasks. Thus, although we can retain terminology that equates development with all prenatal brain formation, and learning with postnatal activity, the constituent processes of development and learning are clearly not mutually exclusive in this (more biologically realistic) *overlapping* model.

Much of the relevance of exploratory growth systems hinges on a belief that the overlapping model has relevance for ANN maturation, whereas most other EANN research assumes that the lockstep model will suffice. The majority of EANN systems build connections first (during a *development* phase driven solely by the genome) and then modify the weights (via *learning* based on environmental inputs). On the other hand, the more elaborate systems presume that environmental stimulations and the resulting neural activations have an important influence upon topology formation, an influence that should not be abstracted away by lumping all external effects into synaptic tuning. Under this view, the maturing ANN gets a leg up on its competitors by absorbing the effects of stimuli (e.g., its training cases) before the topology solidifies. Or possibly the young ANN requires only abstract versions of those stimuli to provide an initial guiding bias, such as the muffled voice of a parent singing to a child in the womb (which apparently affects auditory development). These coarse stimuli may be just the right granularity for the rough patterning of a network, with possibly less ambiguity than one finds in a portfolio of more baroque signals.

Support for an overlapping model comes from very early work on cascade correlation ANNs (Fahlman and Lebiere, 1990). These perform supervised learning using a dynamic hidden layer, which produces new nodes and connections on demand when performance stagnates at a local optimum. This approach fares much better than backpropagation on many tasks. In Downing (2010) a similar dynamic layer, this time in an SOM (Kohonen, 2001), is used to investigate an alternative interpretation of the Baldwin effect (Baldwin, 1896), a theory concerning interactions between learning and evolution that has a strong following among EA researchers (Turney et al., 1997).

Unfortunately, the use of exploratory growth in EANNs seems to generate more questions than answers. Chief among these are the following:

- What is the proper abstraction level at which to model neural network topology formation?

- Should development and learning occur in lockstep or with overlap?

- What are the relative influences of the genome versus the environment during development?

- If the environment supplies signals during development, are these any different than the stimuli that drive learning?

All these questions seem highly problem-dependent, but, as such, are important to consider when exploring EANN solutions for a particular task domain. In general, though, it remains to be seen whether axonal migration—one of the most tangible examples of a trial-and-error search process that underlies emergent intelligence—has any legitimate role in Bio-AI.

12.6   The Model of Evolving Neural Aggregates

Consider an alternative EANN framework, one akin to many of the preceding systems but which adheres to several of the pillars of evolvability while avoiding the computationally impractical and biologically implausible use of direct encodings of individual neurons and synapses. This is the *model of evolving neural aggregates* (MENA).

Beginning at the phenotype level, the basic units of neural networks in MENA are the standard fare: neurons, weighted connections, and neuromodulators. However, MENA only works with *aggregates* of these primitives at the genotype level, as do the genomes of complex biological species, where genes code for class properties of neurons, synapses, and modulators, not for specific instances. These are the three principal aggregates in MENA:

- *Layer*   A collection of neurons with similar properties, a *neuron group*.

- *Link*   A set of weighted neuron-to-neuron connections with similar properties.

- *Modulator*   A signal pathway from a group of neurons to a group of synapses.

These aggregates support a few types of modularity in MENA. First, the contents of each aggregate's specification are encoded by contiguous bits of the chromosome. Second, since a gene codes for the general properties of a layer or link, and each gene maps to at most one such general specification, the genotype-phenotype mapping is very modular. Third, a good deal of this translation process involves the aggregates, not the individual components, and this facilitates (but does not enforce) the production of modular phenotypes: it is very easy to produce a layer with high intraconnectivity but limited communication with other layers, and equally easy to produce one with full connectivity between two layers.

A layer has the following properties:

- *Self tag*   A bit string that determines the types of links that can use the layer's members as presynaptic or postsynaptic neurons.

- *Type*   A general specifier of the type of every neuron in the layer, e.g., CTRNN, leaky integrate and fire.

- *Function*   The activation function, e.g., sigmoid, step, linear.

- *Size*   The number of neurons in the layer. This can be explicitly encoded in the genome or implicitly determined by matching strength during gene regulation.

- *Parameters*   Other general factors associated with the neuron type, e.g., time constants and gains for a CTRNN neuron.

Each link forms a bond (of many synapses) between the neurons of two layers (or within a single layer). Links have several key features:

- *Self tag*   A bit string determining the neuromodulators that can affect the link's connections.

- *Input tag*   A bit string determining the neuron groups that can supply *presynaptic* neurons for the link's connections.

- *Output tag*   A bit string determining the neuron groups that can supply *postsynaptic* neurons for the link's connections.

- *Topology function*   A procedure for connecting the neurons of the presynaptic layer to those of the postsynaptic layer.

- *Weight function*   A procedure for determining the initial weights of each of the link's connections.

- *Learning function*   A procedure for modifying the synaptic strengths of all the link's connections, based on learning rules such as Hebbian, BCM, Oja.

- *Learning rate*   The degree to which recent conditions affect the current synaptic weight.

Layers and links are treated as atomic (i.e., indivisible) units during the evolution of ANNs, before they direct the production of neurons and connections, which then become the focal structural units when the ANN runs. Modulators, however, are treated as atomic units throughout all phases of simulation: they do not produce lower-level components. Their primary properties are as follows:

- *Input tag*   A bit string determining the layers that can serve as a source for this neuromodulatory signal.
- *Output tag*   A bit string determining the links that can function as receivers for this signal.
- *Weight*   A factor multiplied by the total activity of the source neurons to yield the net effect of the neuromodulator on the link's connection weights.

These are just some of the possible attributes that the genome of MENA may encode, and several of them are simplifications. For example, any reference to a function is assumed to be a simple integer index into a list of possible options, but functions with arguments will require the genome to encode them as well. For the purposes of this book, most of these details have little significance. The important points are the following:

- The specification of each aggregate constitutes one (often lengthy) gene, all of whose constituent parts form a contiguous section of the chromosome, thus exhibiting the preceding modularities.
- The genome can expand via the duplication of genes, thus producing copies of phenotypic aggregates and thereby supporting complexification via duplication and differentiation.
- The coarse topology of the phenotype (ANN), in terms of the structural coupling of aggregates, stems from tag matching, which facilitates weak linkage.

This basic framework covers three of the four pillars of evolvability while ignoring the property of exploratory growth, which, as argued earlier, has a precarious position in EANN research. The focus on aggregates instead of individual neurons and synapses has more biological realism and greater potential for scaling up, with the fine-tuning of synapses resulting from a lifetime of environmental exposure rather than from a short period of insulated development.

We can stop at this level of biological realism and facilitated variation by employing the genome of figure 12.19, which encodes each aggregate specification as a gene on a linear chromosome (that permits gene duplication). With respect to the aggregates, this constitutes a fairly direct representation: one gene for each

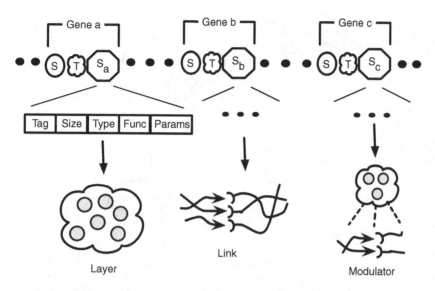

Figure 12.19
A linear chromosome for MENA that directly encodes each aggregate specification (octagon), which is then translated into a set of neurons or connections, or into a neuromodulator. Only the details of one gene, a layer specifier, are shown. S (circle) denotes the start code for a gene; T, the general type of gene—link, layer, or modulator; dark ovals, noncoding genetic material.

aggregate in the phenotype; but at the level of the individual neurons and connections, the genome acts as a generator.

Development of the phenotype ANN from this genotype involves creating the neurons for each layer, determining the links in which a layer participates, producing the connections for each link, and assigning their initial weights. In addition, each modulator gene produces a signal channel from a neuron group to a link, such that activity of the neurons affects the weights in the link. Presumably, the ANN's input and output layers—which must interface with the task domain, a robot, and so on—will stem from fixed, hand-crafted specifications. However, the tags for these special aggregates could still reside on the genome, thus expanding the space of network topologies that evolution can explore.

As shown in figure 12.20, the tags associated with layers, links, and modulators govern the emerging topology. Assuming that the tags are simple bit strings, the Hamming distance may be a sufficient metric for choosing the best presynaptic and postsynaptic groups for a link, and the best layers and links for a modulator. However, as shown in chapter 9, more complex metrics, such as the *streak* model,

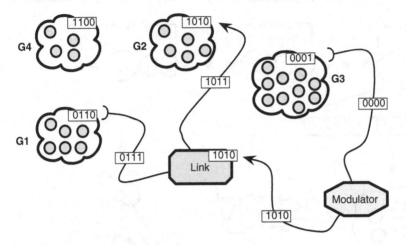

Figure 12.20
Emergent ANN topology as governed by tags (bit patterns) on neural groups (layers), links, and modulators. Tags on each line nearly (or completely) match those on their respective targets, with Hamming distance as the similarity metric. Cupped line ends denote the source or input aggregate; arrows, the output or destination aggregate. For example, the link uses neurons from group G1 as presynaptic inputs and those from G2 as postsynaptic output.

can enhance robustness by reducing the odds that a single mutation will drastically alter matching preferences and thereby change the entire topology.

At a deeper level of network detail, the topology function of each link dictates the connection pattern between its presynaptic and postsynaptic layers, whether all-to-all, one-to-one, random, and so on. Running this along with the weight function yields the initial ANN, ready for deployment.

To provide evolution with even more flexibility while retaining the core notions of aggregates, tags, and gradual complexification, we can elaborate the genome into a full-fledged GRN, with some (functional) genes encoding aggregate specifications and other (regulatory) genes encoding signals that can promote or inhibit other genes. Any of the four GRNs described earlier (see figure 12.18) would suffice, though the functional genes should now encode aggregates, not individual neurons.

Figure 12.21 presents another GRN alternative. Here, each gene consists of a start sequence, a type (regulatory or functional), a regulatory region, and a product. In addition, the model includes an environment consisting of several bit strings; these can bind to regulatory regions to activate or inhibit genes.

The GRN operates as a rule-based system, where each gene encodes a rule. The environment behaves as a signal base (SB), consisting of an initial set of strings along with anything produced by active regulatory genes. The GRN repeatedly

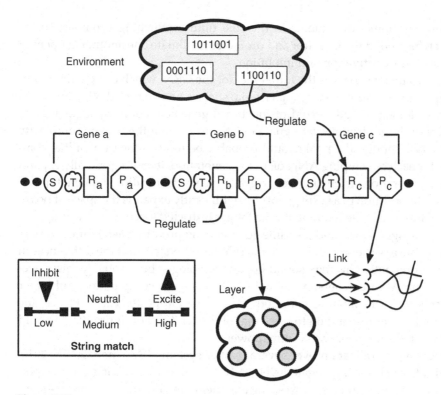

Figure 12.21
One possible GRN for MENA. Genes include a start sequence S, a type indicator T (signal, layer, link, or modulator), a regulatory region (square), and a product (octagon). The environment functions as a preloaded signal base SB, with genetic rules consuming signals that match their regulatory patterns and producing either signals (for the SB) or aggregates for the phenotype. The match degree of signal strings to (the strings of) regulatory sites determines whether the signal inhibits, excites, or has no effect upon the gene.

cycles through the SB, comparing signals to randomly chosen rules, which activate when the signal matches their regulatory region (above some prespecified similarity threshold). An active rule produces one example of its product, which is either a new signal added to the SB, or an aggregate specifier, which supplements the phenotype. In the case of neuron groups (layers), the strength of the regulatory match affects the group size. Signals behave as conserved entities, so after triggering a rule, a signal vanishes.

Signals can also inhibit rules by having a very high Hamming distance (i.e., very weak match) to the regulatory string: at the bit level, the strings are nearly opposites of one another. As depicted in figure 12.21, a low match entails inhibition. A signal inhibits a rule by blocking it for further activation. By convention, an inhibitor remains in the rule until a matching promoter comes along, at which point the inhibitor and promoter annihilate one another, reopening the rule for future activation and inhibition.

The environment serves as the initial fuel for the GRN, which can continue running until it meets various stopping criteria, such as an exhausted SB, a quiescent SB state, or a long period devoid of functional gene activation. By adjusting the initial size of the environmental signal pool, the user can influence the complexity of the ANNs. Though the pool should probably be reset to the same initial composition for each developing ANN of an EA generation, it could gradually expand across the generations to encourage the gradual complexification of phenotypes.

Clearly, the addition of a GRN genotype significantly expands the space of possible neural topologies: development is no longer restricted to a one-to-one mapping from genes to aggregates, and the same genotype can produce vastly different phenotypes because of the stochasticity of the rule-based system and the differences in initial environments. This may sound good to a biologist but probably raises eyebrows among AI researchers: more biology means more computation and larger design spaces. By facilitating variation, we often complicate search. Hence, the choice of which representation to use—that of figures 12.19 or figure 12.21—could depend upon the available computing power.

MENA is a generalization over several earlier systems (Downing, 2007b; 2010; 2012), all of which use aggregate-level ANN encodings. By relying on unsupervised learning (and, in some cases, development) to fine-tune synapses—since the genome does not encode individual weights—these systems exhibit biological realism but perform poorly on EANN benchmarks such as pole balancing. Still, if EANNs are to scale up to more cognitively challenging tasks, the astronomical space of synaptic weights will prohibit effective search using direct-encoded weight vectors. Furthermore, the evolutionary hardwiring of all synapses is clearly counterproductive to survival in dynamic situations (i.e., the majority of real-world problem scenarios). Capabilities for experience-based synaptic modification and modulation seem unavoidable, but these too can greatly complicate evolution's task.

The bottom line is that complex behavior requires complex neural networks, with many diverse layers and regional connection patterns, and various avenues of neuromodulation. A MENA approach permits exploration of this daunting search space by adhering to several key principles of facilitated variation. It can generate a wide diversity of artificial neural anatomies, which despite their computational drawbacks may just come in handy. If ANNs are to fulfill their promise as a substrate for truly general AI systems, I believe that the neural architectures supportive of such intelligence will reside in the far reaches of that anatomy space; and evolutionary search may be our only hope of finding them.

12.7 Picking Prophets

I often think back to a dinner meeting in the late 1990s with a few GOFAI researchers, myself being one of them at the time. I mentioned that I was considering shifting focus to neural networks and the brain, to which one of the diners somberly replied, "Don't." After several years of dabbling with computational neuroscience, I sourly began to view the field as one into which *many good AI people have ventured and disappeared without a trace of publishable results*, and I seriously wondered whether my dinner colleague's blunt statement had been all too prophetic.

There is no doubt that ANNs present a gauntlet of challenges for both engineering and scientific pursuits, but at the same time they provide an extremely versatile substrate for representing information. The two key problems are deciphering the information that they contain (chapters 7 and 13) and getting it in there in the first place (chapters 10, 11, and the current one).

In some domains these knowledge acquisition challenges buckle under to supervised learning via standard backpropagation. Other situations give clear indications of the proper network topology but have no detailed feedback as to the proper response for each input scenario; these are often amenable to standard evolving weight vectors. But some problems simply lack sufficient structure, are very open-ended, or provide only inconsistent feedback from an ever-changing environment. These require the utmost flexibility in a representation and the full repertoire of adaptive mechanisms.

I hope this chapter has convinced readers that there are EANN tools for many jobs. Some have proven their worth, some have clear potential, and others represent very preliminary explorations. Together, they show that all three elements of POE contribute to EANNs, though only rarely in the same system.

To date, few POE EANNs have shown any noteworthy performance on difficult tasks. They may survive and proliferate in ALife environments, where *something* viable normally evolves and outcompetes other species during the coevolution of arcane agents. But channeling the technology into standard problem-solving contexts is nontrivial. Most successful EANNs are straight P systems with direct encodings, or PO systems using generative genomes such as GRNs. Adding learning (the E, for epigenesis) into the system rarely brings big improvements in all but very small networks for simple tasks.

The dilemma is that challenging AI domains, such as robotics, board games, and video games have little support for supervised learning (because of limited and imprecise feedback), which is traditionally easy for an ANN to handle. Instead, they are classic reinforcement learning scenarios, and ANNs to do all RL activities involve both normal and neuromodulatory neurons, along with neuromodulated synapses. These more complex networks have been evolved (from simple primitives) in a few rare cases (Soltoggio et al., 2008), but most RL work with ANNs uses fixed topologies (Niv et al., 2002) or avoids evolution completely (Izhikevich, 2007). Most EANN work simply replaces the lifetime adaptivity of RL with evolutionary adaptivity and thereby avoids learning altogether. In fact, many interesting domains are made for RL but have been successfully conquered by standard direct-encoded P-style EANNs. Adding learning only complicates the search process. In this sense, EANNs in practice have not ventured far from the intentions of the earliest investigators: to replace ANN learning with evolution.

Of course, the goal of AI systems is to solve problems, not to incorporate as many levels of adaptivity as possible. However, to move beyond EANN systems that are specially tailored to particular problems, there is little hope that one level of adaptivity will suffice. The P-style NEAT or SANE approach may produce excellent Othello players, but a general game-playing agent (trainable on several tasks) will probably require more than just evolution to craft an effective net. A legitimate goal for EANN (and AI) research is for some design process (such as evolution) to produce a structure (such as an ANN) that has the basic topology (e.g., network) and primitive components (e.g., neurons) to learn a wide range of capabilities.

Scientists pursuing general AI (more officially, artificial general intelligence, AGI) have the most to gain from this discussion of evolvability and its four pillars, though others exist in the literature. EANNs will continue to make progress on the next generation of race car simulators and action video games, and all without changing the basic tenets of NEAT and SANE. But general intelligence may require a paradigm shift, and EANN workers will have to make one in order to play that game. So in looking for prophets, I tend to turn to innovative folks in the EANN trenches, particularly those attuned to neuroscience and evolvability (Bongard and Pfeifer, 2001; Soltoggio et al., 2008; Reisinger and Miikkulainen, 2007), and I try to forget the advice of that dinner colleague from the previous century.

13   Recognizing Emergent Intelligence

13.1   Bits and Brains

Imagine watching a large dance troupe. Though some global patterns of movement are easily detectable (as when all dancers jump together or pirouette in
sequence), others are more sophisticated. With modern dance, some of the patterns may only be *meaningful* to the choreographer herself, who understands the
timing and interdancer signaling required to attain it.

Now imagine that you are the choreographer for another (competing) dance
troupe. You know nothing of the timing and signaling protocols but respect the
intricacy of the dance. You may not see the global pattern directly but *sense* the
profound aesthetics of the gestalt. However, unless you can dissect and formalize
it, you cannot teach it to your own troupe.

As a start, you will need to verify that interactions are indeed occurring; verify
that the dance is not just the aggregate random movements of 24 individuals. Once
convinced of this, you can proceed to categorize the patterns themselves, hypoth-
esize various interaction protocols that could achieve them, and generally *parse*
what you are now convinced is a legitimate dance.

Now imagine the same task, but with several thousand dancers. Detecting all but
the simplest global and regional movement patterns becomes nearly impossible,
and before you waste a lot of time trying, you would like some confirmation that
patterns actually do exist somewhere in the mass of gyrating bodies.

Neuroscientists face similar challenges in trying to discern behavioral patterns
in neural networks. The simultaneous measurement (on millisecond time scales) of
thousands (millions or billions) of neurons lies a bit beyond current wet lab capabilities, but computational neuroscientists can easily record from large groups of
simulated neurons. And this data, once parsed, gives general indications of an artificial neural network's mechanisms for performing *intelligently* on various tasks.

Information theory provides formal mathematical tools for verifying the *existence* of relations in collections of random variables, whether the locations of

dancers, the per capita incomes of countries, or the activation levels of neurons. It can also help discern (and even explain) individual patterns, but in its basic form, information theory primarily provides the initial confirmation that *something* interesting is happening in a system.

Computational neuroscientists have defined many metrics of *interestingness*, often synonymous with *complexity*. Some of these tie directly to classic information-theoretic notions of communication efficiency, while others have more implicit links to these core concepts. After reviewing several of these fundamentals, this chapter examines a few of the more popular metrics.

These formalisms serve several purposes. They can detect meaningful interactions between small groups of neurons when the behavioral *signature* of each is not simply an on/off state, but a complex millisecond-scale series of membrane potentials or spike times—a topic covered thoroughly by Rieke et al. (1999) but beyond the scope of this book. By abstracting these time series to on/off characterizations over longer time scales (where *on* may mean "firing above its normal frequency"), patterns between large groups of neurons register on the radar of many metrics designed to assess the complexity of large-scale network behavior. These methods permit classifications and comparisons of brains and ANNs with respect to metrics that typically cannot prove the network is performing a complex task, let alone doing it optimally, but rather that intricate, structured interactions are indeed occurring. These same metrics can also be employed as objective functions (fitness functions in evolutionary algorithms) for optimization procedures designed to search for complex ANNs.

13.2   Formalizing Diverse Correlations

Given that the essence of knowledge lies in functional correlations, in *differences that make a difference* (Edelman and Tononi, 2000) in the agent's behavior, then any quantification of these correlations should help codify the knowledge and intelligence of that agent. However, these relations span many diverse domains. For example, visual scenes and acoustic sequences map to neural firing patterns; other neural patterns map to motor sequences; and still other correlations are between different neural ensembles. Unfortunately, the basic mathematical notion of correlation cannot handle these domains in which the states (i.e., different patterns) have no natural ordering (as the real numbers do).

Consider Pearson's correlation metric in equation (13.1), a standard in statistics. To compare two samples (e.g., time series), X and Y, this metric examines corresponding pairs from each series, x_i and y_i (where i is a time stamp) and computes their differences from the respective series' averages, $\overline{X}$ and $\overline{Y}$. The standard deviations of each series are in the denominator. To attain a high positive correlation

($r \approx +1$), the x and y values should *match up* in the sense that whenever x_i is above (below) average, then so too should y_i be above (below) its series' average. A high negative correlation ($r \approx -1$) occurs when the y values are consistently above (below) average when their x counterparts are below (above) it.

$$r = \frac{\sum_{i=1}^{n}(x_i - \overline{X})(y_i - \overline{Y})}{(n-1)\sigma_X\sigma_Y} \tag{13.1}$$

The critical points here are that the series must have an average, and the distance between all data points and the average must be quantifiable. Any domain housing scalar-valued state values will clearly satisfy those criteria. But what about domains in which the states have no natural ordering and hence no meaningful averages nor distances between states? What is the average of a collection of two-dimensional pixel images or of binary vectors of neural states? Even when some notion of average state can be computed, the resulting correlation values may not make much sense.

Imagine trying to assess difference-making differences in a series of mappings from visual scenes through multiple levels of neural patterns to motor sequences. This calls for a string of correlation readings, with each mapping between two domains, none of which admit a straightforward ordering among states. Pearson's metric makes little headway, but information theory succeeds by ignoring averages and orderings and focusing upon one simple type of question: when $x_i = a$, how often is $y_i = b$, where a and b are any of the possible states of X and Y, respectively? This allows information theory to assess correlations across any two domains in which the sets of viable states can be listed (in no particular order). Hence, with information theory, one can compare apples to oranges: visual images to other visual images, to acoustic patterns, to neural states, or to motor sequences.

Figure 13.1 compares time series of patterns across domains of hexagonal and triangular patterns. In the upper series pair, each hexagonal pattern maps to a unique triangular pattern, so the *mutual information* (i.e., information provided by the occurrence of a particular pattern in one domain about the corresponding pattern in the other domain) is high. In the lower series pair, the second and fourth hexagonal patterns map to the same triangular pattern. In addition, the third and fifth hexagonals are identical, but they map to different triangulars. All of this introduces ambiguity into the relation between the domains: knowing the state of one domain may not suffice to predict the partner domain's state. The lower mutual information value reflects this ambiguity.

Mutual information provides a scalar, quantitative measure of interdomain correlation even when the individuals in each domain have no meaningful scalar

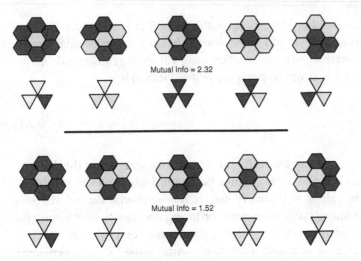

Figure 13.1
Assessing correlations between two domains (hexagonal and triangular) via the information-theoretic metric known as mutual information

(or quantitative) descriptor. It is therefore an important tool in assessing the myriad correlations between and among world, body, and brain states that govern intelligence.

13.3   Information and Probability

The field of information theory began in the late 1940s with the groundbreaking work of Shannon (1948) and receives thorough coverage in several popular textbooks (Cover and Thomas, 1991; MacKay, 2003). What follows is a condensed motivation and introduction to the basic concepts.

Information theory provides mathematical formalisms for assessing the content and transfer of information, thus enabling engineers to design communication protocols, lines, and networks providing the most effective storage and flow of information with the least amount of resources. In science the same concepts help to analyze the information-handling capabilities of signaling systems such as the brain. In both engineering and science typical resources include *wiring* and *message length*.

An important insight of all communication theory is that senders and receivers often exchange relatively small amounts of information, but these signals actually represent (or index) much larger bodies of information. For example, if Arthur and

Betty know the exact contents of each other's bookshelves, then the signal "On the Road, 32" from Arthur would spur Betty to pull down Jack Kerouac's classic and turn to page 32, the contents of which were presumably the complete message that Arthur wished to convey. If Arthur knows the order of books on Betty's shelf, an even simpler message, such as "57, 32" may suffice, where Arthur and Betty have agreed upon the protocol that books are numbered top-to-bottom, left-to-right on Betty's shelf.

Information theory relies heavily upon assumptions of significant background knowledge, such as the agreed information sources available to senders and receivers, or the probability distributions of various signals. These (usually realistic) assumptions greatly enhance coding efficiency and justify the standard use of bits as the basic quantity of information.

Consider a simple military scenario in which two battlefield regiments have very limited communication capabilities over a simple channel. The two basic signals are a short and a long pulse, denoted by 0 and 1, respectively; and the line must be used sparingly, for fear of both enemy detection and general wear-and-tear. It will presumably be used just once each day, at a previously agreed-upon time.

Assume that conventional battlefield wisdom indicates that the only relevant messages in these situations are the five shown in figure 13.2. In this example, each such message constitutes a *symbol*, e.g., A – E. The basic coding theory problem is then, What combinations of short and long pulses should be used for each of the five symbols so as to minimize the total number of pulses sent over the channel in the course of the entire battle?

The probabilities of the underlying situations (e.g., attack, flood, plague) turn out to be of extreme importance in designing efficient codes, with the basic insight that less frequent events should have longer codes. A naive approach ignores these probabilities and simply assigns three bits (pulses) to each symbol,

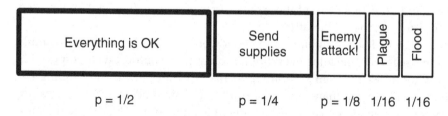

Figure 13.2
Military messages and the probabilities that they will be sent, i.e., that the underlying events will actually occur

since $\lceil \log_2(5) \rceil = 3$. In contexts where many signals will be sent in sequence, the assumption of a common fixed length (k) for all symbols simplifies the decoding process, since the receiver knows to pull out and decode k bits at a time. Thus, there is no need for symbol-ending delimiter signals, which, of course, require extra bits.

However, these fixed-length codes waste bits compared to the most efficient codes, which manage to use variable-length symbol codes without introducing delimiters. This is achieved by *prefix coding*, wherein symbol codes are designed so that the full code for any symbol is never the beginning (prefix) of another symbol. In decoding a message in prefix code, as soon as a variable-length segment matches the code for any symbol, the decoder can immediately pause and record that symbol; upon resuming reading from the channel, the decoder knows that the next signal is the beginning of a new symbol.

When there is no statistical relation between neighboring symbols in a stream, the most efficient coding scheme turns out to be the *Huffman code*, a well-known prefix code designed at the Massachusetts Institute of Technology (Huffman, 1952). Huffman codes are designed by an ingeniously simple algorithm that builds a labeled tree based on the symbol frequencies. In a nutshell, the algorithm clusters symbols based on frequency, with the least frequent symbols clustering earliest and thus ending up at the bottom of the tree. Frequent symbols cluster later, end up near the top of the tree, and thus have shorter codes.

Figure 13.3 presents the Huffman tree for the five military messages, and figure 13.4 details the codes (read directly from the tree as the string of labels from the root to the particular event), the number of bits per code, and the *information content* of each symbol. As shown, the information content of a symbol equals the negative logarithm of its corresponding event's probability, which is also referred to as the *surprisal* of an event. Thus, the occurrence of rare events constitutes more information than that of common events. This information content also equals the number of bits needed to encode the symbol under an optimal (Huffman) protocol, as shown in figure 13.4. Both of these definitions of information content are used in the information theory literature.

Given these codes, the military regiments would normally exchange a simple one-bit message, "0", meaning everything is okay. If one outpost were responsible for receiving messages from several regiments, bundling them, and sending the summary to a high commander, a typical message might be 00110010, indicating that regiments 1 and 2 are okay(0,0), regiment 3 is under attack (110), regiment 4 is okay(0), and regiment 5 needs supplies (10). Thus, five messages require only eight bits.

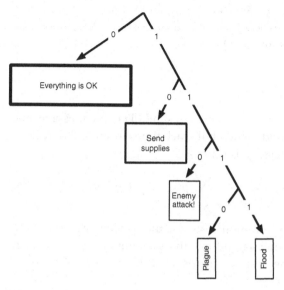

Figure 13.3
Huffman tree (with codes read off along the branches) for the military message example. Note that the most likely messages have the shortest codes and that no code is the prefix of another code.

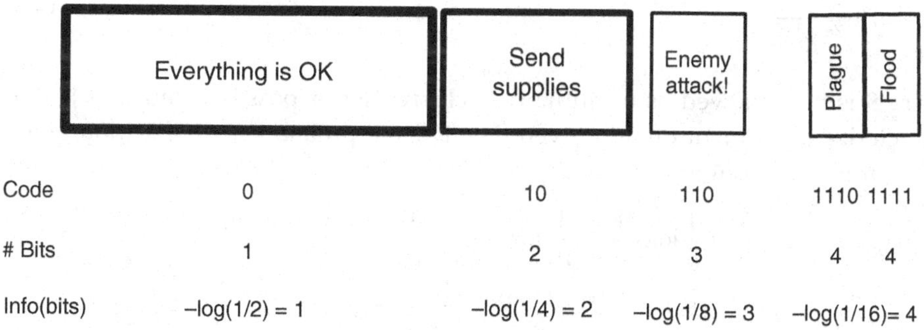

Everything is OK	Send supplies	Enemy attack!	Plague	Flood
Code	0	10	110	1110 1111
# Bits	1	2	3	4 4
Info(bits)	$-\log(1/2) = 1$	$-\log(1/4) = 2$	$-\log(1/8) = 3$	$-\log(1/16) = 4$

Figure 13.4
Information content of each symbol, defined as $-\log_2(p(s_i)) = \log_2(\frac{1}{p(s_i)})$, is shown to be equivalent to the number of bits in the Huffman code for that symbol. Thus, the information in a symbol is directly proportional to its length (under an optimal encoding), which is inversely proportional to the probability of the event underlying the symbol

13.3.1   Entropy

If these military event probabilities remain true throughout the signal-passing scenario (e.g., the war), then one can expect the average length of an N-symbol message (sent by the outpost to the commander) to be

$$\sum_{s_i \in S} N p(s_i) \log_2 \frac{1}{p(s_i)}, \qquad (13.2)$$

where $Np(s_i)$ is the average number of occurrences of symbol i in the message, and $\log_2 \frac{1}{p(s_i)}$ is the number of bits required for symbol i each time it occurs. The average length of a single symbol in this message is thus

$$\frac{1}{N} \sum_{s_i \in S} N p(s_i) \log_2 \frac{1}{p(s_i)} = \sum_{s_i \in S} p(s_i) \log_2 \frac{1}{p(s_i)} = - \sum_{s_i \in S} p(s_i) \log_2 p(s_i). \qquad (13.3)$$

The rightmost expression is the common form for the *Shannon entropy* of the code,[1] which represents both the average bit length of a symbol in any message (assuming optimal coding), and the average surprisal (*self-information*) of a symbol in any message, defined as $-\log_2 p(s_i)$.

More generally, the information-theoretic definition of entropy is the uncertainty in a situation. Mathematically, entropy is a single number (in bit units) that summarizes the probability distribution over a set of mutually exclusive states. In the military example, the uncertainty about the state of a regiment is

$$\sum_{s_i \in S} p(s_i) \log_2 \frac{1}{p(s_i)}, \qquad (13.4)$$

where S is now viewed as a mutually exclusive set of possible states, $\{s_i\}$, that characterize a regiment on any given day. For the probability distribution given earlier, this works out to

$$-\frac{1}{2} \log_2 \frac{1}{2} - \frac{1}{4} \log_2 \frac{1}{4} - \frac{1}{8} \log_2 \frac{1}{8} - \frac{1}{16} \log_2 \frac{1}{16} - \frac{1}{16} \log_2 \frac{1}{16}$$

$$= \frac{1}{2}(1) + \frac{1}{4}(2) + \frac{1}{8}(3) + \frac{1}{16}(4) + \frac{1}{16}(4)$$

$$= 1.875.$$

Thus, the uncertainty of a regiment's state is 1.875 bits, which also represents the average length of a signal stream required to communicate that state to another regiment, the average surprisal associated with any message sent by the regiment, or the amount of uncertainty resolved by a symbol when received.

The maximum possible uncertainty occurs when each regiment state has the same probability: $\frac{1}{5}$. This yields an entropy of $-5\frac{1}{5} \log_2 \frac{1}{5} = \log_2(5) = 2.322$. Such a homogeneous probability distribution would necessitate a different Huffman code, with an average expected symbol length of 2.322 bits.

At the other extreme, minimal uncertainty occurs when one state is known to be true and thus has a probability of 1, while all others have probability zero. This yields an entropy of zero (since $\log_2(1) = 0$ and all other probabilities are zero), and the average Huffman code would be zero as well. Basically, the regiment need not send a message at all if everyone already knows its state.

In general, the entropy of system S is expressed by

$$H(S) = -\sum_{s_i \in S} p(s_i) \log_2 p(s_i),\tag{13.5}$$

where S is modeled as the set $\{s_i\}$ of mutually exclusive states. $H(S)$ then characterizes the uncertainty of the system, with a maximum value (of $\log_2 M$, where M is the size of $\{s_i\}$) occurring when all states are equiprobable, and a minimum value of zero when one state is known to be true and all others false.

13.3.2   Information as Entropy Reduction

Imagine a different scenario in which you are searching for something, such as a bakery, in a large city. Figure 13.5 depicts two possible probability distributions reflecting your uncertainty in the situation. In the high-entropy case, each city block seems equally likely, indicating maximum uncertainty. The general equation for entropy would then produce its maximum value for this $N \times N$ city grid: $\log_2 N^2 = 2\log_2 N$. The low-entropy situation is quite different, since now a good many city blocks have low probabilities while eight others have much higher values. This skewed distribution yields a much lower entropy.

Graphics like figure 13.5 illustrate an abstract connection between the notions of entropy in physics, statistical mechanics, and information theory. In physics the entropy of a system is energy that is not available for work. Imagine a box in which gas molecules are free to move, and the shading in a cell (of figure 13.5) indicates the probability of a random gas molecule residing in that cell (at some frozen moment in time). Under normal conditions, the gas will fill the volume with approximately the same number of molecules per cubic meter, and thus the probability distribution for an individual molecule should be uniform. This represents maximum entropy, since the molecules will not be capable of any work because of their homogeneous distribution.

Next, imagine the molecules pressed into a corner so that only a few of the cubic blocks of the container contain molecules. Here, the probability distribution for a molecule is sharply skewed. Air pressure differences between the filled and empty

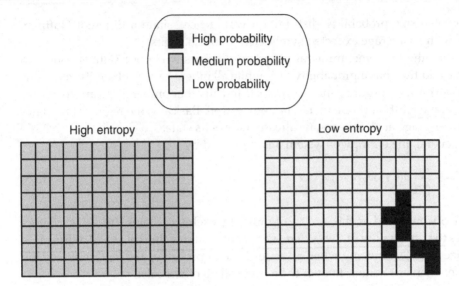

Figure 13.5
High– and low–entropy situations, where each grid represents the complete set of individual states (cells) of a large system. In high–entropy systems all states are equally likely, whereas the probabilities are much less uniform in a low-entropy system; i.e., low entropy indicates less uncertainty, since some states have a much higher likelihood than others.

blocks now produce a situation of low physical entropy, since much of the system's energy is poised to do work. In general, skewed distributions indicate the presence of at least some neighboring blocks with nonequal molecule counts, thus presenting opportunities for pressure-driven work.

In information theory the information content of a message can also be explained as the entropy difference between the situation just prior to and just after the message. For example, if the grids of figure 13.6 represent the bakery-finding problem, then messages might be tips given by various pedestrians as to probable and improbable locations of the bakery. The changing probability distributions from left to right indicate the information content of the messages that link one state of uncertainty to the next. So the information in a message that converts distribution D_1 to D_2 is $H(D_1) - H(D_2)$, the entropy change.

The information content of a single symbol in an alphabet can be understood in the same manner as the bakery-finding problem. Disregarding any intersymbol relations (such as knowing that q is almost always followed by u in English), the receiver of a message assumes that the next symbol will be drawn from the symbol pool with a probability distribution (D_L) that is characteristic of the language (L) used by the sender and receiver. L has an entropy of $H(D_L)$, which represents the expected surprise of any single symbol in the language: a weighted average of surprisals. Upon receiving the actual symbol, s, the receiver's

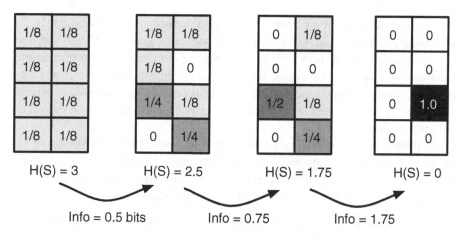

Figure 13.6
Entropy-changing effects of messages

updated probability distribution, D^*, is maximally biased such that $p(s) = 1$ and $p(s^*) = 0 \; \forall s^* \neq s$. Hence, $H(D^*) = 0$. Thus, in situations where exactly one symbol (stochastically drawn from D_L) is expected, the information content of *any* symbol is $H(D_L) - H(D_*) = H(D_L)$. Note that this differs from the self-information or surprisal of the symbol.

13.3.3   Mutual Information

One of the most common problems in analyzing information exchange within the brain, or between the brain and the environment, is detecting patterns of correlated behavior between systems consisting of many components. To evaluate whether actual information has been transmitted from area A to B, one needs to judge whether a complicated pattern of activity in A somehow influences the activity pattern in B.

An analogous situation arises in sports (such as American football), where a coach (and often several assistants) make various sideline gestures (i.e., choose symbols from a large alphabet) in order to signal a formation to their players. In watching a game, a rival scout may record many cases, each consisting of the coach's and assistants' signals along with the resulting formation. In theory, this would allow the rival team to learn the correlation between signals and formations. However, in practice, coaches tend to use several assistants, some of whom serve as decoys. If the decoy changes on each play, the correlations between signals and formations become much more complicated, if not impossible, to detect.

Similarly, when looking at interactions between brain regions A and B, there are many decoys: neurons that happen to fire in A at time t but do not actually have an effect on subsequent firing patterns in B. And the decoys change frequently. So an exact firing pattern (P_A) in A (consisting of a particular group of N highly active neurons) may never reappear exactly, though many P_A variants may exert the same general influence upon B. Detecting these persistent interactions—despite varying patterns, some of which may never exactly repeat themselves—may puzzle the human observer but yield to information theory, in particular, to the notion of mutual information.

Mutual information was originally developed to assess the effectiveness of an information channel, i.e., to quantify the degree to which a sender's messages actually arrive, uncorrupted, at the receiver. From a neuroscience perspective, this can be expressed as slightly different (but mathematically equivalent) questions:

- If the state of the environment constitutes a signal, then how accurately is it transmitted to the brain? For example, does the same environmental state lead to the same or similar brain states?

- If the brain state constitutes a signal (for an observer reading brain-imaging data), then how accurately do these signals reflect the current state of the environment?

- To what degree do changes in the environment cause changes in brain states?

- To what degree do changing brain states reflect environmental changes?

The mutual information between environmental states and brain states provides the same quantitative answer to each of these queries.

Consider the scenario of figure 13.7, where both the brain (X) and perceived world (W) have four possible states, P1–4 and A–D, respectively. The data table displays eight observations, snapshots of co-occurring world and brain states. From this data, the mutual information between the world and brain can be assessed. The mutual information between two systems is commonly denoted $I(X ; W)$, or $I(W ; X)$, since the relation is symmetric. A variety of mathematical expressions are used to define it, including the following:

$$I(W; X) = H(W) + H(X) - H(X, W). \tag{13.6}$$

$$I(W; X) = H(W) - H(W|X). \tag{13.7}$$

$$I(W; X) = H(X) - H(X|W). \tag{13.8}$$

Expression (13.6) defines mutual information as the sum of the entropies (uncertainties) in X and W minus the entropy in the *aggregate system*, i.e., that consisting of all combinations of states in X and W. The symmetric definitions in expressions (13.7) and (13.8) involve the difference between the uncertainty in a system on its own (or in the absence of any other information) and that same system's

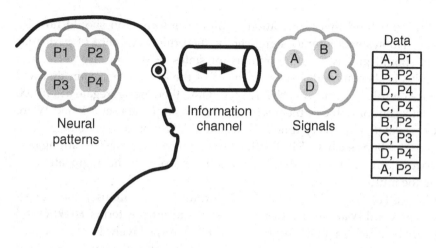

Figure 13.7
Correlations between brain and world states

uncertainty when the other system's state is known. These are referred to as the *marginal* and *conditional* entropies, respectively.

First, to calculate the marginal entropies from the simple data set, use the eight cases to produce frequency distributions for each state. In the case of the world states, each appears exactly twice, so each has a frequency of $\frac{1}{4}$, yielding $H(W) = H(\{\frac{1}{4}, \frac{1}{4}, \frac{1}{4}, \frac{1}{4}\}) = \log_2(4) = 2$, i.e., the maximum uncertainty for a four-state system. However, the brain states have a lower entropy (more biased) distribution, since states P2 and P4 occur more frequently than P1 and P3. Thus, $H(X) = H(\{\frac{1}{8}, \frac{3}{8}, \frac{1}{8}, \frac{3}{8}\}) = 1.811$.

To calculate the entropy of the aggregate system (X, W), the frequencies over all combination states are tallied. Of the 16 possible states, 6 occur in the data set. Two of them, (B, P2) and (D, P4), occur twice, while the other four appear once. Since the nonoccurring states can essentially be ignored in the entropy calculation, $H(X, W) = H(\{\frac{2}{8}, \frac{2}{8}, \frac{1}{8}, \frac{1}{8}, \frac{1}{8}, \frac{1}{8}\}) = 2.5$.

The mutual information between the two systems is

$$I(W; X) = H(W) + H(X) - H(X, W) = 2.0 + 1.811 - 2.5 = 1.311. \tag{13.9}$$

Alternatively, if the correlations between brain and world states had been perfect, then both occurrences of A (C) would have involved P1 (P3) in the data set. Then, $H(X)$ would have also been 2. Furthermore, there would have been exactly four aggregate states (instead of six), all with an equal frequency of $\frac{1}{4}$, yielding $H(X, W) = \log_2(4) = 2$. The mutual information would then be maximized at $2 + 2 - 2 = 2$.

The intuition behind this formulation of mutual information should now be clear. If the two systems interact significantly, then the occurrence of particular states in X should greatly bias the options for states in W, and vice versa. The stronger this bias becomes, the *fewer* the states of the aggregate system that will appear. For example, if X and W each have 100 states, then the aggregate has 10,000 possible states, but only 100 of them will appear in a large data set if X and W are perfectly correlated, e.g., every time X is in state P1, W would have to be in state A, so aggregate states such as (P1, B) (P1, C) would never appear. This strong bias toward particular combination states reduces the entropy of the aggregate, thus increasing the mutual information.

Formulations (13.7) and (13.8) have equally intuitive explanations. If the interactions between X and W are strong, then the conditional entropy terms, $H(W|X)$ and $H(X|W)$, will be much smaller than $H(W)$ and $H(X)$, respectively. In other words, the uncertainty in either system is greatly reduced when the other system's state is known.

13.3.4   Conditional Entropy

The basic definition of conditional entropy is given in equation (13.10). This is the weighted average (over all states x of X) of the remaining uncertainty of W when x is known to be the state of X.

$$H(W|X) = \sum_{x \in X} p(x)H(W|X = x) = \sum_{x \in X} p(x)\{-\sum_{w \in W} p(w|x)\log p(w|x)\}$$

$$= -\sum_{x \in X, w \in W} p(x)p(w|x)\log p(w|x) = -\sum_{x \in X, w \in W} p(x,w)\log p(w|x). \quad (13.10)$$

The mutual information of world and brain (figure 13.7) can now be computed in terms of conditional entropy (instead of the entropy of the aggregate), as shown in equation (13.11), where the four conditional entropies of the world W are computed for each state of the brain. For example, the second term in (13.11) stems from the observation that X is in state P_2 in three of the eight cases, and the frequency distribution of W in those three cases is $\{\frac{1}{3}, \frac{2}{3}, 0, 0\}$, which has the same entropy as $\{\frac{1}{3}, \frac{2}{3}\}$. Note that both formulations (13.9) and (13.12) produce the same result: $I(W; X) = 1.311$.

$$H(W|X) = \frac{1}{8}H(W|X = P_1) + \frac{3}{8}H(W|X = P_2) + \frac{1}{8}H(W|X = P_3) + \frac{3}{8}H(W|X = P_4)$$

$$= \frac{1}{8}H(\{1,0\}) + \frac{3}{8}H(\{\frac{1}{3}, \frac{2}{3}\}) + \frac{1}{8}H(\{1,0\}) + \frac{3}{8}H(\{\frac{1}{3}, \frac{2}{3}\})$$

$$= \frac{1}{8}(0) + \frac{3}{8}(0.918) + \frac{1}{8}(0) + \frac{3}{8}(0.918) = 0.689. \quad (13.11)$$

$$I(W; X) = H(W) - H(W|X) = 2 - 0.689 = 1.311. \quad (13.12)$$

Equating the right-hand sides of expressions (13.6) and (13.7) and simplifying, yields the *chain rule of conditional entropy* in equation (13.13). This says that the remaining uncertainty in W when X is known is the same as the uncertainty in the aggregate system minus the uncertainty of X.

$$H(W|X) = H(X,W) - H(X). \tag{13.13}$$

An equivalent expression is

$$H(X,W) = H(X) + H(W|X). \tag{13.14}$$

When W and X are independent, $H(W|X) = H(W)$ and thus $H(X,W) = H(X) + H(W)$, i.e., the entropy is additive. However, when X and W interact, $H(W|X) < H(W)$ and thus $H(X,W) < H(X) + H(W)$, i.e., the summed individual uncertainties exceed that of the aggregate system. As discussed later, in metrics for neural network *complexity* this superadditivity of individual entropies serves as an indicator of neural interaction.

13.3.5 Venn Diagrams for Information Theory
Venn diagrams often help clarify information-theoretic relations. For example, figure 13.8 illustrates links between mutual information, conditional entropy, and

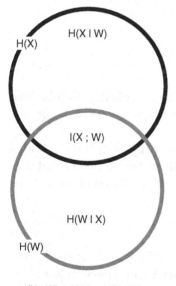

$I(X\,;\,W) = H(X) - H(X\,|\,W)$

Figure 13.8
Venn diagram relating mutual information ($I(X;W) = I(W;X)$) to marginal ($H(W)$ and $H(X)$) and conditional ($H(W|X)$ and $H(X|W)$) entropies

marginal entropy. In these diagrams, space represents uncertainty. Each circle can be viewed as a wide hole or basin. Reduction of uncertainty entails filling the hole, either partly or completely. When two basins, e.g. $H(X)$ and $H(W)$ overlap, then completely filling $H(X)$ will partly fill $H(W)$, with an amount equivalent to the mutual information between X and W, which is often described as *the uncertainty reduction in W when the state of X is known*. That part of $H(W)$ that remains unfilled, given knowledge of X, is $H(W|X)$, *the remaining uncertainty in W given X's state*.

13.3.6   Conditional Mutual Information

Several metrics for neural interaction involve interactions between three or more components (or systems), as discussed later when temporal aspects come into play. Underlying many of these temporal information metrics is conditional mutual information, $I(X;Y|W)$, described as *the information that Y provides about X (or X about Y) when W is known*, and expressed as

$$I(X;Y|W) = H(X|W) + H(Y|W) - H(X,Y|W). \tag{13.15}$$

Via multiple uses of the chain rule of conditional entropy,

$$
\begin{aligned}
I(X;Y|W) &= H(X|W) + H(Y|W) - H(X,Y|W) \\
&= H(X|W) + [H(Y,W) - H(W)] - [H(X,Y,W) - H(W)] \\
&= H(X|W) + H(Y,W) - H(X,Y,W) \\
&= H(X|W) - [H(X,Y,W) - H(Y,W)]. \tag{13.16}
\end{aligned}
$$
$$I(X;Y|W) = H(X|W) - H(X|Y,W). \tag{13.17}$$

Thus, the conditional mutual information can also be described as *the additional reduction in uncertainty of X when both Y and W are known, compared to when just W is known*.

In the Venn diagram of figure 13.9, $I(X;Y|W)$ is most easily described as the area at the intersection of $H(X)$ and $H(Y)$ (i.e., $I(X ; Y)$) minus the portion of that area shared by $H(W)$ (i.e., $I(X ; Y ; W)$). This yields another definition of conditional mutual information:

$$I(X;Y|W) = I(X;Y) - I(X;Y;W), \tag{13.18}$$

where $I(X ; Y ; W)$, the *multivariate mutual information*, is the amount of uncertainty that X can reduce in both Y and W; equivalently, that Y can reduce in X and W; equivalently, that W can reduce in X and Y. Formally,

$$I(X;Y;W) = H(X) + H(Y) + H(W) - H(X,Y) - H(X,W) - H(Y,W) + H(X,Y,W) \tag{13.19}$$

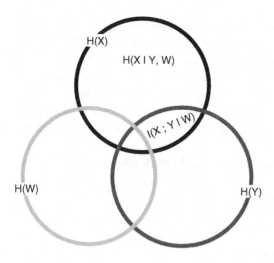

$$I(X\,;\,Y\mid W) = H(X\mid W) - H(X\mid Y,\,W)$$

Figure 13.9
Venn diagram relating conditional mutual information ($I(X;Y|W) = I(Y;X|W)$) to marginal ($H(W)$, $H(X)$, and $H(Y)$), and conditional ($H(X|W)$ and $H(X|Y,W)$) entropies

Table 13.1
Example of conditional mutual information

Observed States			
X	Y	W	$X \cup Y \cup W$
1	0	0	4
0	1	0	2
0	0	1	1
1	1	1	7
0	0	1	1
1	1	1	7
0	1	0	2
1	0	0	4
Entropies			
$H(X)$	$H(Y)$	$H(W)$	$H(X,Y,W)$
1	1	1	2

Note: Numbers in the rightmost column are indices
into the set of all eight possible three-variable states.

Table 13.1 provides a simple data set from which conditional mutual information, $I(X;Y|W)$, can be computed using the standard definition of equation (13.17):

$$I(X;Y|W) = H(X|W) - H(X|Y,W) = 1 - 0 = 1.$$

Here, notice that knowing W does not reduce the uncertainty of X: looking at all rows in which W is 1, note that X has an equal distribution of 1's and 0's in those rows, i.e., maximum entropy. The same is true in the rows where $W = 0$. Hence $H(X|W) = 1$. However, when both Y and W are fixed—to (1, 0), for example—then X has no remaining uncertainty and thus zero entropy. This holds for all four combinations of Y and W, yielding $H(X|Y,W) = 0$. Thus, $I(X;Y|W) = 1$: knowing Y and W reduces the entropy in X by one bit more than the reduction provided by W alone.

In contrast, table 13.2 shows no conditional mutual information because of complete independence of the three variables. Again, notice how independence gives high entropy for the aggregate state (X,Y,W).

13.4   Information Flow through Time

In systems with time-varying state, e.g., neural networks, the actions of one component often influence those of another, such that one can speak of an information transfer between the two. A sizable slice of the system's history can then give the statistical basis for quantifying this transfer in information-theoretic terms. For example, knowledge of the state of component Y at time t can reduce the uncertainty of (i.e., predict) the state of component X at time $t + k$. As shown in figure 13.10, any combination of slices from X's and Y's histories can help predict the future of X.

Table 13.2
Example of total independence, zero conditional mutual information, and zero mutual information for all pairs of variables

Observed States			
X	Y	W	X∪Y∪W
1	0	0	4
0	1	0	2
0	0	1	1
1	1	1	7
0	1	1	3
1	0	1	5
0	0	0	0
1	1	0	6

Entropies			
H(X)	H(Y)	H(W)	H(X,Y,W)
1	1	1	3

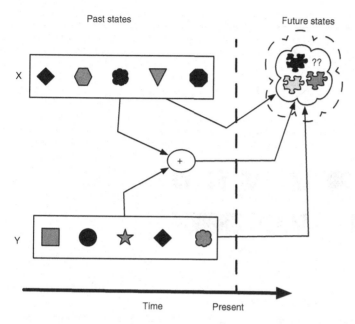

Figure 13.10
Relations used to compute the flow of information. The histories of variables X and Y (alone and in combination) reduce uncertainties of future states of X (dashed–line cloud to solid–line cloud).

There are many different metrics for assessing this movement of information. These serve the important purpose of helping to detect emergent structure in complex systems that, to the naked eye, seem to exhibit little more than random internal behavior.

13.4.1   Entropy Rate

Imagine a time-varying system X, with states $x_0, x_1, \ldots, x_n$. The degree to which X's history predicts its next state constitutes its *entropy rate* (or *information density*), which equation (13.20) codifies as the conditional entropy of the present state conditioned on the history.

$$h_X = H(X_{n+1}|X_n^k) = - \sum_{x_{n+1} \in X} \sum_{x_n^{(k)} \in X^{(k)}} p(x_{n+1}, x_n^{(k)}) \log p(x_{n+1}|x_n^{(k)}), \tag{13.20}$$

where $\{x_n^{(k)} = x_n, x_{n-1}, \ldots, x_{n-k+1}\}$ is the sequence of k states of X that lead up to state x_{n+1}. Thus, h_X represents the average uncertainty in the next state of X, given X's k-state history.

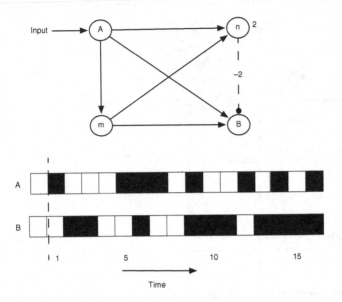

Figure 13.11
(*Top*) ANN in which neuron B detects patterns in the firing history of neuron A. B fires whenever exactly one of A's two previous states was a 1. Each solid arc has a weight of 1 and a delay of one timestep; each neuron has a step-function activation with a threshold of 1, except the inhibitor, n, whose threshold is 2. Neuron m serves as a memory for the previous state of A. The inhibitory link from n to B has a weight of −2 and no delay. (*Bottom*) Firing sequence for neurons A and B driven by a random input series. Black denotes a firing neuron (1); and white, inactivity (0).

It has alternative formulations such as the entropy difference between a $k+1-$ step and a k-step history of X:

$$h_X = H_{X^{k+1}} - H_{X^k},$$ (13.21)

and the limit of X's average entropy per symbol (i.e., time-tagged state) when using longer and longer histories:

$$h_X = \lim_{L \to \infty} \frac{H(X_1, X_2, ..., X_L)}{L}.$$ (13.22)

13.4.2   Transfer Entropy

Now consider two time-varying systems, X and Y. Let $X_{t-1}^{(k)} = \{X_{t-1}, X_{t-2}, ..., X_{t-k}\}$ represent the collection of all possible k-state histories leading up to any of the possible states at time t, X_t; and similarly, let $Y_{t-1}^{(j)} = \{Y_{t-1}, Y_{t-2}, ..., Y_{t-j}\}$ be the j-state histories prior to time t for system Y. Presumably, $j = k$ in most cases.

As introduced by Schreiber (2000), transfer entropy is the information that Y transfers to X, or, technically speaking, the *conditional mutual entropy* between X's current state and Y's history, given (i.e., conditioned on) X's history:

$$T_{Y \to X} = I(X_t, Y_{t-1}^{(j)} | X_{t-1}^{(k)}) = H(X_t | X_{t-1}^{(k)}) - H(X_t | Y_{t-1}^{(j)}, X_{t-1}^{(k)}). \tag{13.23}$$

In other words, this gives the additional reduction in the uncertainty of X_t that Y's j-step history provides, beyond that provided by X's k-step history. The actual calculation involves a joint and several conditional probabilities:

$$T_{Y \to X} = \sum_{x \in X} \sum_{y^{(j)} \in Y^{(j)}} \sum_{x^{(k)} \in X^{(k)}} p(x, y^{(j)}, x^{(k)}) \log \frac{p(x, y^{(j)} | x^{(k)})}{p(x | x^{(k)}) p(y^{(j)} | x^{(k)})}. \tag{13.24}$$

Transfer entropy enables the detection of causality, not just correlation, since unlike mutual information, it is an asymmetric relation: the entropy transferred from Y to X need not equal that going from X to Y.

As a simple example of transfer entropy, consider the four-neuron system of figure 13.11. Here, the firing of neuron A can stimulate neuron B to fire, but only after a one-timestep transmission delay, and only when A has fired exactly once in the past two timesteps; B detects the exclusive OR A's immediate two-step history. Conversely, B has no causal influence upon A.

To detect these relations using only the firing data, we need the asymmetry of transfer entropy. As shown in table 13.4, the mutual information between B's current state and A's two-step history is high: $I(B_t; A_{t-1}^{(2)}) = 0.954$. This signals a correlation, but no causality. However, the transfer entropy from A to B is much higher than that from B to A. This asymmetry reveals the influence of A upon B.

Also note that the first and third conditional entropies of table 13.4 constitute the entropy rates of neurons A and B, respectively, based on a two-step history. Since both are high, we can infer that neither neuron's history plays a big role in predicting its future.

13.5 Information Gain, Loss, and Flow

These three related metrics assess the degree to which information about a variable X (normally assumed to be constant) moves about (i.e., flows) in the system, as reflected in the time-varying states of a variable Y (e.g., the firing rate of a neuron); or equivalently, as changes in the uncertainty (entropy) of X conditioned on changing values of Y. As a simple example, if a subject views a red ball for several seconds, then information flow (from the ball to the neural network) quantifies

Table 13.3
Basis for calculating transfer entropy between neurons A and B in figure 13.11 using equation (13.23). Note: State spaces for columns 3 and 5 are two-element binary vectors that represent two–step histories of variables A and B, respectively

		Firing States		
Time(t)	A_t	$A^{(2)}_{t-1}$	B_t	$B^{(2)}_{t-1}$
0	0	0	0	0
1	1	(0,0)	0	(0,0)
2	0	(1,0)	1	(0,0)
3	0	(0,1)	1	(1,0)
4	0	(0,0)	0	(1,1)
5	1	(0,0)	0	(0,1)
6	1	(1,0)	1	(0,0)
7	1	(1,1)	0	(1,0)
8	0	(1,1)	0	(0,1)
9	1	(0,1)	1	(0,0)
10	0	(1,0)	1	(1,0)
11	0	(0,1)	1	(1,1)
12	1	(0,0)	0	(1,1)
13	0	(1,0)	1	(0,1)
14	1	(0,1)	1	(1,0)
15	0	(1,0)	1	(1,1)
16	1	(0,1)	1	(1,1)

		Marginal Entropies		
	$H(A_t)$	$H(A^{(2)}_{t-1})$	$H(B_t)$	$H(B^{(2)}_{t-1})$
	1.0	1.92	0.95	1.98

		Mutual Information		
$I(A_t;B_t)$	$I(A_t;A^{(2)}_{t-1})$	$I(B_t;B^{(2)}_{t-1})$	$I(A^{(2)}_{t-1};B^{(2)}_{t-1})$	$I(B_t;A^{(2)}_{t-1})$
0.05	0.14	0.07	0.40	**0.954**

	Conditional Entropies		
$H(A_t|A^{(2)}_{t-1})$	$H(A_t|A^{(2)}_{t-1},B^{(2)}_{t-1})$	$H(B_t|B^{(2)}_{t-1})$	$H(B_t|B^{(2)}_{t-1},A^{(2)}_{t-1})$
0.86	0.50	0.88	**0.00**

Transfer Entropies	
$T_{B\to A}$	$T_{A\to B}$
0.36	**0.88**

Table 13.4
Basis for calculating information flow to neuron A from neurons B, C, and D using definitions of information gain and loss (equations (13.26) and (13.27), respectively), and for calculating transfer entropy (equation (13.23))

				Firing States				
Time(t)	A_t	A_{t-1}	B_t	B_{t-1}	C_t	C_{t-1}	D_t	D_{t-1}
0	0	0	0	0	0	0	0	0
1	1	0	1	0	1	0	1	0
2	0	1	0	1	0	1	0	1
3	1	0	1	0	1	0	1	0
4	0	1	0	1	0	1	1	1
5	0	0	0	0	0	0	0	1
6	1	0	0	0	1	0	1	0
7	0	1	0	0	0	1	0	1
8	1	0	0	0	1	0	0	0
9	1	1	0	0	1	1	1	0
10	0	1	0	0	0	1	1	1
11	0	0	0	0	0	0	1	1
12	0	0	1	0	0	0	0	1
13	1	0	1	1	1	0	0	0
14	1	1	1	1	1	1	1	0
15	0	1	0	1	0	1	0	1
16	1	0	1	0	1	0	1	0

				Entropies				
	$H(A_t)$	$H(A_{t-1})$	$H(B_t)$	$H(B_{t-1})$	$H(C_t)$	$H(C_{t-1})$	$H(D_t)$	$H(D_{t-1})$
	1.0	0.99	0.95	0.90	1.0	0.99	0.99	1.0

			Conditional Entropies					
$H(A_t|B_t)$	$H(A_t|B_{t-1})$	$H(A_t|B_{t-1},B_t)$	$H(A_t|C_t)$	$H(A_t|C_{t-1})$	$H(A_t|C_{t-1},C_t)$	$H(A_t|D_t)$	$H(A_t|D_{t-1})$	$H(A_t|D_{t-1},D_t)$
0.79	0.99	0.63	0.0	0.89	0.0	0.89	0.0	0.0

	Transfer Entropy	
$T_{B\rightarrow A}$, $T_{A\rightarrow B}$	$T_{C\rightarrow A}$, $T_{A\rightarrow C}$	$T_{D\rightarrow A}$, $T_{A\rightarrow D}$
0.04 , 0.70	0.0 , 0.0	0.89 , 0.06

		Info Flow = Gain - Loss						
$I_G(A,B)$	$I_L(A,B)$	$I_{flow}(A \rightarrow B)$	$I_G(A,C)$	$I_L(A,C)$	$I_{flow}(A \rightarrow C)$	$I_G(A,D)$	$I_L(A,D)$	$I_{flow}(A \rightarrow D)$
0.36	0.16	**0.20**	0.89	0.0	**0.89**	0.0	0.89	**-0.89**

changes (across the milliseconds) that the network experiences in its uncertainty that the ball is indeed colored red, where uncertainty equates with the entropy of the ball's color conditioned on the state of the network.

The work of Williams (2011) and Williams and Beer (2010) defines information flow from X to Y as the difference between information gain and loss:

$$I_{\text{flow}}(X \rightarrow Y, t) = I_G(X, Y, t) - I_L(X, Y, t). \tag{13.25}$$

This expresses the change in uncertainty of variable X (at time t) given knowledge of another variable, Y, at times t and $t - 1$. Equation (13.26) formalizes information gain as new information (uncertainty reduction) about X_t gleaned from Y_t but excluding that which was already given by Y_{t-1}.

$$I_G(X, Y, t) = I(X_t; Y_t | Y_{t-1}) = H(X_t | Y_{t-1}) - H(X_t | Y_t, Y_{t-1}). \tag{13.26}$$

As expressed in equation (13.27), information loss reflects the increase in uncertainty about X_t when excluding knowledge of Y's immediate past, Y_{t-1}: the difference in uncertainty in X_t when knowing both Y's present and past states versus just its present status.

$$I_L(X, Y, t) = I(X_t; Y_{t-1} | Y_t) = H(X_t | Y_t) - H(X_t | Y_t, Y_{t-1}). \tag{13.27}$$

Figure 13.12 presents a Venn diagram of information gain and loss, showing that each is based upon $H(X_t | Y_t, Y_{t-1})$, the uncertainty remaining in X_t when both Y_t and Y_{t-1} are known.

Figure 13.13, provides a concrete example to illustrate these concepts. It shows the firing patterns of four neurons, whose numeric firing states are listed in table 13.4. Close inspection reveals that A and B have similar firing sequences, and A and C are identical. However, D_{t-1} is always the opposite of A_t. In terms of conditional

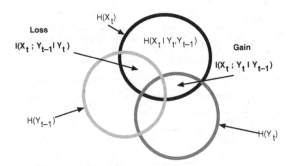

Figure 13.12
Venn diagram portraying information gain and loss

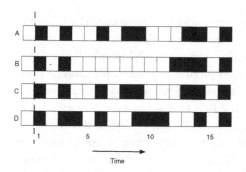

Figure 13.13
Firing diagram for neurons A–D, where black denotes an on state (1), and white denotes an off state (0)

entropy, this means that $H(A_t|C_t)$ and $H(A_t|D_{t-1})$ are both zero; knowing the state of either C_t or D_{t-1} enables the unique determination of A_t. Also, B_t provides more information about A_t than does B_{t-1}, as indicated by $H(A_t|B_t) = 0.79 < 0.99 = H(A_t|B_{t-1})^2$.

Computing the information gain about A due to B

$$I_G(A, B) = H(A_t|B_{t-1}) - H(A_t|B_{t-1}, B_t) = 0.99 - 0.63 = 0.36, \qquad (13.28)$$

whereas the loss is

$$I_L(A, B) = H(A_t|B_t) - H(A_t|B_{t-1}, B_t) = 0.79 - 0.63 = 0.16. \qquad (13.29)$$

Hence,

$$I_{\text{flow}}(A \rightarrow B) = I_G(A, B) - I_L(A, B) = 0.36 - 0.16 = 0.20. \qquad (13.30)$$

Thus, the contribution that knowledge of B_t makes to reducing uncertainty in A_t is 0.20 bits per timestep. However, the flow from A to C is much higher: 0.89 bits per timestep. Even though C_t perfectly predicts A_t, the flow is less than 1, since it measures the information about A_t that is beyond that provided by C_{t-1}, which is small but significant: $I(A_t; C_{t-1}) = 0.11$.

Finally, the flow from A to D is −0.89, because of the large loss term; D_{t-1} strongly predictes A_t, but D_t does not, so transitioning from time t-1 to time t entails a loss of information about A. Note how this definition of flow implies that the system has no memory for previous states: knowledge of D_{t-1} cannot be used to reduce uncertainty in A_t once time t arrives.

Information flow plays a central role in the groundbreaking research by Williams and Beer (2010), which combines information theory with Beer's trademark, a thorough dynamic systems analysis of minimally cognitive, ANN-controlled agents. Together, these techniques generate a detailed picture of the complex interactions between brain, body, and environment in a simple, 1970s-era video game agent, as shown in figure 13.14. After witnessing the first falling object, the agent should only *catch* the second object (i.e., remain underneath it until it lands) if it is smaller than the first. Otherwise, the agent should move away from it. As in much of Beer's other work (Beer and Gallagher, 1992; Beer, 1995; 1996; 2003), the authors evolve the synaptic weights along with neural biases and time constants for fixed-topology CTRNNs. In this and earlier work, the highest-fitness agents behave impressively well, often having success on over 99 percent of the test cases.

Dynamic systems analysis typically focuses on the time-varying states of variables and their interactions, both in a general and specific sense: plots of these variables as a function of one another (or of time) reveal interesting patterns and trends along with particular points at which significant behavioral changes occur. For example, the authors dissect one of the highest-fitness networks (CTRNN-1) and find important relations between a pair of neurons, N1 and N3. As shown in figure 13.15, activity of N1 strongly affects movement, while N3 integrates sensory data to encode (via its activation level) the size of the first falling object (O1). N3 also inhibits N1, such that the larger O1, the more charge that N3 builds up, and thus the more quickly N1 shuts down. Then, as the second object (O2) falls, the

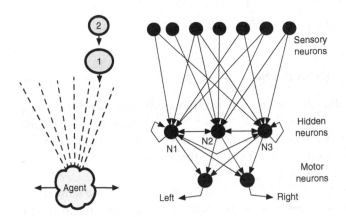

Figure 13.14
(*Left*) Relational categorization task of a simple agent with simple sensing (dashed lines) and horizontal movement capabilities. (*Right*) Fixed-topology CTRNN whose parameters (e.g., synaptic weights, neural biases, and time constants) can be evolved to handle this task.

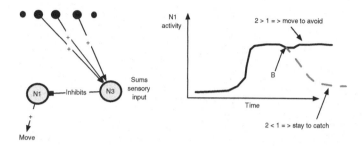

Figure 13.15
(*Left*) Relation between sensors and two neurons in CTRNN-1. (*Right*) The approach of neuron N1 to a
bifurcation point (B) at which the agent decides to stay and catch, or move and avoid the object.

neural system approaches a bifurcation point with a velocity proportional to the
size of O2. For large O2, the bifurcation arrives early and is more likely to precede
the time at which N1 turns off. Thus, the agent keeps moving and avoids O2,
which seems appropriate. The fact that N1 is still active at the bifurcation point
indicates that N3 did not inhibit it too strongly, which means that O1 was small,
probably smaller than O2. However, if O1 is larger than O2, then N1 shuts off
prior to the bifurcation point, and the agent stops moving and catches O2, as it
should.

As pointed out by Williams (2011), the ANN exploits two forms of representation: the activity of N3 represents the size of O1, while the arrival time of the bifurcation point encodes the size of O2. Furthermore, a third representational strategy
arises in another high-fitness network (CTRNN-2), which moves the agent to a
particular lateral location in order to encode the size of O1, thus off-loading memory to the agent-environment coupling (as shown in figure 13.16). None of these
representations are designed into the CTRNNs despite their fixed topologies. They
all emerge from the dynamics of the brain-body-world interaction.

Thus, classic dynamic systems analysis reveals interesting points and trends
in the CTRNN's behavior space that can be coupled into causal explanations of
agent behavior. Information theory complements these accounts by uncovering
pathways of information transmission and their effects upon behavior. It also
enables information tracking at small spatial and temporal scales. Though information theory typically serves to estimate general relations between variables
or subsystems based on time-averaged state correlations, newer metrics such as
information flow can be monitored over time to pinpoint informational transition
points, such as the onset and conclusion of flow between the size of O1 and N3.

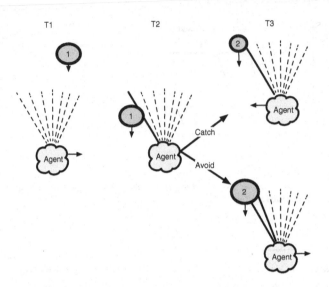

Figure 13.16
Behavior of the agent controlled by CTRNN-2 (Williams, 2011) for three timesteps, T1–T3. (*Left*) Upon detecting the first falling object, it moves right such that only the leftmost sensor detects O1 during the last few steps of its descent. This strategic positioning implicitly stores the size of O1. (*Right*) The agent knows that O2 is larger than O1 (thus initiating rightward fleeing movement) only when it triggers the second leftmost sensor. Otherwise, it moves left to catch.

The authors combine several examples of such flow to reveal a dynamic information matrix uniting brain, body, and environment. For example, information flow from object size to sensory neurons clearly verifies that the informational effects of falling objects begins at central sensors and spreads peripherally for objects directly above the agent. Also, in CTRNN-1 they show that information about the sizes of O1 and O2 flows to several neurons right up to the bifurcation point, wherein all such information is lost but then replaced by binary information indicating which of the two is larger, the crucial determinant of the catch/avoid decision.

The work of Beer and colleagues has provided very deep insights into relations between the behaviors of complex ANNs and simple minimally cognitive agents. Information theory is one of the more recent additions to their extensive dynamic systems toolbox for elucidating these intricate links, which, more than most other simulation results, show the complexity of interactions underlying overt behavior but the relative simplicity of the individual processes that suffice to achieve basic cognition. A mastery of these mathematical and computational tools greatly enhances efforts to recognize and dissect emergent intelligence.

13.6 Quantifying Neural Complexity

With their clear ties to signal transmission, the metrics for information transfer constitute relatively classic applications of information theory to neural networks. However, the concept of mutual information and its extensions also enable the assessment of more abstract concepts, such as emergence and complexity. These, too, have a basis in signal transmission, but as more widespread and general properties of large networks with multidirectional information flow.

Much of the seminal work on this topic, by Tononi, Sporns, and Edelman (often abbreviated TSE) (Tononi et al., 1994, Tononi et al., 1996; Tononi and Sporns, 2004), revolves around two key (and somewhat opposing) network properties: *differentiation* and *integration*. Differentiation (segregation) involves many network regions, each of which can reside in a large number of states. A highly differentiated system can respond to many different external situations when several of these regions occupy one of their many possible states. This dovetails with Edelman's notion of *differences that make a difference*: different external states should promote different neural states, which, in turn, should lead to different actions. It also relates to modularity, since a modular system has isolated regions with significant independence.

Integration is then the degree to which the states of one region affect those of another. An integrated system involves enough cross-talk between regions to facilitate the coordination (in perception, reasoning, or action) required for difficult tasks. From the (quite popular) TSE perspective, a complex neural system involves a mixture of relatively high differentiation and integration: many regions that can occupy many local states, each of which can affect the states of other regions.

Figure 13.17 displays networks exhibiting different levels of complexity. The network at the upper left exhibits high disorder or chaos (and thus low complexity) because of an abundance of differentiation without counterbalancing integration. Each subsystem has enough neurons to produce many local states, but the extreme modularity stemming from three totally isolated components precludes all integration. Thus, each subsystem is completely independent of the others, and the system as a whole can occupy very many global states, most of which show little correlation between regions.

In contrast, the upper-right network has an overabundance of integration because of much lower modularity. Hence, individual regions can have excessive influence upon other areas such that one region may completely determine the state of another. This constitutes a very ordered regime in which each region is so overconstrained by extra-regional signals that it may seldom change state. Or when it does change, so, too, does a lot of the system because of this large web of interaction. A neural system exhibiting this type of order can easily enter a synchronous state wherein all neurons fire and relax together. Again, this represents low complexity.

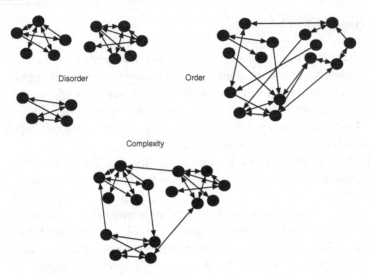

Figure 13.17
Networks displaying different levels of differentiation and integration, which determine a location along the spectrum from disorder to order, with complexity in between. (*Top left*) High differentiation, low integration. (*Top right*) Low differentiation, high integration. (*Bottom*) High differentiation and integration.

The bottom network of figure 13.17 exhibits high complexity via a combination of differentiation and integration. Each module has many neurons and connections (and thus can transition to many local states), and it can do so with some, but not excessive, interregional constraint. In short, the modules have detailed internal dynamics and are weakly interconnected.

Tononi, Edelman, and Sporns liken these three basic network topologies to those of the young, old, and mature brain, respectively. Young brains have intricate but poorly interconnected regions, whereas old brains have less local wiring but retain substantial integration; and mature brains have significant connectivity at both local and global levels. Simulations (discussed by Sporns (2011)) show young ANNS flashing randomly, old ANNs flashing as one unit, and mature ANNs displaying complex waves of correlated activity that form moving blobs of active neurons (resembling the behavior of cellular automata near the edge of chaos).

13.6.1   TSE Complexity

To quantify this combination of differentiation and integration, often referred to as *TSE complexity*, the three authors employ a few different formulations. First, the entropy of a local region aptly quantifies differentiation: the more states a component can occupy, the higher its entropy, and vice versa. However, integration, as formalized in equation (13.31), compares individual to global entropy, just as mutual information compares individual to joint entropy.

$$\text{Int}(X) = \sum_{i=1}^{n} H(x_i) - H(X). \tag{13.31}$$

This generalization of mutual information registers high values when each component occupies many states, but the global system exhibits relatively few (compared to the set of all possible global states) because of dependencies among components. Lower integration occurs when either the local state options are low or the global entropy is high (because of a lack of intercomponent constraint).

This can also be expressed recursively in terms of any bipartition (i.e., a division into a subset and its complement) of the network as

$$\text{Int}(X) = \text{Int}(X_j^k) + \text{Int}(X - X_j^k) + I(X_j^k; X - X_j^k), \tag{13.32}$$

where X_j^k is the jth subset of size k, and $I(X_j^k; X - X_j^k)$ is the mutual information between the subset and its complement. From this, and the fact that mutual information is never negative, it follows that $\text{Int}(S) \geq \text{Int}(s)$ for any subset s of S. This, in turn, entails that $\langle \text{Int}(X_j^{k+1}) \rangle_j \geq \langle \text{Int}(X_j^k) \rangle_j$, where $\langle \rangle_j$ denotes the average over all subsets j of a particular size. In short, TSE integration monotonically increases with increasing subset size, but the absolute increase should decrease to 1 or less as k approaches n.

TSE complexity, equation (13.33), is then the sum (over all size classes) of average mutual information for subsets of that size (compared to their complements).

$$C_N(X) = \sum_{k=1}^{n/2} \langle I(X_j^k, X - X_j^k) \rangle_j. \tag{13.33}$$

Equation (13.34) gives an equivalent definition of TSE complexity in terms of TSE integration instead of mutual information:

$$C_N(X) = \sum_{k=1}^{n} [\frac{k}{n}\text{Int}(X) - \langle \text{Int}(X_j^k) \rangle_j] \tag{13.34}$$

In this model, $\frac{k}{n}\text{Int}(X)$ represents the expected average integration for size k subsets based on the integration value for the entire system.

This formulation highlights an important aspect of TSE complexity: it gives the greatest values to networks with high global integration, $\text{Int}(X)$, but with relatively low integration at many spatial scales. Thus, it rewards a heterogeneous mixture of integration and segregation. A homogeneous network, i.e., one in which the integration at each spatial scale matches the expectation, $\frac{k}{n}\text{Int}(X)$, yields zero TSE complexity. Only when some scales deviate from the expectation does complexity accumulate in this metric. It turns out that deviations at the smaller scales give the most significant contribution, since if $\text{Int}(X)$ is high, it is difficult to achieve much segregation at the highest levels; high interactivity among an entire system entails the same among its larger subsets.

This follows from the observation that the difference between consecutive size-class averages decreases as k increases, thus making it harder to produce large differences between the expected and actual integration values. The smaller size classes have the most potential to contribute to $C_N(X)$ and can do so by having low integration values: high independence or maximum disorder.

$$C_N(X) = \sum_{k=1}^{n} [\langle H(X_j^k) \rangle_j - \frac{k}{n} H(X)]. \tag{13.35}$$

Equation (13.35) provides a third equivalent formulation of TSE complexity, based on entropy differences between subsets and the entire system. Note that, once again, the smaller subsets have the greatest potential to contribute to $C_N(S)$, since large subsets will tend to have entropies similar to the linearly scaled entropy of the entire system. Thus, complexity is highest when small subsets can exist in many different states (i.e., have high differentiation), but because of interactions among subregions (i.e., integration) the system as a whole exhibits reduced entropy: $H(X)$ is low, indicating that X does not evenly sample the complete state space.

13.6.2   Degeneracy and Redundancy

Edelman and Tononi (2000) include *degeneracy* as a central component of their theory of neuronal group selection (the survival-of-the-best-networkers view of brain development; see in chapter 9). Degeneracy is a system property wherein different structural configurations can achieve the same function. It is often contrasted with the property of *redundancy*: maintaining multiple, nearly identical structures, all of which support the same function.

For example, a delivery service exhibits redundancy by having many (interchangeable) trucks, and degeneracy by owning helicopters and employing bicycle and pedestrian couriers. Redundancy helps if one truck breaks down, but degeneracy earns dividends when the principle access road to a community closes for construction. Clearly, both enhance system robustness, but degeneracy will, theoretically, provide a more versatile repertoire of responses to perturbations.

In routing networks, degeneracy entails multiple paths (i.e., structural assemblies of edges) between any two nodes; while in neural networks Tononi et al. (1999)

equate it with many alternative subnetworks that suffice to attain particular output firing patterns. If one subnetwork malfunctions, a substitute preserves the output behavior.

$$R(X;O) = \sum_{j=1}^{n}[I(S_j^1;O)] - I(X;O). \tag{13.36}$$

Redundancy is expressed in equation (13.36) (Tononi et al., 1999) for a system consisting of a collection of input and internal neurons (denoted X), along with a group of output neurons (O); S_j^1 denotes the jth individual neuron in X. Redundancy compares the influences of these single neurons upon O, in terms of mutual information $I(S_j^1;O)$, to the synergistic effects of all neurons in X upon O via $I(X; O)$. High values of $R(X;O)$ entail that each neuron predicts a good deal of the output (high $I(S_j^1;O)$ values), and the contributions of each single neuron to O are similar. Hence, knowing all of X tells us little more about O than knowing just one component of X, and thus $I(X;O)$ is only slightly larger than any $I(S_j^1;O)$. In short, each of the neurons in X performs a similar function by single-handedly coercing O into the same or similar pattern.

In contrast, when $R(X;O)$ is low, the individual neurons have different (possibly completely independent) effects upon O such that (equation (13.36)), their summed individual influences no longer dominate the effect of their aggregate (synergistic) state, $I(X;O)$. Thus, no (or few) individual neurons can strongly determine the complete output state, and few pairs of neurons, when perturbed individually, can force the output layer into the same state.

The formal connection between redundancy and degeneracy becomes clearer after dividing the right side of equation (13.36) by n, to yield equation (13.37):

$$\frac{1}{n}\sum_{j=1}^{n}[I(S_j^1;O)] - \frac{1}{n}I(X;O) = \langle I(S_j^k;O)\rangle_j - \frac{k}{n}I(X;O) \tag{13.37}$$

where $k = 1$ and $\langle I(S_j^k;O)\rangle_j$ denotes the average mutual information between O and each subset of size k, i.e., each individual neuron in X. This average is simply compared to a scaled version of the systemwide mutual information, $I(X;O)$.

Now degeneracy, $D_N(X;O)$, as defined by Tononi et al. (1999), equals the sum of these differences over all subset sizes, as given by equation (13.38):

$$D_N(X;O) = \sum_{k=1}^{n}[\langle I(S_j^k;O)\rangle - \frac{k}{n}I(X;O)] \tag{13.38}$$

Thus, degeneracy is the summed differences between the actual average influence of each subset group upon the output, and the scaled systemwide effect of X upon O. The same logic applies to degeneracy as to redundancy, but now the metric operates on all subsets in the system, not just those of size 1. High degeneracy entails that for many sizes, the subsets of that size have an ability to strongly influence O, and often in a redundant manner: several subsets have the same effect upon O.

Figure 13.18 illustrates redundancy, degeneracy, and their differences. In each of the three networks, neurons (with binary activity levels) in the core have varying degrees of intraconnectivity; whenever one turns on, it can excite all its neighbors, which in turn can excite their neighbors. In assessing $I(X;O)$, redundancy, and degeneracy, assume that a given subset of the core neurons receive external stimulation; then activation spreads throughout X via internal links, and finally neurons in O are stimulated. All calculations of entropy and mutual information (see appendix B) are then based on the equilibrium states of X and O across all the possible trials (i.e., all possible subsets of X neurons that receive external stimulation).

In moving from network A to C, the possibilities for individual neurons to influence large portions of the output state decrease, but when stimulated in isolation (with all other members of X left alone in the off state), they have the same influence upon the probability distribution of output states: when one neuron is stimulated in isolation, it either turns on or off a fixed portion of the output. So both the neuron

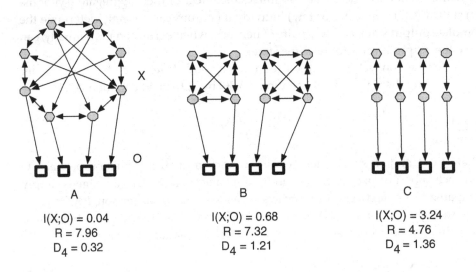

A

$I(X;O) = 0.04$
$R = 7.96$
$D_4 = 0.32$

B

$I(X;O) = 0.68$
$R = 7.32$
$D_4 = 1.21$

C

$I(X;O) = 3.24$
$R = 4.76$
$D_4 = 1.36$

Figure 13.18
Three networks compared with respect to mutual information, $I(X;O)$, between the core (X) and the output layer (O); redundancy, R; and D_4, the contribution of size 4 subsets to the total degeneracy measure, computed as $D_4 = < I(X_j^4;O) > - \frac{4}{8}I(X;O)$.

and the output vector have two equiprobable states, and thus both have entropy = 1. So comparing network A to B and C, redundancy does not decrease because of lack of mutual information between individual neurons and the output state. Rather, it declines because $I(X; O)$ increases: the influence of X upon O goes up, and this stems from the increased entropies of the equilibrium state distributions in X and O. By isolating components of X (i.e., by increasing modularity), there are simply more achievable equilibrium states, and more of the information transferred from X to O involves the synergy of X, not the individual neurons of X.

With degeneracy, the increase in $I(X;O)$ has less effect, since it is scaled by the subset size in equation (13.38). Furthermore, when systems contain more independent components (as in network C), the odds increase that individual neurons in any of the randomly chosen subsets (used in equation (13.38)) end up alone in a component. These neurons have full control of their component and thus single – handedly determine a portion of the output; they can be either on or off with equal probability and thus have an entropy of 1. Conversely, subset neurons that share a component with q others from the subset have only a $\frac{1}{2^{q+1}}$ chance of being off in the equilibrium state; they therefore have less entropy and thus less mutual information with the output layer: a subset cannot transfer more information than it possesses.[3]

In short, modularity (i.e., a greater number of small clusters) decreases redundancy by increasing the mutual information between X and O, but it increases degeneracy by allowing more subsets of neurons to have strong control of the output state. In figure 13.18 the many subsets of X that can produce the same output pattern in O are those with one or more neurons from each of the clusters.

The strong similarity between equations (13.35) and (13.38), along with the fact that high mutual information entails high entropy of the components, indicates a relation between TSE complexity and degeneracy. Indeed, Tononi, Sporns, and Edelman (1994; 1996; 1999) find that neural networks evolved and selected for high degeneracy also exhibit considerable complexity.

The ties between degeneracy, weak linkage, robustness, and thus evolvability are quite strong. Weak linkage supports degeneracy by allowing many different structural configurations to form and then conquer the same functional challenge. This diversity facilitates variation, since a degenerate genotype-phenotype mapping can tolerate variation-producing mutation (to some configurations) by exploiting some of the alternative routes to vital functions. Whereas the duplication-and-differentiation road to variation involves relatively straightforward redundancy, a degenerate approach, with its meshwork of functionally overlapping processes, is much less transparent. Metrics to detect and quantify it are therefore all the more welcome.

13.6.3   Matching Complexity

As discussed earlier, one prime advantage of information theory is its ability to formalize relations between diverse systems, such as the brain, body, and environment. Tononi et al. (1996) exploit this property and extend TSE complexity to the concept of *matching complexity*, which quantifies the ability of neural networks to absorb the statistical essence of environmental signals. Formally, in equation (13.39), it is the difference between the total complexity of the ANN-environment system ($C_N^T(X)$) and both the intrinsic complexity of the ANN ($C_N^I(X)$) and the complexity of the ANN-environment interface (extrinsic complexity, $C_N^E(X)$).

$$C_M(X; S_i) = C_N^T(X) - C_N^I(X) - C_N^E(X). \tag{13.39}$$

Figure 13.19 elaborates this concept. It shows that $C_M(X; S_i)$ isolates those deep internal interactions that stem from the external stimulation itself, not from preexisting internal conditions or from superficial interactions at the network-environment interface. The value can be positive or negative, reflecting the degree to which internal correlations are enhanced or reduced by external signals, respectively. Hence, elevated $C_M(X; S_i)$ indicates the ability of a neural system to adapt in response to exogenous factors. These additional internal correlations mediated by the environment constitute a representation, and thus this metric also captures aspects of information transmission.

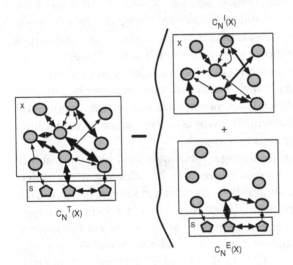

Figure 13.19
Matching complexity as total complexity minus intrinsic and extrinsic complexity. Arrow thickness reflects the level of informational interaction between components, not necessarily synaptic strength. Note that some connections in region X exhibit different strengths on the left and right. These differences contribute to the matching complexity. Based on Tononi et al. (1996).

13.6.4   Effective Information and Information Integration

The symmetry of mutual information makes it an appropriate basis for TSE integration and complexity, since both are inherently bidirectional concepts. But causal metrics, such as information transfer and flow, require the extraction of directionality via conditional entropy across time slices of system behavior. An alternative approach involves large-scale manipulation of a network's structure and careful setting of neural states in one region to help clarify its effects upon other regions.

Tononi and Sporns (2004) introduce the concept of *effective information* as a prelude to a more causal complexity metric known as *minimum information bipartition* (MIB) *complexity*. The effective information from A to B, $EI(A \to B)$, measures the influence of A's states upon B's states under a strict protocol. First, all internal connections within A are disabled and then every possible vector of activation values for A's neurons is imposed, with the effects upon B recorded. This insures that every state of A has the opportunity to affect B. Based on the many pairings of A and B states, the effective information is measured as

$$EI(A \to B) = I(A^{H_{max}}; B), \tag{13.40}$$

where $A^{H}max$ denotes the subsystem A that has evenly sampled all its possible states. Note how this differs from a standard mutual information measurement, wherein the setting of A's state could be modified by active intra-A connections, thus clouding the effect of the original state upon B. Also, this metric focuses on all possible A states, not just those encountered during normal operation of the system. Thus, it embodies the *potential causal influence* of A upon B.

They then define the bidirectional version as the sum of all potential causality:

$$EI(A \leftrightarrow B) = EI(A \to B) + EI(B \to A). \tag{13.41}$$

Tononi and Sporns also note that if a subset S of system X has any bipartition (into a region, A, and its complement $S - A$), where $EI(A \leftrightarrow S - A) = 0$, then S has no possibilities for information integration. So a subset's capacity to integrate information is based on its weakest bipartition. This makes sense in light of equation (13.32), where the integration of S is the combined integration of any bipartition plus its mutual information. In that metric a lack of mutual information indicates that any integration for S as a whole will have to come from within the the two independent pieces. In this newer metric that independence automatically discredits S as an integrator of information.

This leads to the definition of information integration capacity, $\Phi(S)$, of neural group S ($S \subseteq X$) as the minimum information bipartition of S, MIB(S): the bipartition of S, $[A : B]_S$, that minimizes the following expression:

$$\frac{EI(A \leftrightarrow B)}{\min\{H_{\max}(A), H_{\max}(B)\}}, \qquad (13.42)$$

where the denominator normalizes EI to permit a fair comparison between unequal-sized subsets when computing integration capacity for the entire system: $\Phi(X)$. This comparison is a critical factor in the metric, which uses orderings among its *complexes* as the basis for $\Phi(X)$. The complexes of X are defined as

$$\{S_i \subseteq X : \neg \exists S_j \subseteq X \ni S_i \subset S_j \wedge \Phi(S_i) < \Phi(S_j), \qquad (13.43)$$

i.e., the collection of all subsets that are not contained within a better-integrated subset. $\Phi(X)$ is then the value of its highest-capacity subset.

In summary,

$\Phi(S_i) = EI(\text{MIB}(S_i))$ for any $S_i \subseteq X$.

$\Phi(X) = \max \Phi(S_i)$ over all complexes S_i.

$\Phi(X)$ proves useful in analyzing ANN topologies, such as those in figure 13.20. Note that networks containing heterogeneous weight vectors have higher values than their homogeneous counterparts. This reflects a key asymmetry that allows many patterns in one region to invoke unique patterns in another: differences in A that make a difference in B (differentiation or segregation). Homogeneous vectors, on the other hand, constrain B to react similarly to many of A's states, thus reducing the effective information between A and B. Connectivity also plays an important role in $\Phi(X)$, since large connected components have the potential for greater integration capacity despite scaling in the denominator of equation (13.42). So $\Phi(X)$ clearly rewards both differentiation and integration just as $C_N(X)$ does.

While $C_N(X)$ does detect an important feature of emergence—different levels of integration at different spatial scales—it fails to distinguish between systems that are fully interconnected and those that involve several tightly intraconnected modules. $\Phi(X)$ registers that difference, giving the highest marks to the latter configuration. In general, it favors bidirectional interactions (integration) between heterogeneous neural regions (segregation) but is not fooled by, for example, two modules that have no direct ties but are driven by the same third module. In short, it detects causality, not merely correlation.

These properties make $\Phi(X)$ an interesting metric for quantifying the level of consciousness in a neural system (Tononi, 2003; Sporns, 2011). In fact, when applied to various ANNs whose topologies mirror those of brain regions such as the cerebellum, basal ganglia, and thalamocortical loop, the latter achieves much larger $\Phi(X)$ values. Not surprisingly, Edelman and Tononi (2000), among others

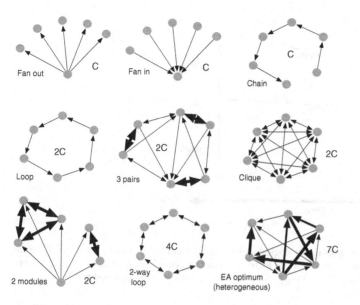

Figure 13.20
Information integration values ($\Phi(X)$) for a sampling of ANNs found by Tononi and Sporns (2004). $C \approx 10$ is a base value of $\Phi(X)$ for simple networks with homogeneous weight vectors; arrow thickness denotes relative synaptic strength. The bottom rightmost network was found by an evolutionary algorithm using $\Phi(X)$ as its fitness function.

(Granger, 2006; Sherman and Guillery, 2006) view this loop as the anatomical center of consciousness.

13.6.5   Entropy and Self-Organization

Brooks and Wiley (1988) reconcile evolution with the second law of thermodynamics by showing that order can increase in a biological system despite a simultaneous rise in entropy. They do so by separating the maximum potential entropy of a system (H_{max}) from its actual observed entropy (H_{obs}) and then focusing on their difference, $\Delta H = H_{max} - H_{obs}$, which represents the amount of unachieved potential disorder, i.e, the *constraint* upon or *order* within the system.

As shown in figure 13.21, both entropies and ΔH increase over time. For example, during ontogeny, the number of cell types in an organism gradually rises, as does the entropy of the distribution of cells over cell types. Similarly, but on an evolutionary time scale, the space of possible genotypes goes up, as does the entropy of the distribution of genotypes over lineages (Brooks and Wiley, 1988). Thus, both potential and actual diversity have evolved higher values, an indication that the second law still holds. Since ΔH also increases, the system as a whole becomes

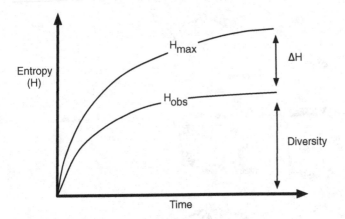

Figure 13.21
Temporal progression of entropies in a biological system (Brooks and Wiley, 1988). Over time, the number of potential options and realized options increases. The difference between the two entropies represents order, which also rises. Both entropy and order can increase in biological systems.

more ordered despite the dual entropy increases. Progress toward greater structure moves forward without violating the second law.

Hierarchical information theory (Brooks and Wiley, 1988) defines self-organizing systems as those in which

- both H_{max} and H_{obs} increase over time,
- ΔH increases as one moves up from lower to higher levels of structural organization.

Hence, an emergent or self-organizing process is one in which both lower and higher levels experience an expansion of their possibilities, but only lower levels retain an ability to explore them, whereas higher levels become more and more constrained. Note how this closely matches the notion of high integration, as defined in equation (13.31), and the expression of TSE complexity, in equation (13.34), since both depend upon high entropy (and low interaction constraint) at the lowest levels to achieve their high values. Thus, one useful sign of emergence in a neural network is increasing order with decreasing resolution, where order is gauged by differences between potential and actual entropy.

As a simple example of this concept, consider the patterns in figure 13.22, one of which is completely random whereas the other is biased to produce blocks of 1's and 0's (of lengths 3–10). The latter pattern is an abstraction of real-world visual stimuli, which tend to have much more correlation between pixels than do random patterns.

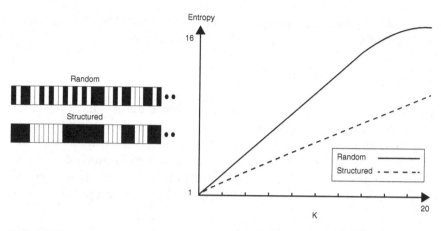

Figure 13.22
(*Left*) Examples of random and structured bit patterns. (*Right*) Entropy as a function of window size (*K*) for a random and a structured pattern, both of length 100,000 bits. The random curve begins to flatten out around 16 because of small-sample bias: larger windows require more data to achieve their maximum possible entropy.

The graph in figure 13.22 plots the entropies of one random and one structured pattern computed over all resolutions from single pixels up to groups of 20 bits. For groups of size K, a sliding window of width K moves along the pattern and counts the frequency of occurrence of each K-bit subpattern, and these frequencies then form the basis of the entropy calculation. The entropy of the random string is analogous to $H_{\max}$ in figure 13.21, and the entropy of the structured string mirrors H_{obs}, though now the x-axis measures spatial resolution, not temporal progression.

Notice that the difference between the two entropy curves increases with decreasing resolution (i.e., moving right on the x-axis). For larger values of K, the structured pattern explores its subpattern space in an increasingly more uneven/biased manner than does the random pattern.

Now imagine a neural network with a hierarchical topology in which the neurons of lower layers process local patterns in a sensory field, while higher levels integrate outputs from lower levels and thus are (at least indirectly) affected by larger segments of the sensory field. As described in relation to matching complexity, a fully functioning neural network should reflect the statistical complexity of its environment. So a neuron whose effective sensory field has a size of K should have an entropy level similar to that of the size-K bins of the original (structured) pattern. Thus, it is not unreasonable to assume that the entropy of neural layers will follow an entropy trajectory similar to the lower curves in figures 13.21 and 13.22, with the x-axis now plotting the level in the neural hierarchy. On the surface, at least, the definition of a self-organized system proposed by Brooks and Wiley (1988) makes sense for an ANN that has adapted to a real-world environment.

13.7 Converging Complexities

As summarized by Melanie Mitchell (2006), there is broad disagreement over definitions of the terms *complexity* and *emergence*. Consequently, there is no consensus on how to measure them. Candidates for quantifying complexity include the length of an algorithm to construct the system, the amount of historical information needed to predict its future behavior, fractal dimension, capacity to perform computations, degree of hierarchical organization, Shannon entropy, and many others.

Assessing the complexity of the brain requires a commitment to a particular perspective. If it's just another organ, then thermodynamic or constructive considerations may be most appropriate. If viewed structurally as a meshwork of neurons, then its fractal dimension or small-world topology (Sporns, 2011) become pivotal. But if viewed as a controller or problem solver, then computational and informational aspects come into play. These last two are arguably the most interesting, given the brain's special contributions to life and intelligence.

From the perspective of emergent intelligence, information theory provides an extremely general currency—one exchanged between brains, bodies, and worlds. The ability to follow this money trail offers much-appreciated leverage when hunting for a network of salient, task-directed activity embedded within a much larger matrix of connections. Basically, neural networks are so full of surprises, so hard to predict, that we need a few general indicators of interestingness as basic assurance that indeed something sophisticated has arisen from these multiple layers of adaptivity. And while some measures, such as entropy and mutual information, give only a coarse indication of saliency—of exploratory behavior and interaction, respectively—others, such as transfer entropy and information flow, permit more fine-grained detection of causal chains. Combined with other tools for dynamic systems analysis, these create an advanced repertoire of quantitative methods for mining well-hidden relations from the avalanche of data produced by a population of neurons.

In contrast to this search for specific interactions, the attempt to assess the complexity level of a neural network is a broad-brush effort, though it involves an exorbitant number of individual neuron and small group calculations[4]. But regardless of whether we actually carry out the computations, the general concept deserves careful attention. Rarely do we find quantitative metrics that serve such a foundational role in the explanation of consciousness (Tononi, 2003) or of the diverse functional roles of brain regions (Edelman and Tononi, 2000). Scientific discussions of natural phenomena tend to feel more comfortable when theories eventually ground out in numbers, not metaphysics.

Sporns (2011) provides an insightful chapter on neural complexity, covering a wide range of metrics, including several examined here. Central to all is his conviction that cognition arises directly from a combination of order and disorder, as manifest in a careful mixture of differentiation and integration. With differentiation comes modularity, and with integration comes intermodular communication and the emergence of hierarchies, which echoes Simon's view of complex systems as *nearly decomposable* hierarchies. An anchor in Simon's definition garners a lot of explanatory leverage because of the strong relation between hierarchy and evolvability, which Simon (1996) summarizes:

> We have shown thus far that complex systems will evolve from simple systems much more rapidly if there are stable intermediate forms than if there are not. The resulting complex forms in the former case will be hierarchic. We have only to turn the argument around to explain the observed predominance of hierarchies among the complex systems nature presents to us. Among possible complex forms, hierarchies are the ones that have the time to evolve. (196)

Another important focus of Sporns (2011) and others (Shanahan, 2008; Strogatz, 2003) is the small-world nature of brains and its contribution to their behavioral complexity. These networks are characterized by a power law distribution over the lengths of connections from any given node: the vast majority of connections are short, but a few are long. In several of these studies, small-world topologies were shown to produce neural behavior exhibiting high TSE complexity.

A modern reincarnation of PolyWorld (Yaeger, 2009) (see chapter 12) also helped unite several of these concepts. In challenging environments the evolved Poly-World agents showed both small-world ANN topologies and TSE behavioral complexity, thus giving synthetic support to both Simon and Sporns.

Power laws, hierarchies, a mixture of order and disorder (akin to an *edge-of-chaos* definition of complexity), and evolution to that edge—all these common signatures of complex systems appear in neural networks, both natural and artificial. Tononi, Sporns, Edelman, Yaeger, Simon, Williams, Beer, and others have shown that cognitive scientists have a lot to gain by studying complex systems in general and their informational properties in particular. With all these concepts seeming to converge in neural systems, maybe more complex-systems experts should take a closer look at the brain.

14 Conclusion

14.1 Circuits in Cycles in Loops

When you watch birds flying in a nearly perfect V formation, or termites building earthen edifices of engineering ingenuity, or fish swimming in synchronized
schools that swirl, part, and merge to confuse attacking predators, it is hard to
imagine that human intelligence has similar origins. Of course, everything organic
shares humble beginnings in deep geological time. But intelligence can appear
much *shallower* than that: it emerges from different starting points and extends
in diverse directions all the time.

These starting points can often cloud our perceptions of and lessen our appreciation for self-organization, since they exhibit stability, but one earned by emergence over other time scales. Each such structure, each model of order, is another
springboard for the endless saga of wild leaps and ruthless selection that move life
forward.

The developmental recipe embodied in our DNA arose over evolutionary time.
From it emerges the immature organism, fully equipped with adaptive tools for
a lifetime in an unpredictable world. Each day, each second, these tools process information and sew it into the self-organized and self-modifying tapestry
of the mind. From this stable, yet ever-changing knowledge network, each new
thought—whether a creative breakthrough or a bland tautologous deduction—
comes about as the system settles into various attractor states, some of which
leave deep tracks, stain the tapestry, while others blow away like dust from a
welcome mat.

The story of natural intelligence is the story of emergence compounded, of circuits within cycles within loops, all continuously marching, but to the beats of
different drummers, some imperceptibly slow. The story of artificial intelligence is
much different. It began with the imposition of logical structure upon high-level
reasoning, experienced impressive early success, fell on some hard times, but then
rallied with a multipronged (but uncoordinated) attack, wielding new tools at both

the high and low levels, from probabilistic enhancements of knowledge and optimized theorem provers on one end to neural networks, swarming agents, and evolutionary algorithms on the other.

Despite their many differences, the high and low roads to AI share a reliance upon one essential process: search. Sophisticated intelligence is simply too hard to model with a myopic, perpetually forward-directed procedure, one that always makes the proper choice for the current situation without hesitation or regret. True intelligence seems to require significant lookahead and backtracking, where the latter may entail simply shifting focus away from predictions of the future or actually undoing the present. Either way, the cognitive system is forced to navigate in a world of possibilities, not merely to follow a breadcrumb trail through life.

However, the preferred approaches to GOFAI search have traditionally involved as much specific knowledge and general heuristics as possible to maintain control over the blossoming tree of possible states. Nature employs a much dumber approach, one closer to trial and error. Random attempts, based on little or no insight into the problem, compete for footholds sculpted by the problem, like rock climbing in the dark.

Some classic (toy) AI problems, such as the 8-puzzle and the Tower of Hanoi, admit to effective heuristics or simple recursive, divide-and-conquer strategies. These serve the educational side of the field, but they can easily create the illusion that difficult cognitive challenges eventually succumb to a nifty rule of thumb or clever partitioning of the problem pie. The ever-expanding set of NP-complete problems testifies to the falsity of that view. Though knowledge-intensive AI and operations research methods often prove reasonably effective on such problems, they continue to elude all polynomial-time, general-case solutions. In short, there are always problem instances for which randomly generating and testing solutions is no worse than any other method. For NP-complete problems, there is no well-marked royal road to a sure conquest.

Many of the processes underlying intelligence map directly to NP-complete or even NP-hard problems, such as Steiner tree and minimal independent set (both discussed earlier). Even the simple abstract Hopfield network has attractor states that can evade all but the exponential time algorithms. So none of these problems of intelligence has a foolproof, *intelligent* solution. Trial-and-error search may work just as well as anything else, and it is almost surely easier to evolve.

Intelligence is a poorly defined concept, so any claims of optimal routes to this hazy destination will appear optimistic at best, misguided and misleading at worst. Nature has found its way to a brain that displays cognitive sophistry, but whether that route and that brain should guide AI research is unclear. In the early days of AI the most convincing demonstrations of automated intelligence were on human-made tasks: chess, physics problem solving, geometry theorem proving, and so on. These endeavors have always had a high status in the intellectual hierarchy. Only later, when human-easy began to be recognized as AI-hard, did AI people begin to appreciate excellence in *nature's game* as a worthy display of intelligence. That game includes the brunt of everyday living, with all its personal and social challenges. If this general intelligence persists as a high-priority goal of AI, then it seems prudent to continue drawing inspiration from nature's solution: both process and result. Thus, we cannot ignore the contribution to intelligence of dumb search filtered by selection.

Of course, Darwinian evolution is the paragon of selected variation, the slow incessant driver of life in all its more or less intelligent forms. Though often described as random variation honed by selection, recent research into evolvability shows a method to the mix-and-match madness of variation as well. Hundreds of millions of years have fine-tuned DNA to exhibit properties such as modularity and weak linkage, and to produce them in phenotypes. These, in turn, support an exploratory growth process that enables organisms to self-configure, thus deconstraining evolutionary search to the point that sophisticated life-forms no longer appear as needles in a haystack. Intelligence has emerged slowly, but once the genetic ball got rolling, random variation gradually morphed into a patterned and pattern-forming process that opened many routes to the complexity of phenotypic form and function.

We see this duo of uninformed generation and ruthless testing everywhere in the hierarchy of organic processes, from the blind casting of microtubular lines in hopes of hooking a chromosome during meiosis, and the unrefereed competition between filopodia to secure contact points on the blastocoel wall during neural crest cell migration, to the haphazard extensions and retractions of axonal growth cones in search of targets and the emergent cooperation between presynaptic neurons to activate postsynaptic targets, with the successful cooperators increasing their influence while synapses from unsynchronized afferents wither away.

At the highest levels of behavior, trial and error also reigns, though with an auxiliary ally: memory. We often attempt things randomly but tend to keep track of our actions so as to minimize repetition, particularly of failures, thus producing a systematic though somewhat brute-force search. Over time, we accumulate so much knowledge about so many different situations that we rarely need to start from scratch and resort to this wild explore-and-bookkeep approach, but without background knowledge it is our only viable alternative.

Reinforcement learning at the psychological level has direct parallels in neuroscience: the trial and error that we see in real life stems from a similar process under the hood. At the neural level those firing combinations that lead to successful overt behavior reap the benefits of an ensuing neuromodulator flood, which increases their future contributions. So as the arms and legs of my rock-climbing children flail in all directions, 5 meters over my head, their brains are similarly working through different activation patterns, only some of which will prove successful (on the rock wall) and be posted on the mental wall.

14.2   Sources and Seeds of Complexity

As discussed earlier, Herbert Simon postulates an important similarity between ants, mice, and humans: they employ simplistic reasoning systems that receive both prodding and constraint from a complex environment that includes the agent's memory. Now we have seen that separating memories/representations from reasoning machinery is no trivial matter in a neural network, since substrate and content seem inseparably intertwined. However, a common categorization posits spreading activation and attractor dynamics as reasoning, while connection-strength matrices constitute memory; and it is not unreasonable to view the former as relatively simple process (of transmitting action potentials) whose complexity skyrockets when stretched in all directions across a synaptic network.

Though placing memory in the environment seems a bit ad hoc, we cannot discount information off-loading in everything from the use of spatial position to encode falling object sizes in Randall Beer's minimally cognitive agents to the human use of external memory aids such as paper and digital media. Given the human propensity to document large chunks of life, our individual off-line representations may soon dwarf their neural counterparts.

A more contentious issue is Simon's view of functional hierarchies as the hallmarks of complex systems: modules within modules within modules. Unfortunately, mammalian brains fail to cleanly segment into functional regions, as do the digestive and circulatory systems; brains do consist of well-demarcated areas but with myriad interconnections. As shown by some of the EANN systems reviewed earlier, effective modularity at one level does not entail modularity in levels above or below it; and the superiority of modularity in general often comes into question. So although hierarchical organization plays a central role in engineering complexity, neural networks (both natural and artificial) do not (or cannot) always gain an advantage from strict decomposition, although Simon's conception of *nearly decomposable* may be another theory-preserving compromise.[1]

As illustrated by EANN research, the benefits of modularity are not universally accessible to a multilayered adaptive system. Unlike the human design process, evolution, development, and learning work bottom up as part of a continuously living/functioning system. Each new innovation must mesh with the existing components, and this severely precludes the independent emergence and fine-tuning of modules followed by their later integration into a functioning whole. In most Bio-AI, as in life, the whole must operate perpetually.

In short, the complexity of both the brain and Bio-AI systems stems directly from their emergent origins. Self-organization produces nice patterns at some level, but this hardly means that the underlying system is meticulously organized and easy to understand. To glean knowledge from these emergent structures, we turn to some of the tools discussed earlier, such as data clustering, principle component analysis, and information theory. Searching for functioning phenotypes is the job of nature and its corresponding Bio-AI tools. Searching for meaning in the emergent constellation is still a matter of statistics.

14.3   Representations

In this book the circuit in the cycle in the loop mirrors the three adaptive levels: the learning neural network within the brain's and body's developmental and self-maintenance processes, all in an evolutionary context. The tools of Bio-AI capture each of these mechanisms, though rarely in the same system and not yet in a manner that fully integrates and capitalizes on the strengths of each method. For example, complete POE systems often do development and learning in lockstep or employ very simple versions of each. Basically, Bio-AI has not yet achieved the technical sophistry to fully combine the best of the three adaptive worlds, nor has it developed a sufficiently stable conceptual basis to determine those biological phenomena that are worth preserving in silico.

Any attempt to automate multiple layers of adaptivity in an AI system is bound to bring representational issues to light. In this book three have primary importance: the correspondence between ANNs and EAs on the one hand and brains and natural evolution on the other; the genotypic representation of the phenotype; and the phenotype's encoding of information about the brain-body-world interaction.

The first of these involves the level of biological commitment in a Bio-AI project. Introductions to artificial neural networks (chapter 6) and evolutionary algorithms (chapter 5) have presented the basic biological structures and mechanisms from which a very thorough abstraction process begins. Little but the critical essence remains, but this has proven quite sufficient in both cases. Programmers introduce more detail with some trepidation, since as illustrated by several of the POE-EANN systems, the functionality often drowns in a sea of computational biology, all interesting but with unclear contributions to AI success.

Second, an aspect of representation that separates Bio-AI from GOFAI is the genotype-phenotype distinction. GOFAI has no use for one representation that generates another, since the field typically banks on intelligent search operators that directly modify semantic-level constructs; a syntactic-level encoding would be superfluous. Bio-AI, however, performs search at this lower level while doing evaluation at the next, an approach clearly motivated by nature's own, and one that capitalizes on a developmental process that allows a random modification of one component to spawn coordinated change in others. Chapter 12 on evolving neural networks has hopefully conveyed the creative power inherent in this approach, where linear genomes encode abstract versions of genetic regulatory networks, which then produce brains for cognitive agents. Not all applications of EAs require such a gap between genotypes and phenotypes, but the intimidating complexity of a multileveled adaptive system renders design search rather problematic if all search operators must operate at the phenotypic level. Such a restriction often shackles the creative potential of evolution.

These first two representational issues receive little attention in GOFAI. Originally, AI researchers cared about the fidelity of their models with respect to the brain, but most have now relinquished the view of mind as a logical theorem prover, particularly in light of considerable convincing evidence (Kahneman, 2011) that many everyday decisions are quite irrational though understandable. GOFAI methods retain their engineering significance, but their scientific implications have diminished. Second, although the original classifier systems (Holland, 1992) employ a bit string genotype to encode a classic GOFAI phenotype, a rule-based system, these are hardly mainstream GOFAI tools. Most expert systems research and applications involves more expressive rule languages (that are difficult to encode and manipulate as bits). This is another reason why the genotype-phenotype distinction has never carried much weight in GOFAI.

However, the ability of a phenotype to represent the world has always had major significance in GOFAI, almost certainly more so than in Bio-AI. In fact, the whole concept of representation has this as its default meaning in GOFAI. Managing the trade-off between epistemological expressibility and computational efficiency is a primary goal of many AI projects, particularly those with a knowledge representation focus. GOFAI systems tend to wear their knowledge on their sleeve, leaving little doubt as to what aspects of the real world are encoded, and how. Conversely, Bio-AI systems in general, and ANNs in particular, keep their encodings all too secretive, forcing researchers to craft elaborate tools just to mine out facts/tendencies, such as "the agent likes to pick up blue blocks." In addition, Bio-AI systems conveniently off-load representational work to the agent's body and surrounding environment (Pfeifer and Scheier, 1999; Pfeifer and Bongard, 2007; Clark, 2011) in a manner that is often emergent and thus completely unpredictable from the starting conditions of experiments.

So explicit representational choices concerning genotypes, phenotypes, and their relations can strongly affect the representations that emerge in the

ANN-body-world trio. The circus of circuits manifest in this trio, combined with three interacting levels of adaptation and tangled neural topologies, produces representations that hardly spell out "all large dogs are dangerous" but yield very explicit big-dog avoidance behavior and thereby succeed at nature's game.

14.4   Nurturing Nature in Our Machines

The spectacle of the circus of circuits makes the hardwiring of knowledge and intelligence very difficult, if not impossible, in Bio-AI systems. This is anathema to GOFAI adherents accustomed to loading large amounts of explicit knowledge into their formal reasoning systems. It is no surprise that GOFAI has largely eschewed both embodiment and emergence (as conveyed by Pfeifer and Bongard (2007)). Basically, the philosophical shift renders the standard GOFAI toolkit largely irrelevant for modeling natural intelligence, so it is understandable that researchers who have spent their careers fine-tuning logical representations and reasoning engines will resist a paradigm shift. But for AI to move forward in pursuit of general intelligence, that shift seems justified.

In this newer AI the main job is nurturing emergence, not force-feeding systems with high-level information. Knowledge representations emerge via complex interactions that we can only vaguely envision, not predict in detail; yet this combination of (neural) structure and dynamics governs elaborate behavior. And as covered thoroughly by Pfeifer and Bongard (2007), the morphology and material composition of the body also plays a huge role in realizing intelligent activity.

The nurturing process should draw considerable inspiration from the theory of facilitated variation, in which evolution found a set of core processes and can now mine success from the astronomical design space defined by their combinations. In Bio-AI the component pieces still need further refinement, but tools such as GRNs, ANNs, Hebbian learning, and neuromodulated reinforcement learning have strong potential as primitives, whose parameters and exact combinations can form the basis for an AI design space, explorable by many standard search techniques, including evolutionary algorithms. From this perspective, much of the past and current work in Bio-AI can be viewed as an extensive exploration phase that will one day become massively exploitative as proven components become the *atoms* in an automated search for *molecules* of cognition. The work of researchers such as Sims (1994), Yaeger (1993), and Bongard (2002) shows that combinations of many complex core processes are both practically feasible in terms of modern-day computing power, and conceptually attainable, since ALife workers are already quite adept at mixing and matching adaptive components in their systems. The core primitives just need more polishing.

It all boils down to search, lots of it. Now, instead of relying on one standard search algorithm and a few insightful heuristics, we build evolutionary search routines that produce thousands of recipes, each of which incorporates weak linkage and exploratory search to grow an agent, which then refines its own understanding of the world via the trial-and-error search of reinforcement learning. This is parallel search to produce parallel search procedures to generate a parallel-searching neural network. Some hardcore EA proponents may even recommend an additional outer search shell that finds the proper parameters or genome representations for the evolutionary algorithm. To paraphrase Dennett (1995), it's search (not necessarily mutating universes) all the way down.

Bio-AI is keeping and nurturing the baby, search, but tossing out the bathwater of formal representation because it's too clean, too perfect, and thus completely insufficient for the adaptivity required in nature's dirty world. Forget about the expressibility and efficiency of a representation, and focus on its flexibility. No internal substrate can capture all the essentials, all the dirt, in the world, but a good one can change quickly and interface seamlessly with its external assistants, the body and the environment (Clark, 2004).

Bio-AI embraces these diverse contributions to intelligence rather than relegating them to distant points of the research world (and separate conferences). However, there is no grand plan as to how everything fits together. There are guiding theories (Bateson, 1979; Maturana and Varela, 1987; Pfeifer and Scheier, 1999), but each implemented system can only address a few aspects of any one of them. And not every AI researcher has access to a world-class robotics lab.

It's a dirty, bumpy, confusing route, this low road to intelligence. But with frequent use, bad roads tend to get paved, straightened, widened, and better connected to the main thoroughfares. We then have to ask who or what is best equipped to drive on them? Very specifically, autonomous vehicle research has progressed to a point where we can legitimately question whether humans should get behind the steering wheel at all. Machines may increase both the safety and efficiency of transportation, and the human way begins to look like the inferior way. By proxy, a Bio-AI approach (based on nature's solution) seems less promising than the traditional AI and engineering methods that underlie today's self-driving cars.

The critical question is whether the human approach falls short in the general case as well. As human champions in a wide variety of disciplines fall to computational agents, few of which have deep Bio-AI roots, it is only natural to wonder what promise nature really holds when it comes to intelligence. Maybe GOFAI methods combined with the ability to examine millions of possible situations (or documents) in seconds is the recipe for success, as IBM's Deep Blue and Watson have demonstrated. Or maybe there really is something truly miraculous about the architecture and emergence of the mammalian brain, something worth incorporating into AI systems.

It is tempting to say that a belief in Bio-AI is a belief in ourselves, but this sells short the ability of our technological minds to best our natural brains on a good many challenges. Rather, the belief in Bio-AI is the understanding that nature has something very unique to offer these technological minds. To ignore that is akin to ignoring the Amazon rain forests in our search for new medicines or the advice of our relatives and the history of our ancestors in planning the future. There is an often-overlooked value in *that which got us here.*

As I sit at my favorite coffee shop, mulling over these final sentences, an unexpected visitor wanders along the meshwork of my metal table, seemingly campaigning for some recognition of its own: an ant. We began this journey with it, so in the service of yet another cycle, we can easily end with it. However, the ant's tale is just as precautionary as encouraging. The ant and the colony do have salient connections to high-level intelligence, but the wise AI researcher knows where the science ends and the fiction begins. As with any natural inspiration, there is an abstraction level below which the contributions to machine intelligence become questionable. That level often varies with the task, but it too deserves recognition.[2]

The challenge, across the many sources of bio-inspiration, lies in finding those sweet spots along the abstraction spectra that can facilitate bio-to-tech transfers of actual value to AI. This book provides some hints via a range of examples. But the search for general modeling principles continues, and in many respects it still feels like climbing walls.

A Evolutionary Algorithm Sidelights

A.1 Essential Components of an Evolutionary Algorithm

Figure A.1 illustrates the basic flow of data structures (typically implemented as objects) in an evolutionary algorithm. The evolutionary process begins with a population of genotypes as shown in the bottom left corner of the diagram. These are normally generated randomly, although some EA applications involve strong biases during initialization because of known or predicted structure in the solutions.

Phenotypes are then derived from the genotypes via the developmental phase. The genotypes themselves are retained for later use in reproduction, just as organisms retain their DNA throughout life.

Fitness testing of the phenotypes then assigns a fitness value to each individual (as depicted in the figure by the numbers in each cloud). This value reflects the ability of the phenotype to solve the problem at hand and then determines the prospects for the individual to pass elements of its genotype into the next generation.

As described in chapter 5, adult selection may weed out inferior (i.e., low-fitness) offspring, allowing admission into the adult pool to only the better performers. Parent selection then chooses individuals whose genotypes (or portions thereof) will be passed on to the next generation. In theory, all adults remain in the adult pool until the next round of adult selection, when all or some of them may get bumped out to make room for promising new adults.

Finally, the genetic material of the chosen parents is recombined and mutated to form a new pool of genotypes, each of which is incorporated into a new object. These objects then begin the next round of development, testing, and selection.

The cycle halts after either a predetermined number of generations or one of the individuals has a fitness value that exceeds a user-defined threshold. The latter is only practical when the user has concrete knowledge of the fitness value that corresponds to an optimal solution. For example, in a difficult traveling salesman

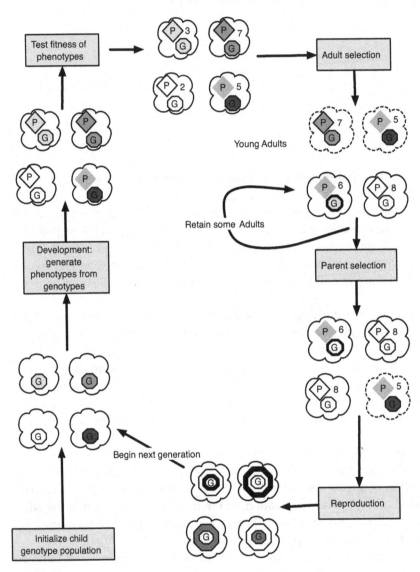

Figure A.1
Flow of individual computing objects (clouds) through an evolutionary algorithm. Objects begin with only a genotype, then acquire a phenotype via development and a fitness value via performance testing. Fitness then biases selection of both adults and parents/mates.

problem the value of an optimal tour may not be known ahead of time, but in a constraint satisfaction problem such as a sudoku the optimal score is easily ascertained without detailed knowledge of the actual solution.

To facilitate the process of figure A.1, an EA designer must include the following basic components:

- Genotype representation
- Phenotype representation
- Translator of genotypes into phenotypes
- Genetic operators for forming new genotypes from existing ones
- Fitness assessment procedure
- Selection strategy for child phenotypes transitioning to adulthood.
- Selection strategy for adult phenotypes transitioning to parenthood.

The first four of these components are often discussed together, since the genotypic level is, by definition, that for which one defines genetic operators, and the choices of genotypes and phenotypes are strongly influenced by the potential ease/difficulty of both the translation mechanisms and genetic operations.

Several of these seven components are general and hence reusable across many different problem domains. Typically, the selection strategies are extremely modular and highly problem-independent. A good EA has a library of a half dozen or so strategies that can be tested to find the best fit for a given situation.

At the other extreme, the fitness assessment procedure is always very problem–dependent, since it embodies the key knowledge as to *what* a solution to a particular problem should do.

Phenotypes also have a tendency to be problem-dependent, since they embody much of the semantics of the domain. However, there is still possibility for code reuse at the phenotypic level. For example, a general phenotypic type such as *permutation of integers 1..N* could be used to represent solutions to both an N-city traveling salesman problem (TSP) and an N-task ordering problem. In this case, the only difference would be in the fitness assessment procedure.

Genotypes are often amenable to reuse, particularly the very low level variants such as bit vectors. For these genotypes, general genetic operators are easily written. As genotypes move to higher levels, i.e., move closer to the phenotypic representations, their generality declines and the need for special-purpose genetic operators often arises.

In general, a good EA provides a solid backbone of code for running the basic evolutionary loop along with a library of reusable phenotypes, genotypes, genotype-to-phenotype translators, and selection strategies. It also includes simple hooks for adding in new representations and translators, preferably as subclasses to existing ones.

Despite this generality, even a good EA will require that the user understand the basic evolutionary procedure in order to tune parameters such as the population size, selection strategy, mutation and crossover rates, and so on. Furthermore, many complex problems require a specialized phenotypic (and possibly genotypic) representation not included in the off-the-shelf EA. In short, it helps to enjoy programming.

A.2   Setting EA Parameters

The astute choice of genotype, phenotype, genetic operators, and fitness function usually makes the difference between problem-solving success and sending the EA off into a hopeless search space with nothing but a broken compass. Thus, at a qualitative level these decisions are pivotal.

At a quantitative level other choices can affect the efficiency of an EA; if not considered seriously, they can also ruin the EA's search. These parameters include population size, stopping criteria, and mutation and crossover rates.

Although EA theoreticians have tried for years to produce useful guidelines for selecting these parameter values, the *no-free-lunch theorem* (D. H. Wolpert and Macready, 1997) has been the most influential, and it effectively says that one *cannot* make general statements about proper parameter settings for search algorithms: they are strongly problem-dependent. So, although an EA guru cannot provide magic values, for population size or mutation or crossover rates, for all problems, a domain expert who has applied EAs to specific problems can identity good parameter settings for EAs in that domain.

For example, if the representational choices have been so difficult that no recombination operator can guarantee high heritability, then a low crossover rate of, say, 0.2 might be a good choice. This indicates that when two parents are chosen for crossover, only 20 percent of the time will they actually be recombined; in the remaining 80 percent they will simply be copied (with possible mutations) into the next generation.

Mutation rates typically come in at least two varieties: per genome and per genome component (e.g., per bit in a bit vector genome). Depending upon the problem, these may vary from as high as 0.05 per component (e.g., 5 percent of all genome components are modified) to 0.01 per individual (e.g., 1 percent of all individuals are mutated in just *one* of their components).

Goldberg (2002) and De Jong (2006) provide some useful general tips for choosing EA parameters. One of the most critical and most general involves the well-known balance between exploration and exploitation (Holland, 1992). De Jong emphasizes a balance of strength between the explorative forces of reproduction and the exploitative powers of selection.

Reproduction explores the search space by creating unique new genomes; as mutation and crossover rates rise, the extent of this exploration increases. Selection mechanisms control exploitation via the degree to which they favor high-fitness over less impressive genotypes. De Jong's key message is that if reproductive exploration is high (e.g., the mutation rate is high), then selection should be highly exploitative as well. On the other hand, low reproductive exploration should be complemented with weak selection.

This makes intuitive sense. If the EA is generating many unique genotypes in each generation, then in any reasonably difficult search space, most of those genotypes will have lower fitness than the parents. Hence, it makes sense to filter them somewhat ruthlessly with a strong selection mechanism, such as tournament selection with low ϵ and high K (see below). But if reproduction produces only a few unique individuals, then weak selection is required to give those new children a fighting chance; otherwise, evolution will stagnate. Note that in both cases, the end result of a reproductive step followed by a selective step should be approximately the same amount of *innovation* in the next generation. The *absolute* mutation and crossover rates are less important than their exploratory strength *relative* to the exploitative degree of selection.

Also, the population size of an EA deserves careful consideration. Normally, a large population size of 100, 1,000, or even 5,00,000 is desirable for hard problems. Unfortunately, computational resources can often restrict practical population sizes to the 1,000–10,000 range, although this is highly problem-dependent. Some phenotypes, such as solutions to 20-city traveling-salesman problems or 20-node map-coloring problems, are easily checked for fitness, and thus large populations can be simulated over many generations in just a few seconds of run time. Other phenotypes, such as electronic circuits or robot controllers, require extensive simulations (if not live runs of real robots) to test fitness. This can greatly reduce the practical population size to values as low as 10 or 20. Higher values may take weeks to run 50 or 100 generations.

The different branches of evolutionary computation (see chapter 5) have varying philosophies on population size. In general, evolutionary strategies (ES) and evolutionary programming (EP) researchers tend to use very small populations of 20 or less; some use a single individual. Genetic algorithm (GA) and genetic programming (GP) researchers frequently prefer large populations of hundreds or thousands. In general, to achieve the full power of parallel stochastic search in tough solution spaces, large populations are necessary. But to find a satisfactory combination of 30 parameters, for example, a small population of only 10 or 20 may be sufficient.

Finally, the stopping criteria for an EA must be determined. One can either set in a known (or estimated) maximum fitness as a threshold and stop simulation when a genotype achieves it, or simply set a predetermined generation limit, G, and run until then. Both are trivial aspects of the EA, and the optimal choice is easily determined via experience in the problem domain. In general, if multiple runs of the same EA (on the same problem) are being performed in order to accurately assess the EA's problem-solving efficiency, then considerable run-time can be saved by cutting a run when fitness reaches a threshold value.

The tuning of EA parameters can often use up a significant fraction of total project time. It is not unusual to hack together an EA in a few days (or hours) but then spend an order of magnitude more time to actually get it to find good solutions. Start early! Be patient and persistent!

A.3   Fitness Assessment

In population biology *fitness* generally refers to an individual's ability to produce offspring (Roughgarden, 1996), whereas EAs use fitness values to *determine* reproductive success. EAs, unlike biologists, cannot watch a population reproduce and then afterward assign fitness to the productive individuals. EAs explicitly restrict access to the next generation, and the fitness evaluation is the first (and main) step in that process.

As discussed earlier, an EA generally has little knowledge about *how* a good solution/hypothesis should be designed but good hints as to *what* a good solution should be or do. Most of that information is implicit in the fitness function. Whereas the phenotypic representation and the genotype-to-phenotype mapping embody knowledge of how to generate a *legal* individual, it is the fitness function that assesses the individual's *goodness* with respect to the problem. So the extent of *how-to-construct* knowledge in the EA is normally restricted to legal phenotypes, not necessarily superior ones.

Fitness assessment involves two steps: performance testing and fitness assignment. In the former, the phenotype is applied to the problem to be solved by the EA. The results of this are then converted into a quantitative fitness value, which then follows the individual around like a college grade point average to open (and close) doors to the future. That is, the selection mechanisms use the fitness value to prioritize and filter individuals.

In the curve-fitting GP example (see chapter 8), a typical performance test goes through each of the n (x_i, y_i) pairs. The value x_i is used as input to F_j, the complete function defined by phenotype j, producing output $\tilde{y}_i$, which is then compared to y_i

to compute a contribution to the total error, E_j. The sum of squared error is typical for this purpose:

$$E_j = \sum_{i=1}^{n} (y_i - \tilde{y}_i)^2. \tag{A.1}$$

The fitness of phenotype j could then be as simple as

$$\text{Fitness}(F_j) = \frac{1}{1 + E_j}. \tag{A.2}$$

In the robot example (chapter 8), the performance test would involve a simulated two-dimensional block world with red and blue objects. Each phenotype (i.e., set of behavioral rules) would be downloaded into a simulated robot that would then move around this world under the control of these rules. Initially, the red and blue objects would be randomly spread about the plane. The robot would then run for a number of timesteps, say, 1,000. Upon completion, the state of the world would be analyzed for certain factors of relevance to a solution. These factors are ideally necessary and sufficient conditions for problem success, but such conditions are not always easy to define. But in this case, the definition of the task is fairly straightforward, so a good final world state should have all red objects surrounded by many blue objects and few blue objects standing alone in the plane.

To quantify this for each red object, count the number of blue objects that are within a short distance d of its center, i.e., *warning blues*. The initialization procedure might insure that no blue objects are within d of any red object. By itself, this warning blues count could be a fairly effective fitness value. In addition, the EA could take into account the isolated blue objects and, say, subtract their count from that of the warning blues. In most cases, it is useful to avoid negative fitness values, so the isolated blues count might be divided by a scaling constant, or all negative results could simply be mapped to zero fitness.

For a traveling salesman problem, the phenotype would probably encode the proper sequence of cities to visit: first Dallas, then Denver, then Portland, and so on. The performance test would then *walk through* that route and tally up the total distance. A standard fitness value would then be the inverse of this distance.

If an EA were used to design a good classifier for a machine-learning system, then the performance test would involve testing each phenotype classifier on a training set of (input, desired output) pairs. The classification error, E, of a phenotype—i.e., number of the classifier outputs that differ from the desired outputs—could then determine fitness, as in the curve-fitting example above.

Error is also a natural fitness metric for scheduling problems. Assuming that the EA must design a good assignment of college exams to discrete time slots and classrooms, the phenotype would consist of times and rooms for each exam. Then, given a list of all students, professors, and the courses that each takes/gives, the performance test would measure error as the number of violations of hard (e.g., no student or professor can have two exams at the same time) and soft (e.g., no student should have two exams on the same day) constraints.

In general, the exact fitness value matters very little, but the *relative* fitnesses of hypotheses must reflect the relative utility of their solutions. If one individual receives a fitness of 10 while another receives 2, then in the eyes of the system designer, the former should be approximately five times *better*. Failure to achieve appropriate fitness spacing among individuals can cause evolutionary search to sputter. However, as discussed below there are certain phases of the search when it is desirable to artificially widen or narrow the effective fitness gaps.

A.4   Natural and Artificial Selection

Whereas fitness assessment should provide an objective, unbiased measure of an individual's quality, selection strategies often introduce stochasticity into the processes of survival and mating. Just as in nature, where the strongest and fastest may accidentally fall off a cliff, meet an oncoming train, or simply have a couple of off nights during mating season, there is no guarantee with EAs that the most fit individuals will pass on the most genes or that the least fit individuals will not reproduce at all. This is generally an advantage, since high-fitness individuals may have hidden weaknesses or represent merely local optima, while low-fitness genotypes may encode useful traits that have not yet been complemented with enough other traits to show their true utility. In short, an EA normally profits by keeping some options open, by maintaining a good balance between exploration and exploitation.

A.4.1   Fitness Variance

In theoretical evolutionary biology, classic results by Sewall Wright and Ronald Fisher led to the *fundamental theorem of natural selection* (Roughgarden, 1996; Futuyma, 1986), which states that the rate of evolution (measured as the rate of change of the composition of the gene pool) is directly proportional to the *variance* of individual fitness. So evolution requires fitness variation, and the greater the variation, the faster the population evolves (and adapts to its environment).

Noting that fitness in biology refers to reproductive efficacy, the fundamental theorem implies that evolution can only occur when individuals have different reproductive rates. If each individual produces one offspring (or each mating pair produces two), then the gene pool will stagnate.

The same principle applies to evolutionary algorithms: If all individuals have the same fitness, then they will all have approximately the same likelihood of passing on their genes, and the population will not change much from one generation to the next. Hence, for evolution to proceed, a sizable fitness variance should be maintained.

Unfortunately, EAs often suffer from a fitness variance that is either too high or too low. In the early generations of an EA run, most of the individuals tend to have (very) low fitness, while a few will have more respectable values simply by virtue of doing a few things correctly (often by accident). For example, a robot that always picks up blue blocks and then deposits them anytime it sees other blocks, whether blue or red, will score better than one that never picks up anything. This results in a very high fitness variance, since most of the individuals will have zero or near-zero fitness, while a select few will have values reflecting decent performance but nothing exceptional.

The problem is that the low-fitness individuals will normally have almost no chance of reproducing, while the respectable genotypes will quickly dominate the population. Hence, the EA will *prematurely converge* to a homogeneous population of mediocre individuals, a local optimum. This homogeneity implies very low fitness variance, so evolution will grind to a halt, and the EA's best-found solution will be far from optimal. To remedy this situation, the fitness variance should be artificially reduced during the early stages of an EA run.

In contrast, near the end of a run, the population will often converge on an area of the fitness landscape that indeed holds an optimal or near-optimal solution, but since all individuals are quite good, there will be little fitness variance and thus no driving force for continued improvement. The population will essentially run out of useful hints regarding the goal. In these cases of *late stagnation*, the fitness variance should be artificially increased so that the cream of the cream gets a slight but significant reproductive edge.

These artificial modifications to fitness variance are achieved by selection mechanisms, which often scale fitness values before picking and pruning parents. In effect, this alters the fitness landscape, as shown in figure A.2.

A.4.2 Selection Pressure

In EAs the concept of *selection pressure* ties directly to the scenario of figure A.2. In effect, it is the degree to which the real fitness variance translates into differences in reproductive success. A high selection pressure will give a strong reproductive advantage to individuals that are only slightly better than their peers, while a low selection pressure will treat individuals more evenly despite fitness differences. So the example of figure A.2 reflects an initially low selection pressure, with a high

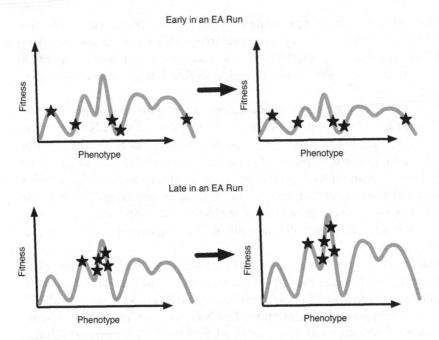

Figure A.2
Virtual modifications to the fitness landscape performed by the fitness scaling that often precedes parent selection. Stars represent individuals. In early stages of evolution fitness variance is artificially decreased, which implicitly compresses the fitness landscape. In late stages variance is artificially increased, leading to an implicit stretch of the landscape.

selection pressure toward the end. This strategy of selection pressure change is often ideal for an EA.

A.5   Selection Strategies for Evolutionary Algorithms

In the classic EA reference literature (De Jong, 2006; Banzhaf et al., 1998; M. Mitchell, 1996), the terminology of selection varies a bit, but the basic concept of a selection mechanism or strategy is usually the same. It involves the filtering and biasing of individuals jockeying to *mature* and *reproduce*.

Referring to the evolutionary algorithm cycle, a selection strategy must be defined for both *adult selection* and *mate selection* (*parent selection*). Adult selection typically involves the immediate removal of a (possibly empty) subset of child genotypes, along with the filtering of some (or all) parents from the previous generation. The remaining children and previous adults then constitute the new adult population. Mate selection consists of repeated choices of adult genotypes, which are then used to produce the next generation of genotypes. Mate selection often involves the scaling of fitness values as the basis for the stochastic choice of parents.

A.5.1   Adult Selection

As shown in figure A.3, there are three main protocols for adult selection:

- *A-I—Full generational replacement* All adults from the previous generation are removed (i.e., die), and all children gain free entrance to the adult pool. Thus, selection pressure on juveniles is completely absent.

- *A-II—Overproduction replacement* All previous adults die, but m (the maximum size of the adult pool) is smaller than n (the number of children). Hence, the children must compete among themselves for the m adult spots, so selection pressure is significant. This is also known as (μ, λ) selection, where μ and λ are sizes of the adult and child pools, respectively.

- *A-III—Generational mixing* The m adults from the previous generation do not necessarily die; they and the n children compete for the m adult spots in the next generation. Here, selection pressure on juveniles is extremely high, since they are competing with some of the best individuals that have evolved so far, regardless of their age. This is also known as $(\mu + \lambda)$ selection, where the plus indicates the mixing of adults and children during competition.

A.5.2   Parent Selection

For parent/mate selection, a host of mechanisms are available. These vary in the degree to which individuals compete locally or globally and by the type of fitness scaling that may occur prior to filtering. With *local* selection mechanisms, such as tournament selection, individuals participate in a competition with only a small subset of the population, with the winner moving immediately to the mating pool. With *global* selection mechanisms, an individual implicitly competes with every member of the adult population.

Global Selection Mechanisms   Many of the global selection mechanisms involve a *roulette wheel* on which each adult is alloted a sector whose size is proportional to the adult's fitness. In asexual reproductive modes, if n children are to be produced, the wheel is spun n times, with the winning parent on each spin sending a (possibly mutated) copy of its genotype into the next generation. In sexual reproductive schemes, pairs of wheel spins are employed, with the two winner parents passing on their genotypes, which may be recombined and mutated, normally yielding two children. Repeating this process $n/2$ times yields the complete child population.

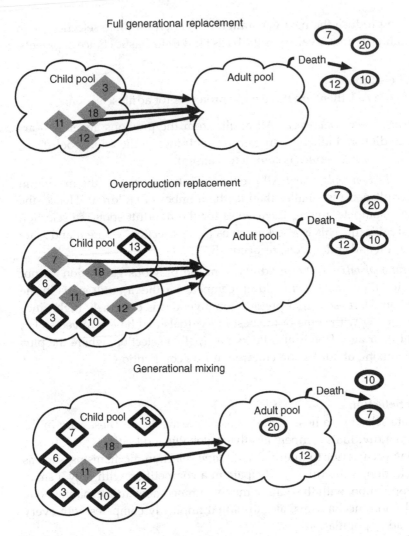

Figure A.3
Three basic protocols for adult selection. (*Top*) Complete turnover of adults, with $m = n$. (*Middle*) Complete turnover of adults, with $n > m$. (*Bottom*) Adults and children compete equally for the m adult spots. Numbers denote fitness; unshaded diamonds, children; circles, adults who lose out in the competition, arrows from diamonds, children who win acceptance into the adult pool.

Naturally, fitness values must be normalized in order to divide up the area of the roulette wheel. In addition, almost all global selection mechanisms scale the fitness values prior to normalization. These scaling techniques are typically the defining feature of the selection mechanism.

The following overview of basic selection mechanisms uses the concept of *expected value*: the expected number of times that the parent will reproduce

(Holland, 1992). The original fitness values are scaled into these expected values prior to roulette wheel normalization. I also closely follow the description of selection mechanisms given by M. Mitchell (1996).

The roulette wheel metaphor is best explained by a simple example. Assume a population of four individuals with the following fitness values, respectively: 4, 3, 2, 1. To convert these to space on the roulette wheel, simply divide each by the sum total of fitness, 10. By *stacking* the resulting fractions, the individuals each get a portion of the [0, 1) number line. The subranges for each are [0 0.4), [0.4, 0.7), [0.7, 0.9), [0.9, 1.0). These are equivalent to sectors on a roulette wheel. Selecting a parent becomes a simple weighted stochastic process wherein a random fraction, F, in the range [0, 1) is generated. The subrange within which F falls determines the chosen parent. Clearly, individuals with higher fitness have a greater chance of selection.

The classic selection mechanism is *fitness-proportionate*, in which fitness values are scaled by the average population fitness. Of course, dividing m fitness values by their average and then normalizing is equivalent to simply normalizing the original m values. Hence, this mechanism merely scales fitnesses so that they sum to 1 and thus properly fill up the roulette wheel; it does not modify their relations to one another (figure A.4). In other words, it does not alter the selective advantages/disadvantages inherent in the original values and thus does not implicitly change the fitness landscape.

EA researchers often describe the roulette wheel as a unique feature of fitness-proportionate selection, but it applies equally well to many global selection mechanisms, since most work by normalizing all expected values and then using randomly generated fractions to simulate the spinning of the wheel and choice of a parent.

Sigma scaling selection successfully modifies the selection pressure inherent in the raw fitness values by using the population's fitness variance as a scaling factor.

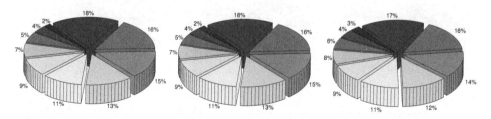

Figure A.4
Selection wheels using (*left*) unscaled but normalized fitness values, (*middle*) fitness-proportionate scaling and normalization, and (*right*) sigma scaling and normalization. The original fitness values in all three cases are 1, 2, 3, 4, 5, 6, 7, 8, 9, 10. Note the similarity of all three wheels this set of fitness values.

Hence, unless this variance is 0 (in which case all fitnesses scale to expected values of 1.0), the conversion is

$$\text{ExpVal}(i,g) = 1 + \frac{f(i) - \bar{f}(g)}{2\sigma(g)}, \qquad\qquad (A.3)$$

where g is the generation number of the EA, $f(i)$ is the fitness of individual i, $\bar{f}(g)$ is the population's fitness average in generation g, and $\sigma(g)$ is the standard deviation of population fitness.

This has the dual effect of damping the selection pressure when a few individuals are much better (or worse) than the rest of the population (since such cases have a high $\sigma(g)$) and increasing the selection pressure when the population has homogeneous fitness (thus low $\sigma(g)$). This helps avoid the problems of early and late stages of EA runs. Figures A.5 and A.6 illustrate these effects, showing the clear advantage of sigma scaling over fitness-proportionate scaling in effectively modifying the selection pressure inherent in the original fitness values.

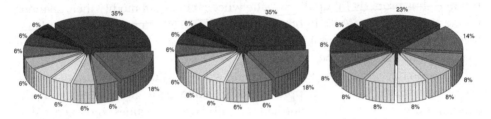

Figure A.5
Selection wheels using (*left*) unscaled but normalized fitness values, (*middle*) fitness-proportionate scaling and normalization, and (*right*) sigma scaling and normalization. The original fitness values in all three cases are 1, 1, 1, 1, 1, 1, 1, 1, 3, 6, which are typical of an early generation of an EA run. Note the more even distribution (i.e., lower selection pressure) for sigma scaling.

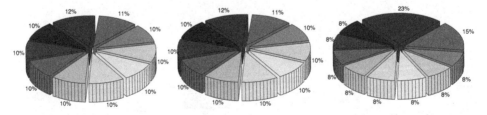

Figure A.6
Selection wheels using (*left*) unscaled but normalized fitness values, (*middle*) fitness-proportionate scaling and normalization, and (*right*) sigma scaling and normalization. The original fitness values in all three cases are 8, 8, 8, 8, 8, 8, 8, 8, 9, 10, which are typical of the later stages of an EA run. Note the more skewed distribution (i.e., higher selection pressure) for sigma scaling.

Boltzmann selection is based on simulated annealing, in which the degree of randomness in decision making is directly proportional to a temperature variable. This is based on the physical property that molecules in a heated mixture exhibit more random movement (than under cooler conditions, especially when the mixture transitions from solid to liquid or from liquid to gas). With Boltzmann selection, higher (lower) heat entails a more (less) random choice of the next parent and hence less (more) selection pressure, since superior individuals have less (more) of a guarantee of passing on their genes, and inferior solutions have a better (worse) chance of getting lucky and reproducing.

The scaling equation for the Boltzmann selector is

$$\text{ExpVal}(i, g) = \frac{e^{f(i)/T}}{\langle e^{f(i)/T} \rangle_g},$$

(A.4)

where g is the generation, T is temperature, $f(i)$ is the original fitness of individual i, and $\langle e^{f(i)/T} \rangle_g$ is the population average of the fitness exponential during g.

As shown in figure A.7, as temperature falls, the odds of choosing the best-fit individual increase dramatically as the top individual garners more and more area on the roulette wheel. Ideally, a Boltzmann selector uses a temperature that gradually decreases throughout the EA run, such that selection pressure gradually increases. This ameliorates both premature convergence and late stagnation.

Rank selection ignores the absolute differences between fitness values and scales them according to their relative ordering. Hence, if the best-fit individual in the population has a fitness of 10, and second place has 3, then the 10 will achieve no more roulette wheel area than would a 3.1. This type of scaling also helps adjust selection pressure. It decreases the advantage of *lucky starters* during early stages of a run, and it tends to add some spacing between individuals when population fitness becomes homogeneous. This, too, helps combat the problems of premature convergence and late stagnation.

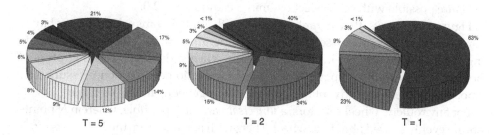

Figure A.7
Increasing selection pressure by decreasing temperature in the Boltzmann selector from (*left*) 5° to (*middle*) 2° to (*right*) 1°. The original ten fitness values are 1, 2, . . ., 10.

The basic scaling equation for rank selection is

$$\text{ExpVal}(i,g) = \min + (\max - \min)\frac{\text{rank}(i,g)-1}{N-1}, \tag{A.5}$$

where N is the popuation size, g is the generation, and min and max are the expected values of the least- and best-fit individuals, respectively. rank(i, g) is the rank of the ith individual during generation g, with the least-fit individual having rank 1 and the best fit having rank N.

M. Mitchell (1996) points out the basic constraints $\max \geq 0$ and $\sum_i \text{ExpVal}(i,g) = N$, from which it is straightforward to prove that $1 \leq \max \leq 2$ and $\min = 2- \max$:

$$\sum_i \text{ExpVal}(i,g) = \sum_{i=1}^{N} \min + (\max - \min)\frac{rank(i,g)-1}{N-1}$$

$$= N \cdot \min + (\max - \min)\sum_{i=1}^{N} \frac{i}{N-1}. \tag{A.6}$$

Then,

$$\sum_{i=1}^{N} \frac{i}{N-1} = \frac{1}{N-1}\sum_{i=1}^{N} i = \frac{1}{N-1} \cdot \frac{(N-1)N}{2} = \frac{N}{2}. \tag{A.7}$$

Hence,

$$N \cdot \min + (\max - \min) \cdot \frac{N}{2} = N. \tag{A.8}$$

Solving the above equation for max yields $\max = 2 - \min$. Since min is the expected value of the worst-fit individual, it should lie within the range [0, 1], the worst-fit individual should get at most one copy of its genome in the next generation. Thus, $1 \leq \max \leq 2$.

As a typical setting, min = 0.5 and max = 1.5, or for the maximum selection pressure possible with rank selection, min = 0 and max = 2.0.

Figure A.8 shows that a rank selector with scaling range [0.5, 1.5] behaves similarly to a Boltzmann selector with temperature = 10. Although the shading schemes are reversed in the two roulette wheels, the corresponding proportions match up well, and both exhibit a clear reduction of selection pressure (i.e., smoothing of selective advantages) for the heterogeneous fitness scenario.

For any roulette wheel selectors, a *universal* variant is possible, wherein N pointers are evenly spaced about the wheel. Then, with just one spin, the N parents for $\frac{N}{2}$ pairings are chosen. This approach removes the random possibility of good individuals dominating the mating to a much greater degree than sanctioned by the scaled fitness distribution. The next generation thus *holds true* to that distribution.

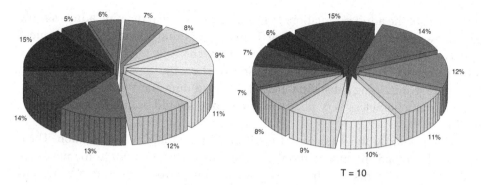

T = 10

Figure A.8
(*Left*) Normalized, rank-scaled fitness values for ten phenotypes with original fitnesses 1, 2, . . . , 10. The scaling range is [0.5, 1.5]. (*Right*) The same fitnesses scaled by a Boltzmann selector using a temperature of 10. Both give a fairly even partitioning of the roulette wheel despite the high fitness variance. Hence, both exhibit low selection pressure for this wide fitness distribution.

Finally, the term *uniform selection* refers to situations where, in effect, there is no selection pressure: all individuals have exactly the same chance of being chosen. In *deterministic uniform selection*, each parent gets to produce the same number of children, while *stochastic uniform selection* involves a roulette wheel on which all adults have equal area. EAs that use uniform selection at one level, such as during mate/parent selection, must typically use some form of fitness-based selection elsewhere.

Local Selection Mechanisms   The classic local mechanism is *tournament selection*, wherein random groups of K adults are chosen and their fitnesses compared. With a probability of $1 - \epsilon$, the best fit of the K is chosen for the next mating, while the choice is random with a probability of ϵ. The parameter ϵ is a user-defined value. N tournaments are thus required to determine all mating pairs.

Notice that this procedure allows poor genotypes to slip through to the next generation if either they get grouped with even worse individuals in a small tournament, or they lose a tournament but sneak through via the ϵ-ruling. In general, selection holds truer to the original (global) fitness distribution when K is large and ϵ is small. In other words, selection pressure is a function of K and ϵ, with low K and high ϵ exhibiting low pressure, and high K and low ϵ giving much stricter selection.

A.5.3   Supplements to Selection

Although traditionally defined as a selection mechanism in its own right, *truncation* involves the immediate removal from mating consideration of an often sizable fraction, F, of the adult population. In the classic case, the remaining adults simply produce an equal number of offspring. However, nothing prevents the subsequent application of any global (or local) selection mechanism to choose among these adults. Truncation is quite useful in situations where the genetic operators have a hard time maintaining substantial heritability, and thus many children from good parents are nonetheless very poor performers and should be culled from the population.

Elitism is simply the retention of the best E individuals in a generation. These are simply copied directly, without mutation or recombination, to the next generation. This prevents EAs from *losing* a superior performer before a better one comes along. Elitism is very important when population diversity is high and thus the odds of recreating a similar good individual on each round are lower than when the population has converged upon a good region of the search space. Again, elitism can easily be added to any local or global selection mechanism. A typical value for E is 1 or 2 individuals, or a small fraction (1% to 5%) of the total population. Higher fractions can often lead to premature convergence.

B Neural Network Sidelights

B.1 Neural Networks as Complex Mappings

In most ANN applications the user possesses a set of data instances, each of which has a value for n different attributes plus a *classification*. She wishes to train the ANN with this data such that in the future, when given a new unclassified data instance, the ANN can predict the proper class. Hence, she wants the ANN to learn the proper *mapping* from input attribute sets to classifications.

If the data set is linearly separable, then a single multi-input neuron (traditionally known as a *perceptron*) can achieve the proper classification. However, most real-world data sets are not linearly separable, and thus they require a multilayered ANN to realize a proper mapping.

Note that the form of the output neuron's activation function will not help to *bend* the border line to separate data. Rather, a more complex function such as a sigmoid only adds a *gray region* on each side of the border line, an area in which positive and negative instances are difficult to differentiate. This is illustrated in figure B.1. Clearly, any net_z value that pushes the sigmoid upward (toward the black region or beyond) is a better candidate for concept membership than one that maps to the base of the sigmoid slope (light gray), but a precise border for concept membership, as in the case of a step function, is absent.

Most ANN applications use some version of the backpropagation algorithm to train their weights. This often produces good mappings, but ones whose internal workings evade simple human analysis. The weight vectors reveal little of the ANN's functionality without painstaking dissection and reconstruction, which do not, however, yield useful modules of the type an engineer or mathematician would devise.

To illustrate the types of mappings that ANNs can achieve, the following example will break a linearly inseparable problem into meaningful modules and show how an ANN can be constructed to represent each module and achieve the desired mapping as well.

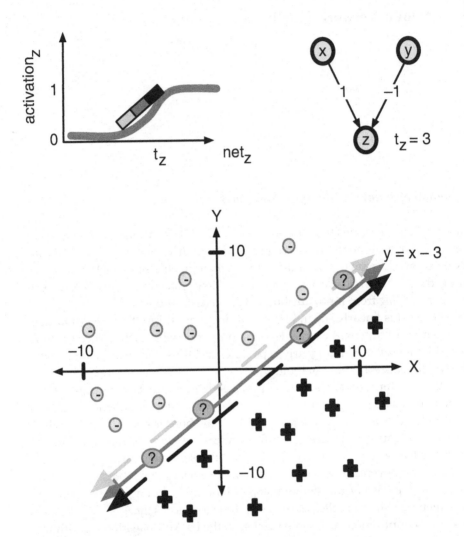

Figure B.1
Imprecision of class membership decisions incurred by the use of a sigmoidal activation function in a
detector neuron. (*Top left*) Sigmoidal curve with bar paralleling the upward slope. Shading represent
strength of class membership (light gray, low; black, high) for a given net_z value. (*Top right*) Detector
neuron, z, with an unspecified activation function but a threshold of 3. (*Bottom*) Crisp separation (solid
gray line) versus fuzzy differentiation (region between dashed gray and black lines) produced in the
Cartesian plane as a result of a step activation function versus a sigmoidal activation function for z. For
the sigmoid, a threshold of 3 is the net_z value that maps to the middle of the upward-sloping curve (*top
left*). Crosses denote positive class instances; circled dashes, negative instances.

The problem is simple. Given the set of data points shown in figure B.2, design a neural network that can map any pair of x-y coordinates in that set into a Boolean value corresponding to the correct classification of the coordinates.

Clearly, a single straight line cannot separate the data into positive and negative exemplars, but a *group of lines* can. As shown at the top of the figure, three lines suffice to segregate the data points such that no region contains both positive and negative instances. Given these lines, a set of neurons is easily crafted to detect the regions carved out by the lines.

The three lines are

$L1 : y = x,$

$L2 : y = -x + 5,$

$L3 : y = -4x + 30.$

The three regions housing the positive instances are

$R1 :$ above $L1$ and below L2,

$R2 :$ above $L1$ and above L3,

$R3 :$ below $L1$, above L2, and below L3.

Now consider the three large areas above each line. These can be expressed as follows:

Above L1 : $y > x \Leftrightarrow y - x > 0,$

Above L2 : $y > -x + 5 \Leftrightarrow y + x > 5,$

Above L3 : $y > -4x + 30 \Leftrightarrow y + 4x > 30.$

The right sides of each bidirectional implication correspond to the firing conditions for simple neurons with step function activations (figure B.2, bottom).

To slightly simplify the example, assume that the lines were devised such that no data point lies directly on L1, L2, or L3.[1] Hence, all points are either above or below a line. The mathematical expressions for the areas below each line are the same as for those above it, but with the inequality reversed (i.e., $>$ becomes $<$). However, the "below" areas need not be represented explicitly; since the network will be testing for area membership, the fact that a point is not above a given line indicates that it must be below it.

To wire up neural groups to detect membership in the three regions (R1, R2, and R3), simply use the three "above" detectors shown at the bottom of figure B.2 as the second layer of the ANN. The regions are then defined as the intersections of these areas by connecting these A neurons (and weighting them appropriately) to a third layer of (R) neurons corresponding to the three regions: R1, R2, and R3. Figure B.3 shows a fragment of the network for region 3.

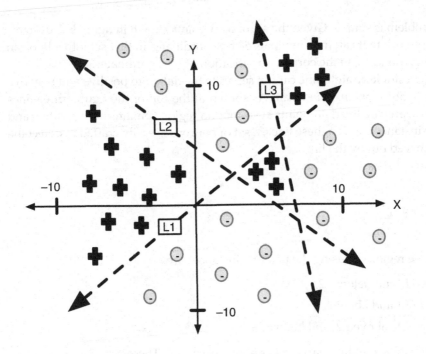

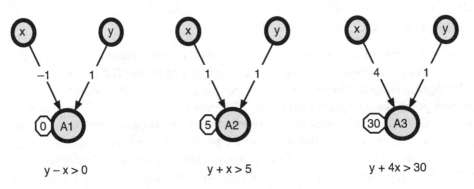

$$y - x > 0 \qquad\qquad y + x > 5 \qquad\qquad y + 4x > 30$$

Figure B.2

(*Top*) Data set defined by two features: x and y coordinates. Together, the lines L1-L3 successfully segregate the positive from the negative instances. (*Bottom*) Detector neurons for three important areas of
the grid: A1, above L1; A2, above L2; A3, above L3. Each neuron uses a binary step activation function
that fires a 1 whenever the sum of weighted inputs is equal to or exceeds the threshold (octagon next
to neuron).

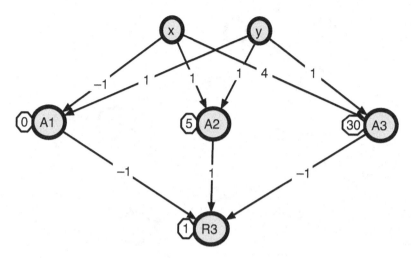

Figure B.3
Network for detecting points in region R3, defined as below L1, above L2, and below L3. Note that each "below" condition entails a negative weight from the detector neuron for the corresponding "above" relation. Since each of the second-level nodes uses a binary step activation function (outputting 0 or 1), the incoming weights plus threshold of R3 give it a Boolean AND functionality with respect to the three conditions for R3 membership.

The full network (see figure B.4) accounts for each of the three regions (R1–R3) via its third layer (or second hidden layer), which includes a copy of the fragment from figure B.3. The final (output) layer houses a single neuron with Boolean OR functionality, which responds when an x-y pair lies within any of the three regions.

The network represents a simple solution to a non-linearly-separable data set, and thus one that requires more than a single perceptron. In theory, all the logic in the two hidden layers could be compressed into a single larger hidden layer. This would work well but probably be harder to understand. The modularity of our current network, with separate area and region detectors, reflects a straightforward comprehensible mathematical decomposition of the problem. Most human engineers would design something similar. This example shows that complex mappings can be achieved with ANNs, often in reasonably elegant ways.

However, in using ANNs for complex problems, engineers and scientists typically do not have the ability, desire, or resources to design neural networks by hand. They will specify parameters, such as the sizes of the different layers, the learning rule(s), activation functions, and so on, and hope that the adaptive algorithm can tune the ANN to the proper mapping after having seen the data set (typically many times). These adaptive procedures, such as evolutionary algorithms or ANN learning procedures—whether supervised (as would be appropriate for the current example, since the class is available for each instance), unsupervised,

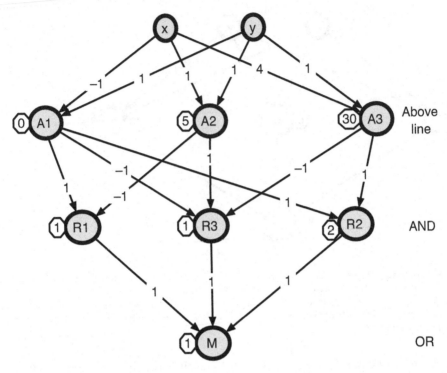

Figure B.4
Complete neural network for detecting the positive instances of the figure B.2 data set. From top to
bottom, the four layers have the following functionality: input the x-y coordinates of a data point;
detect which of the lines (L1–L3) a point is above (otherwise it is below that line); detect membership in
regions defined by conjunctions of above-line and below-line relations; and detect membership in any
of the three regions, which entails membership in the class of positive instances of the concept.

or reinforced—tend to produce networks whose nodes do *not* have such elegant
modular functionality. Their overall behavior can approach perfection, but a look
under the hood leaves most engineers in confusion.

Structured incremental learning approaches that, for example, train one layer at
a time on specific aspects of the data set can improve the prospects of a modular
solution. But if the human designer already has a deep understanding of the data
set, its most salient features, and how the data should be partitioned, then she may
not need an ANN to solve her problem in the first place. The whole point of most
ANN applications is to take a poorly understood data set and allow adaptive ANN
methods to determine a mapping that summarizes the data while also generalizing
to previously unseen cases.

This section illustrates that ANNs have the full potential to represent simple
modularly decomposable solutions for complex problems. Whether that potential

is realized in any given application, however, is difficult to guarantee, because of the stochastic nature of adaptive techniques for training neural networks, combined with the vast search space of weight vectors, many of which yield similar performance.

B.2 $k-m$ Predictability

To quantify the capacity of a distributed representation, the notion of k-m predictability provides some assistance. Consider an n-node binary network that will store patterns containing exactly k 1's, or k-*one patterns*. There are $\binom{n}{k}$ possible patterns of this type. For a given task, assume that the network needs to store a subset of these patterns, Ψ.

In its role as a content-addressable memory, the network must perform pattern completion. When presented with a partial pattern containing $k - m$ 1's, the network must uniquely determine the locations of the remaining m 1's. One prerequisite for this network (or, in fact, any system) to properly predict the missing bits is that no pair of k-one patterns in Ψ contain the same $k - m$ 1's for any possible $k - m$ locations. If two patterns did contain the same $k - m$ ones, then since the completed patterns are different, they must differ in at least one spot in their remaining m ones. This means that when given the shared $k - m$ ones, the system cannot uniquely predict the missing m ones.

A network is k-m *predictive* for a pattern set Ψ if and only if all sets of $k - m$ ones can uniquely predict the remaining m ones. Figure B.5 illustrates that no neural network (or any system) can be k-m predictive for the complete set of $\binom{n}{k}$ patterns when $k = \frac{n}{2}$ and $m = \frac{k}{2}$. In fact, even when $m = 1$, the predictive relations are far from unique.

More generally, any n-neuron network can represent at most $\binom{n}{k-m}$ possible subpatterns of $k - m$ ones. In any k-one pattern exactly $\binom{k}{k-m}$ of these $k - m$ one subpatterns will appear (i.e., each k-one pattern *uses up* this may $k - m$ subpatterns), and each will predict the other m 1's. However, as mentioned, no such $k - m$ subpattern can appear in more than one pattern of Ψ, since that would destroy the uniqueness of the predictions. Hence, if Ψ is k-m predictable by the network, then the maximum number of k-one patterns in Ψ is

$$P_{n,k,m} = \frac{\binom{n}{k-m}}{\binom{k}{k-m}} = \frac{\frac{n!}{(k-m)!(n-k+m)!}}{\frac{k!}{(k-m)!m!}} = \frac{n!\,m!}{k!\,(n-k+m)!}. \tag{B.1}$$

It is informative to express this as the fraction of the total number of k-one patterns, which indicates the degree to which the *promise* of a distributed

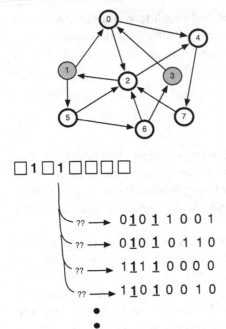

☐ 1 ☐ 1 ☐ ☐ ☐ ☐

?? ⟶ 0 1 0 1 1 0 0 1

?? ⟶ 0 1 0 1 0 1 1 0

?? ⟶ 1 1 1 1 0 0 0 0

?? ⟶ 1 1 0 1 0 0 1 0

•
•
•

Figure B.5
Assume that each of the $\binom{8}{4}$ eight-bit patterns that contain exactly four 1's can be stored in an eight-neuron network. Then given a partial pattern with two 1's, there are $\binom{6}{2} = 15$ possible completions: no unique completion exists. Even if three of the four 1's are given, there are still five possible completions. Thus, the network cannot store all these patterns in a manner that supports perfect content-addressable memory and pattern completion.

representation is fulfilled with respect to a required k-m predictability. This *k-m fufillment fraction* for an *n*-element distributed representation, $F_{n,k,m}$, is expressed as

$$F_{n,k,m} = \frac{\frac{n!m!}{k!(n-k+m)!}}{\binom{n}{k}} = \frac{\frac{n!m!}{k!(n-k+m)!}}{\frac{n!}{k!(n-k)!}} = \frac{(n-k)!\,m!}{(n-k+m)!}. \tag{B.2}$$

Table B.1 displays the maximum capacity (as given by equation B.1) of a 20-node distributed representation that must achieve perfect *k-m* predictability for a range of *k* and *m* values. Note that the capacity peaks for $k = \frac{n}{2}$ but that it drops off sharply for increasing values of *m*. Although it is not obvious from this table, a complete plot for all *k* values reveals that for $\frac{n}{4} \le k \le \frac{n}{2}$, the capacity remains within approximately the same order of magnitude.

Table B.1 also shows the corresponding fulfillment fractions (as given by equation B.2) for the 20-neuron network. Note that near the *k* value that gives optimal network capacity, $\frac{n}{2} = 10$, the fraction is only 9 percent for $m = 1$. However, as

Table B.1
Maximum capacity of a 20-node network for perfect k-m predictability.

			k		
m	1	5	10	15	19
1	1 (0.05)	969 (0.06)	**16,796 (0.09)**	2,584 (0.17)	10 **(0.5)**
2	0 (0)	114 (0.01)	**2,799 (0.02)**	738 (0.05)	7 **(0.33)**
3	0 (0)	19 (0.001)	**646 (0.003)**	276 (0.02)	5 **(0.25)**
4	0 (0)	4 (0.0003)	185 (0.001)	123 (0.008)	4 (0.2)
5	0 (0)	1 (0.00006)	62 (0.0003)	62 (0.004)	3 (0.16)

Note: Fulfillment fractions in parentheses. $n = 20$ in equations (B.1) and (B.2).

stored patterns become longer ($k > 10$), fulfillment approaches perfection even though network capacity decreases dramatically. In other words, pattern lengths that reduce capacity do have the advantage of having partial patterns that are easier to unambiguously complete.

Note that although the word *network* appears in this section, the concept of k-m predictability is completely independent of the underlying representational substrate. Any pattern collection that contains distinct patterns A and B that contain an identical subpattern (S), of length $k - m$, cannot be k-m predictable in any representation, since knowledge of S does not permit a unique prediction of either A or B. The ambiguity is a property of the data set, not the representational form or the reasoning machinery.

Regardless of this representation independence, the k-m predictability notion highlights a fundamental problem with distributed representations: common subpatterns cause interference (and thus ambiguity) during pattern completion.

In analyzing the memory capacity of a neural network, the data set (or types of data sets) to be learned need to be taken into account. k-m predictability gives an upper bound on the size of a data set that can be learned and have all of its $k - m$ length subpatterns successfully completed, but the actual capacity can fall well short of this bound if patterns share large subpatterns.

This touches on an important relation between memory capacity and sparsification. Although a pattern set with similar members yields reduced memory capacity according to k-m predictability analysis, in practice, an organism may not need to distinguish these elements. In fact, it may perform best by mapping partial versions of any of the similar source patterns into the same target pattern, i.e., by sparsifying and thus generalizing. As k-m predictability declines with highly overlapping patterns, sparsification becomes inevitable.

As the k-m predictability concept indicates, one cannot look exclusively at a neural network and ascertain its capacity. The data set, type of encoding (i.e., the value of k), and projected retrieval demands (i.e., m) require careful consideration.

B.3   Combinatorics of Expansion Recoding

Assume that a competitive layer, L_c, detects complex combinations of active upstream neurons (i.e., dense distributed codes) from layer L_a, which houses n_a neurons. Furthermore, assume that the competition level of L_c is such that no more than k neurons are active at any time. For L_c to detect all possible combinations, it must produce unique codes for all 2^{n_a} possible input vectors; and it can produce at most $\binom{n_c}{k}$ codes, where n_c neurons reside in L_c, and exactly k neurons are on at any one time in L_c. Thus,

$$2^{n_a} \leq \binom{n_c}{k} = \frac{n_c!}{k!\,(n_c - k)!}. \tag{B.3}$$

In equation B.3 the right-hand expression is maximized when $k = \frac{n_c}{2}$, but even this is much smaller than 2^{n_c}. And as n_c increases, the ratio of $\binom{n_c}{k}$ to 2^{n_c} decreases. For example, when $n_c = 50$, the ratio is 0.11, but when $n_c = 100,000$, it is 0.002. Furthermore, if L_c is truly a competitive layer, then k will be much less than $\frac{n_c}{2}$, which reduces $\binom{n_c}{k}$. This implies that n_c needs to be quite a bit larger than n_a. For example, if $n_c = 50$ and $k = 5$, then $\frac{\binom{n_c}{5}}{2^{n_c}} = 0.00000000188$. Thus, n_c would need to be over 100 million times larger than n_a.

The size demands upon n_c are relaxed somewhat by appealing to the need for L_c to generalize across L_a patterns, i.e., a group of k L_c neurons should detect a good many similar patterns from L_a. They are further reduced by the fact that in a neural network even a dense distributed code does not represent all 2^{n_a} patterns, but merely $\binom{n_a}{k*}$, where $k* > k$. This changes the relation to

$$\binom{n_a}{k*} \leq \binom{n_c}{k}. \tag{B.4}$$

But for a relatively small k and a value of $k*$ close to $\frac{n_a}{2}$, the ratio of $\binom{n_a}{k}$ to $\binom{n_a}{k*}$ is still quite small. For example, $\frac{\binom{50}{5}}{\binom{50}{25}} = 0.0000000167$. It therefore appears that n_c needs to be considerably larger than n_a in order to faithfully represent the dense codes of L_a with the sparser codes of L_c.

B.4   Interference Reduction via Sparse Coding

Consider three-layered networks wherein a dense-coding layer (A_d) feeds into a sparsely coded competitive layer (A_s), which feeds into a sparsely coded associative layer, which in turn feeds back to the top layer. The advantage of this topology for completing densely distributed codes (in the first layer) is far from obvious. The following is an argument based on interference levels in dense versus sparse-coded networks. The calculations show that when subpatterns (each denoting a *concept*) in a densely coded associative layer have overlap, i.e., share active neurons, then an association between concepts in that layer causes more interference than if each concept were linked in a sparsely coded associative network.

Assume that the network will store C different concepts. These will have dense representations in A_d and single-neuron representations in A_s. In A_d assume that B neurons are needed to encode each concept. Furthermore, assume that each neuron of A_d will, on average, participate in p different concepts. Thus, if a neuron (x) is known to be part of a particular concept, c_1, then, for another concept, c_2, the probability that x also participates in it is $\frac{p-1}{C-1}$.

First, consider the number of neurons needed in each layer. In A_s that number is simply $n_s = C$, one neuron for each concept. In A_d a total of $n_d = \frac{BC}{p}$ neurons are needed. The maximum number of connections, including self-links, between neurons in A_s and A_d is C^2 and $\frac{C^2 B^2}{p^2}$, respectively.

To bidirectionally link any two concepts in A_s requires exactly two connections, so the total fraction of synaptic space required to make a conceptual association is roughly $F_s = \frac{2}{C^2}$.

In layer A_d the calculation is a bit more complicated. Here, each concept consists of B neurons, and any concept will, on average, share $\frac{B(p-1)}{(C-1)}$ neurons with any other concept. To bidirectionally link all the neurons of one concept to another will require approximately $2B^2$ synapses, but this number double-counts the connections between the shared neurons. The exact number is thus

$$2(B^2 - \frac{B^2(p-1)^2}{(C-1)^2}) = 2B^2(1 - \frac{(p-1)^2}{(C-1)^2}). \tag{B.5}$$

The fraction of synaptic space used by a conceptual association in A_d is therefore

$$F_d = \frac{2B^2(1 - \frac{(p-1)^2}{(C-1)^2})}{\frac{B^2 C^2}{p^2}} = \frac{2p^2}{C^2}(1 - \frac{(p-1)^2}{(C-1)^2}). \tag{B.6}$$

In cases where p is much smaller than C, $F_d \approx \frac{2p^2}{C^2}$, and thus $F_d \approx p^2 F_s$. When p is very large, such as $p = \frac{C}{2}$, then $F_d \approx \frac{C^2}{4} F_s$. So, when concepts/subpatterns have overlap (i.e., $p > 1$) each conceptual association in the densely coded layer fills up the synaptic space more quickly than does each corresponding two-node association in a sparsely coded layer. Thus, A_d with subpattern overlap will experience more interference than does A_a, and it therefore behooves a neural system to store conceptual links at sparser levels.

Another complication of storing multiple concepts with overlapping activation patterns in one densely coded layer involves conflicts between excitation and inhibition. For example, assume concepts c_1 and c_2 share a set of neurons, S, but are mutually incompatible—e.g. they may represent concepts such as *liberal* and *conservative*—then auto-association among the neurons of either pattern will excite S, which will thus partially excite the competing concept. These types of relations call for a mixture of inhibition and excitation that may not be feasible within a single layer, particularly since real neurons tend to either excite or inhibit all their immediate downstream neighbors, not both. By spreading the information among different associative layers (some weak and some strong) with intervening competitive layers, the brain allows concepts to share neurons yet still compete for activation. Only the competition winners then get their patterns completed by feedback from the sparsely coded, strongly associative layer.

B.5   Weight Instabilities

The proof that combinations of positive and negative output values from neurons can still lead to instabilities of weights is described by Dayan and Abbott (2001, 286–287). What follows is a summary of that proof.

To assess the change of an entire weight vector, in this case, the vector of all input weights to node v, W_v, we can look at the geometric length of that vector, which is simply

$$| W_v | = \sqrt{\sum_{i=1}^{n} w_i^2}. \tag{B.7}$$

We are interested in whether that length continually increases, decreases, or fluctuates, in other words, in $\frac{d|W_v|}{dt}$. To simply the analysis, we can look at $\frac{d|W_v|^2}{dt}$, which will have the same sign as $\frac{d|W_v|}{dt}$. Then,

$$\frac{d | W_v |^2}{dt} = \frac{d(\sum_{i=1}^{n} w_i^2)}{dt} = \frac{dw_1^2}{dt} + \cdots + \frac{dw_n^2}{dt} = 2w_1 \frac{dw_1}{dt} + \cdots + 2w_n \frac{dw_n}{dt}$$

$$= 2W_v \bullet \frac{dW_v}{dt}. \tag{B.8}$$

So the key relation is

$$\frac{d | W_v |^2}{dt} = 2W_v \bullet \frac{dW_v}{dt}. \tag{B.9}$$

Next, assume that learning involves the classic Hebbian learning rule:

$$\triangle w_i = \lambda u_i v, \tag{B.10}$$

where λ is a positive real number (often less than 1) representing the learning rate. This equation captures the proportionality between weight change and the correlation between the two neural outputs. Rewriting equation (B.10) to account for all input connections to v yields

$$\frac{dW_v}{dt} = \lambda U v, \tag{B.11}$$

where U is the vector composed of $\{u_i : i = 1..n\}$: the outputs of all upstream neighbors of v.

Combining equations (B.9) and (B.11), we get

$$\frac{d \mid W_v \mid^2}{dt} = 2W_v \bullet \lambda U v. \tag{B.12}$$

Now assume that the postsynaptic output is simply the sum of the weighted inputs:

$$v = W_v \bullet U. \tag{B.13}$$

Then,

$$\frac{d \mid W_v \mid^2}{dt} = 2\lambda v W_v \bullet U = 2\lambda v^2. \tag{B.14}$$

Thus, except for the trivial case when $v = 0$,

$$\frac{d \mid W_v \mid^2}{dt} > 0. \tag{B.15}$$

Thus, the length of the weight vector is always increasing, which generally means that the positive weights keep getting larger and the negative weights keep getting smaller (i.e., becoming larger negative numbers). This instability of the weight vector typically leads to a saturation of the network, with many nodes always firing hard and others never firing at all.

It is important to remember that the final step of equation (B.14) hinges on the assumption in equation (B.13) that v is the sum of its weighted inputs, net_v. This is equivalent to assuming the identity transfer function, which, of course, is not always the case.

However, $\frac{d|W_v|^2}{dt}$ remains non-negative as long as v and $W_v \bullet U = net_v$ never have the opposite sign. This is the case for transfer functions with thresholds at $net_v = 0$ and which produce positive outputs above that threshold and nonpositive outputs below it. Sigmoidal transfer functions used in ANNs often have this property, although it may be violated in a small neighborhood below the threshold, where small (i.e., near zero) negative net_v values yield small positive outputs.

At any rate, the general argument should be clear: the use of both negative and positive values for u_i and v does not, on its own, prevent the positive feedbacks that lead to instability (i.e., infinite growth in the positive or negative direction) of the ANN's weights.

Furthermore, if $W_v \bullet U$ and v have the same sign for long periods of time, $\triangle W_v$ can easily become dominated by the positive feedback dynamics. As net_v moves further toward the extremes of the activation function (i.e., *saturates*), this relation is more likely. Hence, it is wise to design networks that can avoid saturation but instead allow neurons to work near the thresholds of their activation functions, where the signs of net_v and v often vary.

Similar problems arise with learning rules such as BCM, which is unstable when θ_v is held constant but becomes stable when θ_v is permitted to grow faster than does the average value of v (Dayan and Abbott, 2001, 288–289).

B.6   Redundancy and Degeneracy

In network A (figure 13.18) any single neuron, when active, can light up the entire core, thus producing 1111 as output. Furthermore, if any neuron has an inactive equilibrium state, this entails that none of the other neurons was stimulated, and hence the output vector is 0000. So the binary state of any single neuron gives perfect information as to the complete output state (of which there will only be two, 0000 and 1111, for network A).

When testing the effects of a single neuron x_j on O, there are only two equiprobable perturbation patterns: stimulate and do not. These produce two equiprobable outcomes in O: 1111 and 0000, respectively. Thus, $H(x_j) = H(O) = H(x_j, O) = H(\{\frac{1}{2}, \frac{1}{2}\}) = 1.0$. Thus, $I(\frac{1}{j}; O) = 1 \forall j$. This yields a redundancy value of $8 - 0.04 = 7.96$. Note that $I(X; O)$ has the low value of 0.04 because 255 of the 256 (2^8) possible stimulation vectors applied to X will produce the same equilibrium state: 11111111, while only one stimulation vector (00000000) produces the equilibrium state 00000000. Consequently, in 255 of 256 cases, the output vector is 1111, and 0000 in the remaining case. Hence, $H(X) = H(O) = H(X, O) = H(\{\frac{255}{256}, \frac{1}{256}\}) = 0.04$, and thus $I(X; O) = 0.04$.

To compute one of the eight components of degeneracy, D_4, note that for any size 4 subset of X, there will be 15 of the 16 possible stimulation patterns that give all four subset members an equilibrium state of 1, while the final stimulation pattern leaves all four off. And any pattern that turns them all on will turn on all outputs, while any that leaves them all off yields 0000 as output. Thus, $I(S; O) = H(\{\frac{15}{16}, \frac{1}{16}\}) = 0.34$ for each of the $\binom{8}{4} = 70$ four-neuron subsets of X. Thus, $< I(X_j^4; O) > -\frac{4}{8}I(X; O) = 0.34 - \frac{0.04}{2} = 0.32$.

Note that the key difference between redundancy and D_4 stems from the different entropies associated with singleton versus size 4 subsets of X. With subsets of size $K = 2$ and up, the $H(S^k)$ value (and hence $I(S^k; O)$) decreases from $H(\{\frac{3}{4}, \frac{1}{4}\}) = 0.81$ to $H(\{\frac{7}{8}, \frac{1}{8}\}) = 0.54$ to $H(\{\frac{15}{16}, \frac{1}{16}\}) = 0.34$. Hence, larger subsets contribute less and less to degeneracy in network A.

In network B (figure 13.18) the partitioning of X reduces the ability of individual neurons to completely control the output pattern, thus decreasing redundancy. The mutual information between X and O is higher, since $H(X)$ and $H(O)$ increase; they can end up in a wider variety of equilibrium states. That trend continues in network C (figure 13.18), where $I(S_j^1; O) = 1$ $\forall j$ is still true, but now $I(X; O)$ is much higher, bringing R down. The calculation of D_4 is more involved, since the four-neuron subsets can now span the four connected components of X in different ways: one in each of the four components; two in two of the components; and two in one component and one in two other components. The 70 four-neuron subsets are distributed over these three classes with 16, 6, and 48 exemplars, respectively.

In the first class, with one neuron in each component, there is no interaction between the externally stimulated neurons, and thus their perturbed state is their equilibrium state. Hence $H(s_j) = H(\{\frac{1}{2}, \frac{1}{2}\}) = 1$. Because of the lack of interaction, the four components are completely independent, and thus $H(S^4) = 4H(s_j) = 4$. Since the four output neurons are directly determined by the states of these four components, $H(O) = H(S^4; O) = 4$ as well. So $I(S^4; O) = H(S^4) + H(O) - H(S^4; O) = 4 + 4 - 4 = 4$.

In the second class, each pair of neurons interacts: turning on either of them will excite the other. Thus, the probability of either component being on in the equilibrium state is $\frac{3}{4}$. The other two, unstimulated, components remain off during all trials. Thus, the entropy of the distribution of states of the subset is $H(S^4) = 2H(\{\frac{3}{4}, \frac{1}{4}\}) = 2(0.81) = 1.62$. By similar reasoning, $I(S^4; O) = 1.62$ also.

In the third case, three components are in use. Two involve independent neurons, while one contains two interacting neurons. Again, the independence of these components insures that their individual entropies can be summed to yield $H(S^4)$. The component entropies are $H(\{\frac{3}{4}, \frac{1}{4}\})$ and two instances of $H(\{\frac{1}{2}, \frac{1}{2}\})$, yielding $H(S^4) = 0.81 + 1 + 1 = 2.81$.

In comparing these three cases, notice that the more finely the subset neurons are distributed over the independent components, the greater their total influence upon the output. The system has high degeneracy precisely because these independent multineuron components allow many different subsets to span many components and control the output. If one such subset fails to function, i.e., if some of its neurons become inoperable, then another subset can step in to achieve the same output pattern.

The weighted average over these three classes produces the total average for size 4 subsets:

$$< I(S^4; O) >= \frac{16}{70}(4) + \frac{6}{70}(1.62) + \frac{48}{70}(2.81) = 2.98. \tag{B.16}$$

Hence,

$$D_4 = 2.98 - \frac{4}{8}I(X; O) = 2.98 - \frac{3.24}{2} = 1.36. \tag{B.17}$$

Notes

Chapter 1

1. The POE acronym (Tyrrell et al., 2003) was originally used in the field of evolutionary hardware (EHW), with *phylogeny* referring to the evolutionary processes that modify gene and genome pools, *ontogeny* covering the development of the phenotype as governed by the genotype, and *epigenesis* denoting all phenotypic change controlled primarily by the environment. Hence, learning is a key element of epigenesis. In biology, ontogeny includes epigenesis, which concerns auxiliary biochemical processes (not directly orchestrated by the DNA) that contribute to the translation of genotypes into immature phenotypes. Assuming a general definition of *epigenesis* as "the earliest process during which the emerging phenotype can begin to differ from the design encoded in the genotype," the biological and ALife/EHW semantics coincide. In the latter, ontogenetic processes rarely receive influence from anything but the genotype and a hardcoded (though commonly stochastic) developmental routine, thus insuring that two identical genotypes normally produce similar phenotypes. Only after ontogeny, during the agent's lifetime, does the environment effect changes in the phenotype. Throughout this book, I employ the ALife/EHW definition of these three terms.

2. In AI the informal distinction between a *neat* and a *scruffy* is between a researcher who uses logic and other formal systems to achieve intelligent behavior, and one who plays fast and loose with formalisms but produces intelligence from nifty coding, originally dubbed *hacking*.

3. Though I can remember myself, as a naive young graduate student, wanting to *prove* the necessity claim.

Chapter 2

1. Unlike Deacon, authors describing these classic types of emergence rarely bring up notions of self; they tend to stick to the hard-science descriptions.

Chapter 3

1. As an implementation note, all states share the list of original points, so a separate copy is not needed in each state. However, separate versions of chosen edges, remaining edges, and Steiner points are necessary, thus adding to the computational load of best-first search.

Chapter 6

1. Some authors use the opposite notation: w_{ij} is the weight on the arc from neuron i to neuron j.
2. In reality, they do not even have a state variable.

Chapter 7

1. Of course, if w_y is a negative value, then the derivation would involve a switch of the inequality, but the slope and location of the border line would remain the same.
2. A justification that sparse coding reduces interference appears in appendix B.

Chapter 9

1. Afek et al. show that when $M \geq 34$, success of the algorithm is, for all intents and purposes, guaranteed.
2. Slight modifications to the algorithm—the most significant of which is the addition of the random decision to return a node to the undifferentiated state—enable it to solve other NP-complete graph problems such as vertex coloring (Garey and Johnson, 1979).

Chapter 10

1. The summation in equation (10.13) is the standard integration function (see chapter 6).

2. Competitive learning networks often use a linear activation function, so a higher weighted sum of inputs yields a correspondingly higher activation.

Chapter 12

1. NEAT uses direct encoding, so genotypes and phenotypes are isomorphic with respect to equivalence testing: two phenotypes are equal if and only if their corresponding genotypes are equal.

2. Note that the use of relative addresses in these codes puts them somewhere in between direct encodings and expanded indirect encodings, admittedly a gray zone that confounds any attempts at precise classification.

Chapter 13

1. This has many similarities to the thermodynamic definition of entropy, a measure of the energy in a system that cannot be used to perform work, though this is a controversial comparison. In this book the term *entropy* refers to the Shannon entropy unless otherwise stated.

2. Although the formal defintions of information gain, loss, and flow given by Williams and Beer (2010) assume that variable A in these examples (the source of flowing information) should be constant throughout *one* run of the system, the information associated with A has little import if A is constant across *all* runs, since $H(A) = 0$, as do all conditional entropies involving A. Hence, getting a useful reading of information flow requires the experimenter to vary A's values across runs. Since analyzing multiple runs is beyond the scope of this chapter, I have allowed A to vary within a single run in order to convey the essence of these calculations.

3. This highlights an important factor in assessing most of these complexity measures: structural connectivity can only tell a small part of the story. The ability of S (a subset of one or more neurons) to influence other parts of the network is indeed reflected in neural topology, but the full information-theoretic degree of influence strongly depends upon $H(S)$. This causal potential reaches highest when $H(S)$ is large, when S can inhabit many states with reasonably equal probability. Conversely, an S that normally resides in a restricted set of its complete state space has little entropy to transfer and thus little influence.

4. The exponential explosion of the TSE complexity metrics is obvious from the summations over all subsets found in most of the formulas.

Chapter 14

1. It is perhaps unsurprising that Simon won a Nobel Prize in economics for his theory of *satisficing*: finding satisfactory but not necessarily optimal solutions.

2. For example, chapter 10 discusses the hippocampus, a popular focus of neuroscientists. Of the many details of this region discovered in the past few decades, which ones can actually contribute to AI? I speculate that the intricacies of individual neurons have little relevance, nor does the 3% to 5% intraconnectivity (recurrence) of area CA3. But the fact that CA3 has significant recurrence (the most of any brain area) and lies within a series of areas (EC, BG, CA3, CA1, subiculum) that interconnect into a loop is a topological configuration that could help guide ANN design, as does the sandwiching of a competitive layer between a densely and a sparsely coding associative layer (see chapter 7). I am banking on that level of neuroanatomy as a more fitting resolution for AI research.

Appendix B

1. Given any two-dimensional data set, one can always find such lines; there is just no guarantee that exactly three (or any other fixed number of) lines will suffice.

References

Ackley, D. H., and M. L. Littman (1992). Interactions between learning and evolution. In C. G. Langton, C. Taylor, J. D. Farmer, and S. Rasmussen (Eds.), *Artificial Life II*, 487–509. Reading, MA: Addison-Wesley.

Afek, Y., N. Alon, O. Barad, E. Hornstein, N. Barkai, and Z. Bar-Joseph (2011). A biological solution to a fundamental distributed computing problem. *Science 331*, 183–185.

AI Challenge: Ants. http://ants.aichallenge.org. Accessed June 20, 2014.

Allman, J. (1999). *Evolving Brains*. New York: W. H. Freeman.

Andersen, P., R. Morris, D. Amaral, T. Bliss, and J. O'Keefe (Eds.) (2007). *The Hippocampus Book*. New York: Oxford University Press.

Arbib, M. (Ed.) (2003). *The Handbook of Brain Theory and Neural Networks*. Cambridge, MA: MIT Press.

Artola, A., S. Bröcher, and W. Singer (1990). Different voltage-dependent thresholds for inducing long-term depression and long-term potentiation in slices of rat visual cortex. *Nature 347*(6288), 69–72.

Astor, J., and C. Adami (2000). A developmental model for the evolution of artificial neural networks. *Artificial Life 6*(3), 189–218.

Bak, P., C. Tang, and K. Wiesenfeld (1988). Self-organized criticality. *Physica Review A 38*(1), 364–375.

Bala, J., K. DeJong, J. Huang, H. Vafaie, and H. Wechsler (1996). Using learning to facilitate the evolution of features for recognizing visual concepts. *Evolutionary Computation 4*(3), 297–311.

Baldwin, J. M. (1896). A new factor in evolution. *American Naturalist 30*, 441–451.

Ball, P. (2004). *Critical Mass: How One Thing Leads to Another*. New York: Farrar, Straus and Giroux.

Banzhaf, W., P. Nordin, R. E. Keller, and F. D. Francone (1998). *Genetic Programming—An Introduction: On the Automatic Evolution of Computer Programs and Its Applications*. San Francisco: Morgan Kaufmann.

Barto, A. G. (1995). Adaptive critics and the basal ganglia. In J. Houk, J. Davis, and D. Beiser (Eds.), *Models of Information Processing in the Basal Ganglia*, 215–232. Cambridge, MA: MIT Press.

Bateson, G. (1979). *Mind and Nature: A Necessary Unity*. New York: Hampton Press.

Bear, M., B. Conners, and M. Paradiso (2001). *Neuroscience: Exploring the Brain* (2d ed.). Baltimore: Lippincott Williams and Wilkins.

Bedau, M. A. (2008). Is weak emergence just in the mind? *Minds and Machines 18*(4), 443–459.

Beer, R. (1995). On the dynamics of small continuous-time recurrent neural networks. *Adaptive Behavior* 3(4), 469–509.

Beer, R. (1996). Toward the evolution of dynamical neural networks for minimally cognitive behavior. In *From Animals to Animats 4: Proceedings of the 4th International Conference on Simulation of Adaptive Behavior*, 421–429.

Beer, R. (2003). The dynamics of active categorical perception in an evolved model agent. *Adaptive Behavior* 11(4), 209–243.

Beer, R., and J. Gallagher (1992). Evolving dynamical neural networks for adaptive behavior. *Adaptive Behavior* 1(1), 91–122.

Bejan, A. and J. P. Zane (2012). *Design in Nature: How the Constructal Law Governs Evolution in Biology, Physics, Technology, and Social Organization*. New York: Doubleday.

Bellman, R. E. (1957). *Dynamic Programming*. Princeton, NJ: Princeton University Press.

Bentley, P. (Ed.) (1999). *Evolutionary Design by Computers*. San Francisco: Morgan Kaufmann.

Bentley, P., and D. Corne (Eds.) (2001). *Creative Evolutionary Systems*. San Francisco: Morgan Kaufmann.

Bergquist, H., and B. Kallen (1953). On the development of neuromeres to migration areas in the vertebrate cerebral tube. *Acta Anatomica (Basel) 18*(1), 65–73.

Bern, M. W., and R. L. Graham (1989). The shortest-network problem. *Scientific American 260*(1), 84–89.

Bienenstock, E., L. Cooper, and P. Munro (1982). Theory for the development of neuron selectivity: Orientation specificity and binocular interaction in visual cortex. *Journal of Neuroscience 2*, 32–48.

Boden, M. (Ed.) (1990). *The Philosophy of Artificial Intelligence*. New York: Oxford University Press.

Boers, E., and I. Sprinkhuisen-Kuyper (2001). Combined biological metaphors. In M. Patel, V. Honavar, and K. Balakrishnan (Eds.), *Advances in the Evolutionary Synthesis of Intelligent Agents*, 153–183. Cambridge, MA: MIT Press.

Bonabeau, E., M. Dorigo, and G. Theraulaz (1999). *Swarm Intelligence: From Natural to Artificial Systems*. New York: Oxford University Press.

Bongard, J. (2002). Evolving modular genetic regulatory networks. In *Proceedings of the IEEE 2002 Congress on Evolutionary Computation (CEC-2002)*, 1872–1877.

Bongard, J., and R. Pfeifer (2001). Repeated structure and dissociation of genotypic and phenotypic complexity in artificial ontogeny. In *Proceedings of the Genetic and Evolutionary Computation Conference (GECCO-2001)*, 829–836.

Brooks, D. R., and E. O. Wiley (1988). *Evolution as Entropy: Toward a Unified Theory of Biology*. Chicago: University of Chicago Press.

Brooks, R. A. (1999). *Cambrian Intelligence: The Early History of the New AI*. Cambridge, MA: MIT Press.

Burgess, N., and J. O'Keefe (2003). Hippocampus: Spatial models. In M. Arbib (Ed.), *The Handbook of Brain Theory and Neural Networks*, 539–543. Cambridge, MA: MIT Press.

Buzsáki, G. (2006). *Rhythms of the Brain*. New York: Oxford University Press.

Calabretta, R., S. Nolfi, D. Parisi, and G. Wagner (2000). Duplication of modules facilitates the evolution of functional specialization. *Artificial Life 6*(1), 69–84.

Callan, R. (1999). *The Essence of Neural Networks*. London: Prentice Hall.

Camazine, S., J.-L. Deneubourg, N. Franks, J. Sneyd, G. Theraulaz, and E. Bonabeau (2001). *Self-Organization in Biological Systems*. Princeton, NJ: Princeton University Press.

Chalmers, D. (1990). The evolution of learning: An experiment in genetic connectionism. In D. Touretzky, J. Elman, T. Sejnowski, and G. Hinton (Eds.), *Connectionist Models: Proceedings of the 1990 Summer School*, 81–90. San Francisco: Morgan Kaufmann.

Cherniak, C., M. Changizi, and D. W. Kang (1999). Large-scale optimization of neuron arbors. *Physical Review E 59*(5), 6001–6009.

Churchland, P. (1999). *The Engine of Reason, the Seat of the Soul*. Cambridge, MA: MIT Press.

Clark, A. (1997). *Being There: Putting Brain, Body, and World Together Again*. Cambridge, MA: MIT Press.

Clark, A. (2001). *Mindware: An Introduction to the Philosophy of Cognitive Science*. Cambridge, MA: MIT Press.

Clark, A. (2004). *Natural-born Cyborgs: Minds, Technologies and the Future of Human Intelligence*. New York: Oxford University Press.

Clark, A. (2011). *Supersizing the Mind: Embodiment, Action and Cognitive Extension*. New York: Oxford University Press.

Clune, J., B. Beckmann, P. McKinley, and C. Ofria (2010). Investigating whether hyperNEAT produces modular neural networks. In *Proceedings of the Genetic and Evolutionary Computation Conference (GECCO-2010)*, 635–642.

Clune, J., J.-B. Mouret, and H. Lipson (2013). The evolutionary origins of modularity. *Proceedings of the Royal Society B 280*(1755), 20122863.

Cover, T., and J. Thomas (1991). *Elements of Information Theory*. New York: Wiley.

Darwin, C. (1859). *On the Origin of Species by Means of Natural Selction*. London: John Murray.

Dayan, P., and L. Abbott (2001). *Theoretical Neuroscience: Computational and Mathematical Modeling of Neural Systems*. Cambridge, MA: MIT Press.

De Jong, K. (2006). *Evolutionary Computation: A Unified Approach*. Cambridge, MA: MIT Press.

Deacon, T. (1998). *The Symbolic Species: The Co-evolution of Language and the Brain*. New York: W.W. Norton.

Deacon, T. (2003). The hierarchic logic of emergence: Untangling the interdependence of evolution and self-organization. In B. Weber and D. Depew (Eds.), *Evolution and Learning: The Baldwin Effect Reconsidered*, 273–308. Cambridge, MA: MIT Press.

Deacon, T. (2012). *Incomplete Nature: How Mind Emerged from Matter*. New York: W.W. Norton.

Dennett, D. C. (1995). *Darwin's Dangerous Idea*. New York: Simon and Schuster.

Dorigo, M., V. Maniezzo, and A. Colorni (1996). Ant system: Optimization by a colony of cooperating agents. *IEEE Transactions on Systems, Man, and Cybernetics—Part B 26*(1), 29–41.

Downing, K. L. (2001). Reinforced genetic programming. *Genetic Programming and Evolvable Machines 2*(3), 259–288.

Downing, K. (2004). Artificial life and natural intelligence. In *Proceedings of the 6th Genetic and Evolutionary Computation Conference (GECCO-2004)*, 81–92.

Downing, K. L. (2007a). Neuroscientific implications for situated and embodied artificial intelligence. *Connection Science 19*(1), 75–104.

Downing, K. L. (2007b). Supplementing evolutionary developmental systems with abstract models of neurogenesis. In *Proceedings of the 9th Genetic and Evolutionary Computation Conference (GECCO-2007)*, 990–996.

Downing, K. L. (2009). Predictive models in the brain. *Connection Science 21*(1), 39–74.

Downing, K. L. (2010). The Baldwin effect in developing neural networks. In *Proceedings of the 12th Genetic and Evolutionary Computation Conference (GECCO-2010)*, 555–562.

Downing, K. L. (2012). Heterochronous neural Baldwinism. In *Proceedings of the 13th International Conference on the Simulation and Synthesis of Living Systems*, 37–44.

Downing, K. L. (2013). Neural predictive mechanisms and their role in cognitive incrementalism. *New Ideas in Psychology 31*, 340–250.

Doya, K. (1999). What are the computations of the cerebellum, the basal ganglia, and the cerebral cortex? *Neural Networks 12*, 961–974.

Dragol, G., K. Harris, and G. Buzsáki (2003). Place representation within hippocampal networks is modified by long-term potentiation. *Neuron 39*(5), 843–853.

Dreyfus, H. (2002). Intelligence without representation: Merleau Ponty's critique of mental representation. *Phenomenology and the Cognitive Sciences 1*(4), 367–383.

Durbin, R., and D. Willshaw (1987). An analogue approach to the traveling salesman problem using an elastic net method. *Nature 326*(6114), 689–691.

Dürr, P., C. Mattiussi, and D. Floreano (2010). Genetic representation and evolvability of modular neural controllers. *IEEE Computational Intelligence Magazine 5*(3), 11–19.

Edelman, G. (1987). *Neural Darwinism: The Theory of Neuronal Group Selection*. New York: Basic Books.

Edelman, G., and G. Tononi (2000). *A Universe of Consciousness*. New York: Basic Books.

Eggenberger, P. (1997a). Creation of neural networks based on developmental and evolutionary principles. In *Proceedings of the International Conference on Artificial Neural Networks (ICANN-97)*, 337–342.

Eggenberger, P. (1997b). Evolving morphologies of simulated 3-D organisms based on differential gene expression. In *Proceedings of the 4th European Conference on Artificial Life (ECAL-97)*, 205–213.

Eiben, A., and J. Smith (2007). *Introduction to Evolutionary Computing*. London: Springer.

Fahlman, S., and C. Lebiere (1990, feb). The cascade-correlation learning architecture. Technical Report CMU-CS-90-100, Carnegie Mellon University.

Federici, D. and K. L. Downing (2006). Evolution and development of a multicellular organism: Scalability, resilience and neutral complexification. *Artificial Life 12*(3), 381–409.

Fellous, J.-M., and R. E. Suri (2003). Dopamine, roles of. In M. Arbib (Ed.), *The Handbook of Brain Theory and Neural Networks*, 361–365. Cambridge, MA: MIT Press.

Finlay, B., and R. Darlington (1995). Linked regularities in the development and evolution of mammalian brains. *Science 268*, 1578–1584.

Floreano, D., P. Dürr, and C. Mattiussi (2008). Neuroevolution: From architectures to learning. *Evolutionary Intelligence 1*(1), 47–62.

Floreano, D., and C. Mattiussi (2008). *Bio-Inspired Artificial Intelligence.* Cambridge, MA: MIT Press.

Fogel, D. (2002). *Blondie24: Playing at the Edge of AI.* San Francisco: Morgan Kaufmann.

Fogel, L. (1966). *Artificial Intelligence through Simulated Evolution.* New York: Wiley.

Forbus, K. D. (1997). Qualitative reasoning. In A. Tucker (Ed.), *The Computer Science and Engineering Handbook*, 715–733. Boca Raton, FL: CRC Press.

Fregnac, Y. (2003). Hebbian synaptic plasticity. In M. Arbib (Ed.), *The Handbook of Brain Theory and Neural Networks*, 515–522. Cambridge, MA: MIT Press.

Fuster, J. (2003). *Cortex and Mind: Unifying Cognition.* Oxford: Oxford University Press.

Futuyma, D. (1986). *Evolutionary Biology.* Sunderland, MA: Sinauer Associates.

Garey, M., and D. Johnson (1979). *Computers and Intractability: A Guide to the Theory of NP-Completeness.* San Francisco: W.H. Freeman.

Gauci, J., and K. O. Stanley (2010). Autonomous evolution of topographic regularities in artificial neural networks. *Neural Computation 22*(7), 1860–1898.

Gerhart, J. C., and M. W. Kirschner (2007). The theory of facilitated variation. *Proceedings of the National Academy of Sciences 104*, 8582–8589.

Gerstner, W., and W. Kistler (2002). *Spiking Neuron Models: Single Neurons, Populations, Plasticity.* Cambridge: Cambridge University Press.

Gilbert, E., and H. Pollak (1968). Steiner minimal trees. *SIAM Journal on Applied Mathematics 16*(1), 1–29.

Gluck, M., and C. Myers (2000). *Gateway to Memory: An Introduction to Neural Network Modeling of the Hippocampus and Learning.* Cambridge, MA: MIT Press.

Goldberg, D. (1989). *Genetic Algorithms in Search, Optimization and Machine Learning.* Reading, MA: Addison-Wesley.

——— (2002). *The Design of Innovation.* Norwell, MA: Kluwer.

Gomez, F., and R. Miikkulainen (1997). Incremental evolution of complex general behavior. *Adaptive Behavior 5*(3-4), 317–342.

Gould, S. J. (1989). *Wonderful Life: The Burgess Shale and the Nature of History.* New York: W.W. Norton.

Granger, R. (2006). Engines of the brain: The computational instruction set of human cognition. *Artificial Intelligence Magazine 27*(2), 15–32.

Graybiel, A. M., and E. Saka (2004). The basal ganglia and the control of action. In M. S. Gazzaniga (Ed.), *The Cognitive Neurosciences III*, 495–510. Cambridge, MA: MIT Press.

Grossberg, S. (1976a). Adaptive pattern classification and universal recoding: I. Parallel development and coding of neural feature detectors. *Biological Cybernetics 23*, 121–134.

Grossberg, S. (1976b). Adaptive pattern classification and universal recoding: II. Feedback, expectations, olfaction, and illusions. *Biological Cybernetics 23*, 187–202.

Gruau, F. (1994). Genetic micro programming of neural networks. In K. E. Kinnear Jr. (Ed.), *Advances in Genetic Programming*, 495–518. Cambridge, MA: MIT Press.

Gruau, F., and D. Whitley (1993). Adding learning to the cellular development of neural networks. *Evolutionary Computation 1*(3), 213–233.

Hart, P., N. Nilsson, and B. Raphael (1968). A formal basis for the heuristic determination of minimum cost paths. *IEEE Transactions on Systems Science and Cybernetics* 4(2), 100–107.

Harvey, I. (1993). *The Artificial Evolution of Adaptive Behavior.* Ph.D. diss., Sussex University, Sussex, England.

Haugeland, J. (Ed.) (1997). *Mind Design II.* Cambridge, MA: MIT Press.

Hawkins, J. (2004). *On Intelligence.* New York: Henry Holt.

Haykin, S. (1999). *Neural Networks: A Comprehensive Foundation.* Upper Saddle River, NJ: Prentice Hall.

Hebb, D. (1949). *The Organization of Behavior.* New York: Wiley.

Hocker, F. (2011). *Vasa: A Swedish Warship.* Stockholm: Medströms.

Hodgkin, A., and A. Huxley (1952). A quantitative description of membrane current and its application to conduction and excitation in a nerve. *Journal of Physiology 117*, 500–544.

Hofstadter, D. R. (1979). *Godel, Escher, Bach: An Eternal Golden Braid.* New York: Basic Books.

Holland, J. H. (1992). *Adaptation in Natural and Artificial Systems* (2d ed.). Cambridge, MA: MIT Press.

Holland, J. H. (1995). *Hidden Order: How Adaptation Builds Complexity.* Reading, MA: Addison-Wesley.

Holland, J. H. (2012). *Signals and Boundaries: Building Blocks for Complex Adaptive Systems.* Cambridge, MA: MIT Press.

Holldobler, B., and E. O. Wilson (2009). *The Superorganism: the Beauty, Elegance and Strangeness of Insect Societies.* New York: W.W. Norton.

Hopfield, J. (1982). Neural networks and physical systems with emergent collective computational abilities. *Proceedings of the National Academy of Sciences 79*, 2554–2558.

Houk, J. (1995). Information processing in modular circuits linking basal ganglia and cerebral cortex. In J. Houk, J. Davis, and D. Beiser (Eds.), *Models of Information Processing in the Basal Ganglia*, 3–9. Cambridge, MA: MIT Press.

Houk, J., J. Adams, and A. G. Barto (1995). A model of how the basal ganglia generate and use neural signals that predict reinforcement. In J. Houk, J. Davis, and D. Beiser (Eds.), *Models of Information Processing in the Basal Ganglia*, 249–270. Cambridge, MA: MIT Press.

Hoverstad, B. (2011). Noise and the evolution of neural network modularity. *Artificial Life 17*, 33–50.

Hubel, D. (1995). *Eye, Brain, and Vision.* New York: Scientific American Library.

Huffman, D. (1952). A method for the construction of minimum-redundancy codes. *Proceedings of the I.R.E.*, 1098–1102.

Husbands, P., I. Harvey, D. Cliff, and G. Miller (1994). The use of genetic algorithms for the development of sensorimotor control systems. In *Proceedings of From Perception to Action Conference*, 110–121.

Izhikevich, E. (2003). Simple model of spiking neurons. *IEEE Transactions on Neural Networks 14(6)*, 1569–1572.

Izhikevich, E. (2007). Solving the distal reward problem through linkage of STDP and dopamine signaling. *Cerebral Cortex 17*, 2443–2452.

Joel, D., Y. Niv, and E. Ruppin (2002). Actor-critic models of the basal ganglia: New anatomical and computational perspectives. *Neural Networks 15*, 535–547.

Kahneman, D. (2011). *Thinking Fast and Slow*. New York: Farrar, Straus and Giroux.

Kandel, E., J. Schwartz, and T. Jessell (2000). *Principles of Neural Science*. New York: McGraw-Hill.

Kashtan, N., and U. Alon (2005). Spontaneous evolution of modularity and network motifs. *Proceedings of the National Academy of Sciences 102*, 13773–13779.

Kauffman, S. (1993). *The Origins of Order*. New York: Oxford University Press.

Kaufmann, M., P. Manolios, and J. S. Moore (2011). *Computer-Aided Reasoning: An Approach* (3d ed.). Boston: Kluwer.

Khan, G., J. Miller, and D. Halliday (2011). Evolution of Cartesian genetic programs for development of learning neural architecture. *Evolutionary Computation 19*(3), 469–523.

Khan, M., A. Ahmad, G. Khan, and J. Miller (2013). Fast learning neural networks using Cartesian genetic programming. *Neurocomputing 121*, 274–289.

Kimura, M. (1983). *The Neutral Theory of Molecular Evolution*. New York: Cambridge University Press.

Kirschner, M. W., and J. C. Gerhart (1998). Evolvability. *Proceedings of the National Academy of Sciences 95*, 8420–8427.

Kirschner, M. W., and J. C. Gerhart (2005). *The Plausibility of Life: Resolving Darwin's Dilemma*. New Haven: Yale University Press.

Kirsh, D. (1991). Today the earwig, tomorrow man? *Artificial Intelligence 47*(1-3), 161–184.

Kitano, H. (1990). Designing neural networks using genetic algorithms with graph generation system. *Complex Systems 4*(4), 461–476.

Kohonen, T. (2001). *Self-Organizing Maps*. Berlin: Springer.

Koza, J. R. (1992). *Genetic Programming: On the Programming of Computers by Natural Selection*. Cambridge, MA: MIT Press.

Koza, J. R. (1994). *Genetic Programming II: Automatic Discovery of Reusable Programs*. Cambridge, MA: MIT Press.

Koza, J. R. (2003). *Genetic Programming IV: Routine Human-Competitive Machine Intelligence*. Norwell, MA: Kluwer.

Koza, J. R., David Andre, F. H. Bennett III, and M. Keane (1999). *Genetic Programming III: Darwinian Invention and Problem Solving*. San Francisco: Morgan Kaufmann.

Lakoff, G., and R. Núñez (2000). *Where Mathematics Comes From*. New York: Basic Books.

Lamarck, J. B. (1914). Of the influence of the environment on the activities and habits of animals, and the influence of the activities and habits of these living bodies in modifying their organization and structure. *Zoological Philosophy*, 106–127.

Langton, C. G. (1989). Artificial life. In C. G. Langton (Ed.), *Artificial Life: Proceedings of an Interdisciplinary Workshop on the Synthesis and Simulation of Living Systems*, 1–49. Reading, MA: Addison-Wesley.

LeDoux, J. (2002). *Synaptic Self: How Our Brains Become Who We Are*. London: Penguin Books.

Lindenmayer, A., and P. Prusinkiewicz (1989). Developmental models of multicellular organisms: A computer graphics perspective. In C. G. Langton (Ed.), *Artificial Life: Proceedings of an Interdisciplinary Workshop on the Synthesis and Simulation of Living Systems*, 221–249. Reading, MA: Addison-Wesley.

Lisman, J., and A. Redish (2009). Prediction, sequences and the hippocampus. *Philosophical Transactions of the Royal Society of London 364*(1521), 1193–1201.

Llinás, R. R. (2001). *i of the vortex*. Cambridge, MA: MIT Press.

Lones, M. A. (2003). *Enzyme Genetic Programming*. Ph.D. diss., University of York, York, England.

Lones, M. A., and A. M. Tyrrell (2002). Biomimetic representation with genetic programming enzyme. *Genetic Programming and Evolvable Machines 3*(2), 193–217.

Luke, S. (2013). *Essentials of Metaheuristics* (2d ed.). Lulu. Available at http://cs.gmu.edu/~sean/book/metaheuristics/.

MacKay, D. (2003). *Information Theory, Inference, and Learning Algorithms*. New York: Cambridge University Press.

Maturana, H. R., and F. J. Varela (1987). *The Tree of Knowledge: A New Look at the Biological Roots of Human Understanding*. Boston: Shambhala/New Science Library.

Marr, D. (1969). A theory of cerebellar cortex. *Journal of Physiology 202*, 437–470.

Marr, D. (1971). Simple memory: A theory for archicortex. *Proceedings of the Royal Society of London, Series B 262*(841), 23–81.

Mattiussi, C., and D. Floreano (2007). Analog genetic encoding for the evolution of circuits and networks. *IEEE Transactions on Evolutionary Computation 11*(5), 596–607.

Mattiussi, C., D. Marbach, P. Dürr, and D. Floreano (2008). The age of analog networks. *AI Magazine 29*(3), 63–76.

Mayley, G. (1996). Landscapes, learning costs and genetic assimilation. *Evolutionary Computation 4*(3), 213–234.

McClelland, J., B. McNaughton, and R. C. O'Reilly (1994). Why there are complementary learning systems in the hippocampus and neocortex: Insights from the successes and failures of connectionist models of learning and memory. Technical Report PDP.CNS.94.1, Carnegie Mellon University.

Mehta, M. (2001). Neural dynamics of predictive coding. *Neuroscientist 7*(6), 490–495.

Mehta, M., A. Lee, and M. Wilson (2002). Role of experience and oscillations in transforming a rate code into a temporal code. *Nature 417*(6890), 741–746.

Merriam-Webster's Collegiate Dictionary (2003). 11th ed. Springfield, MA: Merriam-Webster.

Michalewicz, Z., and D. B. Fogel (2004). *How to Solve It: Modern Heuristics*. London: Springer.

Miller, G. F., P. M. Todd, and S. U. Hegde (1989). Designing neural networks using genetic algorithms. In *Proceedings of the 3d International Conference on Genetic Algorithms*, 379–384.

Miller, J., and W. Banzhaf (2003). Evolving the program for a cell: From French flags to Boolean circuits. In S. Kumar and P. Bentley (Eds.), *On Growth, Form and Computers*, 278–301. San Diego, CA: Elsevier.

Miller, J., D. Job, and V. Vassilev (2000). Principles in the evolutionary design of digital circuits—Part 1. *Genetic Programming and Evolvable Machines 1*(1-2), 7–35.

Mitchell, M. (1996). *An Introduction to Genetic Algorithms*. Cambridge, MA: MIT Press.

Mitchell, M. (2006). Complex systems: Network thinking. *Artificial Intelligence 170*, 1194–1212.

Mitchell, M. (2009). *Complexity: A Guided Tour*. New York: Oxford University Press.

Mitchell, T. (1997). *Machine Learning*. Columbus, OH: WCB/McGraw-Hill.

Montana, D. J. (1995). Strongly typed genetic programming. *Evolutionary Computation* 3(2), 199–230.

Montana, D. J., and L. D. Davis (1989). Training feedforward networks using genetic algorithms. In *Proceedings of the 11th International Joint Conference on Artificial Intelligence*, 762–767.

Moravec, H. (1999). *Robot: Mere Machine to Transcendent Mind*. New York: Oxford University Press.

Moriarty, D. E., and R. Miikkulainen (1997). Forming neural networks through efficient and adaptive coevolution. *Evolutionary Computation* 5(4), 373–399.

Moskewicz, M. W., C. F. Madigan, Y. Zhao, L. Zhang, and S. Malik (2001). Chaff: Engineering an efficient SAT solver. In *Proceedings of the 38th Annual Design Automation Conference* (DAC'01), 530–535.

Mostow, D. (1983). Machine transformation of advice into a heuristic search procedure. In Michalski, R., J. Carbonell, and T. Mitchell (Eds.), *Machine Learning: An Artificial Intelligence Approach*, 367–403. San Francisco: Morgan Kaufmann.

Mountcastle, V. (1998). *Perceptual Neuroscience: The Cerebral Cortex*. Cambridge, MA: Harvard University Press.

Nakajima, K., M. A. Maier, P. Kirkwood, and R. Lemon (2000). Striking differences in transmission of corticospinal excitation to upper limb motoneurons in two primate species. *Journal of Neurophysiology 84*, 698–709.

Newell, A., and H. Simon (1972). *Human Problem Solving*. Englewood Cliffs, NJ: Prentice Hall.

Niv, Y., J. Daphna, I. Meilijson, and E. Ruppin (2002). Evolution of reinforcement learning in uncertain environments: A simple explanation for complex foraging behaviors. *Adaptive Behavior 10(5)*, 5–24.

Nolfi, S., and D. Floreano (2000). *Evolutionary Robotics: The Biology, Intelligence, and Technology of Self-Organizing Machines*. Cambridge, MA: The MIT Press.

Nolfi, S., S. Miglino, and D. Parisi (1994). Phenotypic plasticity in evolving neural networks. In *Proceedings of From Perception to Action Conference*, 146–157.

Ohno, S. (1970). *Evolution by Gene Duplication*. Berlin: Springer.

Oja, E. (1982). A simplified neuron model as a principal component analyzer. *Mathematical Biology 15*, 267–273.

O'Reilly, R. C. (1996). Biologically plausible error-driven learning using local activation differences: The generalized recirculation algorithm. *Neural Computation 8(5)*, 895–938.

O'Reilly, R. C., and Y. Munakata (2000). *Computational Explorations in Cognitive Neuroscience*. Cambridge, MA: MIT Press.

Pfeifer, R., and J. Bongard (2007). *How the Body Shapes the Way We Think: A New View of Intelligence*. Cambridge, MA: MIT Press.

Pfeifer, R., and C. Scheier (1999). *Understanding Intelligence*. Cambridge, MA: MIT Press.

Pinker, S. (1997). *How the Mind Works*. London: Penguin Books.

Prescott, T., K. Gurney, and P. Redgrave (2003). Basal ganglia. In M. Arbib (Ed.), *The Handbook of Brain Theory and Neural Networks*, 147–151. Cambridge, MA: MIT Press.

Puelles, L., and J. Rubenstein (1993). Expression patterns of homeobox and other putative regulatory genes in the embryonic mouse forebrain suggest a neuromeric organization. *Trends in Neuroscience 16*, 472–479.

Radcliffe, N. (1990). *Genetic neural networks on MIMD computers*. Ph.D. diss., University of Edinburgh, Edinburgh, Scotland.

Reil, T. (1999). Dynamics of gene expression in an artificial genome: Implications for biological and artificial ontogeny. In *Proceedings of the 5th European Conference on Artificial Life (ECAL-99)*, 457–466.

Reisinger, J., and R. Miikkulainen (2007). Acquiring evolvability through adaptive representations. In *Proceedings of the 9th Genetic and Evolutionary Computation Conference (GECCO-2007)*, 1045–1052.

Reisinger, J., K. O. Stanley, and R. Miikkulainen (2005). Towards an empirical measure of evolvability. In *Genetic and Evolutionary Computation Conference (GECCO-2005) Workshop Program*, Washington, D.C., 257–264.

Rich, E. (1983). *Artificial Intelligence*. New York: McGraw-Hill.

Rieke, F., D. Warland, R. de Ruyter van Steveninck, and W. Bialek (1999). *Spikes: Exploring the neural code*. Cambridge, MA: MIT Press.

Risi, S., and K. O. Stanley (2010). Indirectly encoding neural plasticity as a pattern of local rules. In *From Animals to Animats 11: Proceedings of the 11th International Conference on Simulation of Adaptive Behavior*, 533–543.

Rodriguez, A., J. Whitson, and R. Granger (2004). Derivation and analysis of basic computational operations of thalamocortical circuits. *Journal of Cognitive Neuroscience 16*(5), 856–877.

Rolls, E., and A. Treves (1998). *Neural Networks and Brain Function*. New York: Oxford University Press.

Roughgarden, J. (1996). *Theory of Population Genetics and Evolutionary Ecology: An Introduction*. Upper Saddle River, NJ: Prentice Hall.

Rumelhart, D., G. Hinton, and R. Williams (1986). Learning representations by back-propagating errors. *Nature 323*(6088), 533–536.

Russell, S., and P. Norvig (2009). *Artificial Intelligence: A Modern Approach* (3d ed.). Essex, England: Pearson.

Sanes, D., T. Reh, and W. Harris (2011). *Development of the Nervous System* (3d ed.). Burlington, MA: Academic Press.

Schreiber, T. (2000). Measuring information transfer. *Physical Review Letters 85*(2), 461–464.

Schultz, W. (1998). Predictive reward signal of dopamine neurons. *Journal of Neurophysiology 80*, 1–27.

Schwefel, H. (1995). *Evolution and Optimum Seeking*. New York: Wiley.

Shanahan, M. (2008). Dynamical complexity in small-world networks of spiking neurons. *Physical Review E 78*(4), 041924.

Shannon, C. (1948). A mathematical theory of communication. *Bell System Technical Journal 27*, 379–423.

Shasha, D. (2003). Puzzling adventures: Short taps. *Scientific American 289*(2), 93.

Sherman, S., and R. Guillery (2006). *Exploring the Thalamus and Its Role in Cortical Function*. Cambridge, MA: MIT Press.

Shors, T. J. (2009). Saving new brain cells. *Scientific American 300*(3), 40–48.

Simon, H. A. (1996). *The Sciences of the Artificial*. Cambridge, MA: MIT Press.

Sims, K. (1994). Evolving 3-D morphology and behavior by competition. In R. Brooks and P. Maes (Eds.), *Artificial Life IV*, 28–39. Cambridge, MA: MIT Press.

Soltoggio, A., J. Bullinaria, C. Mattiussi, P. Dürr, and D. Floreano (2008). Evolutionary advantages of neuromodulated plasticity in dynamic, reward-based scenarios. In S. Bullock, J. Noble, R. Watson, and M. Bedau (Eds.), *Artificial Life XI*, 569–576. Cambridge, MA: MIT Press.

Soltoggio, A., and J. Steil (2012). Solving the distal reward problem with rare correlations. *Neural Computation 24*(4), 940–978.

Song, S., K. Miller, and L. Abbott (2000). Competitive Hebbian learning through spike-timing-dependent synaptic plasticity. *Nature Neuroscience 3*(9), 919–926.

Spears, W., D. Spears, J. Hamann, and R. Heil (2004). Distributed, physics-based control of swarm vehicles. *Autonomous Robots 17*, 137–162.

Sperry, R. (1991). In defense of mentalism and emergent interaction. *Journal of Mind and Behavior 300*(12), 221–245.

Sporns, O. (2011). *Networks of the Brain*. Cambridge, MA: MIT Press.

Squire, L., and S. Zola (1996). Structure and function of declarative and nondeclarative memory systems. *Genetic Programming and Evolvable Machines 93*, 13515–13522.

Stanley, K. O. (2007). Compositional pattern producing networks: A novel abstraction of development. *Genetic Programming and Evolvable Machines 8*(2), 131–162.

Stanley, K. O., B. D. Bryant, and R. Miikkulainen (2005). Evolving neural network agents in the NERO video game. In *Proceedings of the IEEE 2005 Symposium on Computational Intelligence and Games (CIG'05)*, 182–189.

Stanley, K. O., D. D'Ambrosio, and J. Gauci (2009). A hypercube-based encoding for evolving large-scale neural networks. *Artificial Life 15*(2), 189–212.

Stanley, K. O., and R. Miikkulainen (2002). Evolving neural networks through augmenting topologies. *Evolutionary Computation 10*(2), 99–127.

——— (2003). A taxonomy for artificial embryogeny. *Artificial Life 9*(2), 93–130.

Steels, L. (2003). Intelligence with representation. *Philosophical Transactions: Mathematical, Physical and Engineering Sciences 361*(1811), 2381–2395.

Strick, P. L. (2004). Basal ganglia and cerebellar circuits with the cerebral cortex. In M. S. Gazzaniga (Ed.), *The Cognitive Neurosciences III*, 453–461. Cambridge, MA: MIT Press.

Striedter, G. F. (2005). *Principles of Brain Evolution*. Sunderland, MA: Sinauer Associates.

Strogatz, S. (2003). *Sync: How Order Emerges from Chaos in the Universe, Nature, and Daily Life*. New York: Hyperion.

Sutton, R. S., and A. G. Barto (1998). *Reinforcement Learning: An Introduction*. Cambridge, MA: MIT Press.

Tesauro, G. (1995). Temporal difference learning and TD-Gammon. *Communications of the ACM 38*(3), 58–68.

Tononi, G. (2003). An information integration theory of consciousness. *BMC Neuroscience 5*(1), 42.

Tononi, G., and O. Sporns (2004). Measuring information integration. *BMC Neuroscience 4*(1), 31.

Tononi, G., O. Sporns, and G. Edelman (1994). A measure for brain complexity: Relating functional segregation and integration in the nervous system. *Proceedings of the National Academy of Sciences 91*, 5033–5037.

——— (1996). A complexity measure for selective matching of signals by the brain. *Proceedings of the National Academy of Sciences 93*, 3422–3427.

—— (1999). Measures of degeneracy and redundancy in biological networks. *Proceedings of the National Academy of Sciences 96*, 3257–3262.

Turner, A., and J. Miller (2013). Cartesian genetic programming encoded artificial neural networks: A comparison using three benchmarks. In *Proceedings of the Genetic and Evolutionary Computation Conference (GECCO-2013)*, 1005–1012.

Turney, P., D. Whitley, and R. W. Anderson (1997). Evolution, learning, and instinct: 100 years of the Baldwin effect. *Evolutionary Computation 4*(3), iv–viii.

Tyrrell, A., E. Sanchez, D. Floreano, G. Tempesti, D. Mange, J. Moreno, J. Rosenberg, and A. Villa (2003). Poetic tissue: An integrated architecture for bio-inspired hardware. In *Proceedings of the 5th International Conference on Evolvable Systems*, 129–140.

Urzelai, J., and D. Floreano (2001). Evolution of adaptive synapses: Robots with fast adaptive behavior in new environments. *Evolutionary Computation 9*, 495–524.

van Ooyen, A. (Ed.) (2003). *Modeling Neural Development*. Cambridge, MA: MIT Press.

van Rossum, M. (2001). A novel spike distance. *Neural Computation 13*, 751–763.

Varela, F. J., H. R. Maturana, and R. Uribe (1974). Autopoiesis: The organization of living systems, its characterization, and a model. *Biosystems 5*, 187–196.

Wallenstein, G., H. Eichenbaum, and M. Hasselmo (1998). The hippocampus as an associator of discontiguous events. *Trends in Neuroscience 21*(8), 317–323.

Watkins, C., and P. Dayan (1992). Q-learning. *Machine Learning 8*, 279–292.

Weber, B., and D. Depew (Eds.) (2003). *Evolution and Learning: The Baldwin Effect Reconsidered.* Cambridge, MA: MIT Press.

Werbos, P. (1974). *Beyond Regression: New Tools for Prediction and Analysis in the Behavioral Sciences.* Ph.D. diss., Harvard University, Cambridge, Massachusetts.

West-Eberhard, M. J. (2003). *Developmental Plasticity and Evolution.* New York: Oxford University Press.

Whitley, D. (1995). Genetic algorithms and neural networks. In J. Periaux and G. Winter (Eds.), *Genetic Algorithms in Engineering and Computer Science.* New York: Wiley.

Williams, P. (2011). *Information dynamics: Its theory and application to embodied cognitive systems.* Ph.D. diss., Indiana University, Bloomington, Indiana.

Williams, P., and R. Beer (2010). Information dynamics of evolved agents. In *From Animals to Animats 11: Proceedings of the 11th International Conference on Simulation of Adaptive Behavior*, 38–49.

Winter, P. (1987). Steiner problems in networks: A survey. *Networks 17*(2), 129–167.

Winter, P., and M. Zachariasen (1997). Euclidean Steiner minimum trees: An improved exact algorithm. *Networks 30*, 149–166.

Wolpert, D. H., and W. G. Macready (1997). No free lunch theorems for optimization. *IEEE Transactions on Evolutionary Computation 1*(1), 67–82.

Wolpert, D. M., R. C. Miall, and M. Kawato (1998). Internal models in the cerebellum. *Trends in Cognitive Sciences 2*(9), 338–347.

Wolpert, L., R. Beddington, T. Jessell, P. Lawrence, E. Meyerowitz, and J. Smith (2002). *Principles of Development* (2d ed.). New York: Oxford University Press.

Wu, A. S., and R. Lindsay (1995). Empirical studies of the genetic algorithm with non-coding segments. *Evolutionary Computation 3*, 121–148.

Yaeger, L. (1993). Computational genetics, physiology, metabolism, neural systems, learning, vision, and behavior or PolyWorld: Life in a new context. In C. G. Langton (Ed.), *Artificial Life III*, 263–298. Reading, MA: Addison-Wesley.

Yaeger, L. (2009). How evolution guides complexity. *Human Frontier Science Program (HFSP) Journal 3*(5), 328–339.

Yao, X. (1999). Evolving artificial neural networks. *Proceedings of the IEEE 87*(9), 1423–1447.

Yu, T., and J. Miller (2006). Through the interaction of neutral and adaptive mutations, evolutionary search finds a way. *Artificial Life 12*(4), 525–551.

Zhang, L., and S. Malik (2002). The quest for efficient Boolean satisfiability solvers. In *Proceedings of the 14th International Conference on Computer Aided Verification* (CAV'02), 17–36.

Index

Printed in the United States
by Baker & Taylor Publisher Services